S0-BRK-396

RISK-BASED BRIDGE ENGINEERING

Risk-based Bridge Engineering

Edited by

Khaled M. Mahmoud
Bridge Technology Consulting (BTC), New York City, USA

CRC Press is an imprint of the
Taylor & Francis Group, an **informa** business
A BALKEMA BOOK

Cover illustrations:

Front cover: Forth Road Bridge (Forth Rail Bridge in background), Scotland
Back cover: Forth Road Bridge, Scotland

Cover design: Khaled M. Mahmoud, BTC, New York City, USA

CRC Press/Balkema is an imprint of the Taylor & Francis Group, an informa business

Typeset by MPS Limited, Chennai, India

Published by: CRC Press/Balkema
Schipholweg 107C, 2316 XC Leiden, The Netherlands
e-mail: Pub.NL@taylorandfrancis.com
www.crcpress.com – www.taylorandfrancis.com

ISBN: 978-0-367-41673-7 (Hardback)
ISBN: 978-0-367-81564-6 (eBook)

Table of contents

Bridge performance & evaluation

Railway & light rail bridges

Bridge analysis & design

Preface

Risk-based engineering is essential for the efficient asset management and safe operation of bridges. A risk-based asset management strategy couples risk management, standard work, reliability-based inspection and structural analysis, and condition-based maintenance to properly apply resources based on process criticality. This ensures that proper controls are put in place and reliability analysis is used to ensure continuous improvement. An effective risk-based management system includes an enterprise asset management or resource solution that properly catalogues asset attribute data, a functional hierarchy, criticality analysis, risk and failure analysis, control plans, reliability analysis and continuous improvement. Such efforts include periodic inspections, condition evaluations and prioritizing repairs accordingly.

On August 26–27, 2019, bridge engineers from around the world convened at the 10th New York City Bridge Conference to discuss issues of construction, design, inspection, monitoring, preservation and rehabilitation of bridges. This volume documents their contributions to the safe operation of bridge assets.

Suspension bridges are strategic structures with a significant value to the economy and environment enabling social and business connectivity. Therefore the assessment of the residual strength and service life of their main cables is of paramount importance. Only conducting internal inspection, sampling, and testing of wire specimens from different panels along the cables can provide the inputs needed for the strength evaluation. These elements of evaluation mitigate the risk of missing deterioration and provide reliable results. The proceedings lead off with a paper by Mahmoud et al on "Risk-based evaluation of main suspension cables of the Forth Road Bridge in Scotland." The paper provides a summary of the risk-based evaluation of the ongoing investigation of main suspension cables at the Forth Road Bridge in Scotland, utilizing the BTC method. In Denmark, the new 4 km long Storstrøm Bridge between Masnedø and Falster will replace the existing bridge from 1937, which is in poor condition and does not have the capacity to carry the increased railway freight traffic resulting from the future opening of the Fehmarn Belt Connection. Also, the existing bridge is the only section with a single railway track, between Stockholm and Germany. MacAulay and Stoklund Larsen present details of the project in "New cable stayed bridge across Storstrømmen in Denmark." The Bosphorus Bridge is an iconic suspension bridge located in the heart of Istanbul, Turkey serving as a vital link of the city transportation network with an average daily traffic of around 190,000 vehicle/day. Recently, the bridge has undergone a major retrofit work, which also included replacement of the entire suspender system by changing the inclined orientation of the hangers to vertical. In "Hanger cable fatigue life assessment of a major suspension bridge," Durukan and Soyöz aim to evaluate the fatigue performance of the newly installed hanger cables under traffic loads. Three-dimensional finite element model of the bridge is established and the model is verified by using the ambient vibration measurements taken from the bridge. Suspension bridge main cable degradation is a serious worldwide problem. The level of degradation resulted in supplementing or replacement of the main cables, such as the Tancarville and Aquitaine Bridges in France. Dehumidification of main cables has been introduced over 20 years ago, with the hope of reducing main cable degradation. The success of the technique is debatable, as wire cracking and breakage are still observed in dehumidified cables. In "Optimizing main cable dehumidification systems," Bloomstine and Melén present recommendations for design optimization and optimization of systems in service, as well as recommendations for follow-up and inspection of systems in service.

The Honshu-Shikoku Bridge Expressway (HSBE) consists of three routes (total length of 172.9km) between Honshu and Shikoku. It plays an important role as a part of the arterial high-standard highway network. The HSBE maintains 17 long-span bridges located over straits where environment is very harsh. These include the Akashi Kaikyo Bridge, the longest suspension bridge in the world, the Seto-Ohashi Bridges, the longest highway-railway combined bridge link, and the Kurushima Kaikyo Bridges, the unique three consecutive suspension bridges. In "Asset management of Honshu-Shikoku Bridges based on Preventive Maintenance," Uchino and Toyama describe the HSBE's approach for the asset management of this complex network of long span bridges. In Ukraine, there are 16,149 bridges on public roads out of which 35% are surveyed and the survey data are introduced in a database. Since 2004, the works on the development of the Analytical Expert Bridge Management System of Ukraine have been carried out. The system was implemented in 2006. The purpose of the system is to issue the recommendations on the bridges operation strategy taking into account their technical state and funds for bridges repairs. In "Implementation of a bridge management system in the Ukraine," Bodnar and Koval outline the experience of implementing the asset management system. By the example of two bridges over the river Rhine in Duisburg-Neuenkamp and in Leverkusen, Schumm presents "Replacement strategies of existing highway bridges in Germany." Both bridges are welded cable stayed bridges, under traffic since circa 1970–1965 and are constructed as welded steel girder bridges with an orthotropic bridge deck. Recent damage to the steel structure of the bridge deck reduced the service lives of both bridges significantly. Therefore the bridges needed to be replaced within a short time, while maintaining the traffic during all construction stages.

Built in 1882, the New York, Ontario & Western Railway's three-span pin-connected through truss bridge over the East Branch of the Delaware River near Fish's Eddy, New York experienced two collapses in its short 15-year life. The first occurred as a result of a derailed caboose striking the end post of the north span. The second took place while a lightly loaded train was crossing the bridge, collapsing the middle span; the cause of this failure was never determined. In "Two collapses of the Ontario & Western Railway's bridge at Fish's Eddy," Mazurek and Tarhini present a historical review that includes examination of the relevant aspects of the specifications used in the design of the structure. Analysis results demonstrate that the hangers were excessively loaded, and that a hanger failure may have initiated the second collapse event. The environmental and climatic conditions are generally reviewed in the early phase of bridge feasibility study. In such review, flood flow rate of the stream is obtained and bridge's spanning, vertical clearance and type of piers are determined. Piers placed in the stream, are exposed to negative effects of water during their service life. The most important problems that occur around the bridge piers are depositions and scouring. Due to these cyclic effects, bridge structures may experience stability problems and collapse. In order to decrease these negative effects, stream characteristics around the piers, flood flow rate and flow velocity should be investigated. In "Redesign of collapsed river bridges due to flood and scour," Laçin et al. investigate twelve bridges located in Ordu, Turkey. Those bridges were damaged and collapsed due to the flood that occurred in July 2018. The investigation covers the hydraulic effects, which caused the bridges to collapse. The paper also presents the design principles of the new twelve bridges and field precautions to prevent depositions and scouring of piers. On August 14, 2018, the Morandi Bridge, over the Polcevera River, in Genoa, Italy, collapsed, claiming the lives of 43 people. The cable-stayed bridge was named after its designer, Riccardo Morandi. The cables linking the towers to the deck were covered in concrete. On the morning of the incident, Genoa was hit by a fierce thunderstorm. Just before noon, a 91-meter section of the highway collapsed, falling almost 45-meter to the ground. In "Explicit collapse analysis of the Morandi Bridge using the Applied Element Method," Malomo et al. explore the influence of several parameters, including deterioration of cables, and loading effects on the collapse mechanism. The observed and predicted debris were also compared to assess the possible collapse mechanism of the bridge.

The Walter Taylor Bridge over the Brisbane River at Indooroopilly in Queensland Australia, completed in 1936, is a rare example of the Florianopolis type, originated by David B Steinman. His bridge, however, and those by others, used eye bar suspension chains, while the cable at Indooroopilly consists of wire ropes salvaged from the tie-back cables used in construction of the Sydney Harbour Bridge. The bolted clamps that connect the stiffening trusses to the wire rope cables were developed specifically for the bridge at Indooroopilly, and these two features make the bridge unique. When it was opened the bridge had the second longest span in Australia, only exceeded by the then new bridge over Sydney Harbour. Design provenance is clouded by uncertainties, although it is now generally accepted that R. J. McWilliam designed the pylons and W. J. Doak was responsible for the suspended steelwork. At the time, Doak was Bridge Engineer for Queensland Railways, which may explain the reluctance to publically acknowledge his involvement at the time. The bridge is well maintained and remains in service. In his paper, "The Walter Taylor Bridge - Florianopolis Australis," Rothwell presents the history and details of this landmark bridge. South Capitol Street was a primary corridor in Major Pierre L'Enfant's 1791 Plan (*L'Enfant Plan*) of Washington D.C., which developed South, East and North Capitol streets to extend directly from the U.S. Capitol, and become prominent gateways to the city's Monumental Core. Underscoring this historic plan is the replacement of the existing Frederick Douglass Memorial Bridge with a new multiple arch bridge that will transform the South Capitol Street corridor with a landmark entry route to the District's federal area. In their paper, Butler and Porter present the details of "Design and construction of the new Frederick Douglass Memorial Bridge." The ultimate design goal is to improve connectivity, traffic mobility, safety and operational characteristics in the project corridor while ensuring compliance with all environmental requirements related to the protection of all natural and historic resources. The original design of the Tacoma Narrows Bridge developed by Washington State engineer Clark Eldridge included a suspension bridge with a center span of 2,600 feet (792 m), two side spans of 1,300 feet (396 m) each, trusses and cables 39 feet (11.9 m) center to center, stiffening trusses 22 feet (6.7 m) deep, and two travel lanes and sidewalks. However, when the Washington State Toll Bridge Authority (WSTBA) requested federal assistance from the Public Works Administration (PWA), the PWA agreed to a grant of 45% of the construction cost on the condition that the WSTBA hire Leon Moisseiff of New York for the design of the superstructure. Moisseiff told the PWA that his design would reduce estimated construction cost from $11 million to $7 million. The foundation design was to be performed by another New York firm, Moran, Proctor, and Freeman. Moisseiff increased the center span length to 2,800 feet (853.4 m) and reduced the side span lengths to 1,100 feet (335.3 m). The 25-foot deep (7.6 m) stiffening trusses were replaced by 8-foot-deep plate girders. On July 1, 1940, the bridge was opened to traffic, the third longest bridge in the world after the Golden Gate and George Washington Bridges. On November 7, 1940, torsional oscillations caused the failure of the bridge. The wind speed was about 42 miles per hour (67.2 km/hour). In his paper, "The failure of the Tacoma Narrows Bridge," Gandhi covers the construction and destruction of the original bridge and the people connected to it.

In Turkey, which has a high seismic risk, it is required to design bridges so as to ensure a sufficient level of safety, as similar to other types of structural systems. In addition, it is of paramount importance to evaluate the seismic safety of existing bridge structural systems. In this context, currently it is often required to resort to deformation based performance evaluation methods in addition to the conventional strength based performance methods. In "Strength and deformation based performance evaluation of existing bridges," Namlı et al. investigate the seismic performance of bridges. The number of curved bridge structures constructed in the United States has steadily risen over the past several decades. As of 2004, over one-third of all steel superstructure bridges constructed were curved. The popularity of curved bridges experienced a boom since a curved bridge can offer the designer solutions to complicated geometrical limitations or site irregularities as compared to traditional straight bridges. Additionally, as the use of high-performance steel has become more prevalent, engineers have become able to design more complicated structures as the girder can handle greater loads. A previous alternative to constructing a bridge using a curved girder section was to use a chorded structure composed of a series of straight girder sections oriented in a curve to produce a curved bridge. However, using curved girder sections provides aesthetic as well as cost benefits over these traditional chorded structures. In "Effect of thermal loading on the performance of horizontally curved I-girder bridges," William focuses on the response of horizontally curved steel I-girder bridges to changing thermal conditions. The expected continuation or acceleration of climate change could induce additional stresses that increase the risk of failure of critical bridge components. Bridge structures are designed based on historical weather records, where the climatic patterns reflect the local climate in which the bridge is located. In "Structural performance of continuous slab-on-steel girders bridge subjected to extreme climate loads," Mohammed et al. investigate the impact of extreme ambient temperatures and very high thermal gradients on the safety and serviceability of multi-span slab-on-steel girder bridges. A three-span slab-on-girder bridge designed according to the Canadian Highway Bridge Design Code and simulated by a 3-D nonlinear finite element model is investigated for the effect of average ambient and differential temperatures as predicted by climate change models. The results show increases in the moments in critical sections of the bridge superstructure and very large longitudinal and lateral displacements. More investigations are required to further investigate the safety of the bridge elements and connections. The use of heat-straightening to repair damaged steel members dates back to the 1930s with low-grade steels. For many years, heat-straightening of damaged steel bridge girders has been more of an art than a science. Several safety concerns have historically limited its validity as a repair technique. When applied appropriately, heat-straightening, along with mechanical techniques including pressing or jacking, could offer an economical and a viable alternative to replacement of damaged steel girders. Zatar and Nguyen provide details of "Application of heat straightening repair of impacted highway steel bridge girders." Fracture Critical members are steel tension components whose failure is expected to result in collapse of the bridge. It is required to inspect fracture-critical bridges using "arms-length" approach, which is costly and time consuming. Structural health monitoring can be used as alternative approach for inspection providing both accuracy and economy. In "Fracture detection in steel girder bridges using self-powered wireless sensors," Abedin et al. investigate the feasibility of using a handful of self-powered wireless sensors for continuous monitoring and detection of fracture in steel plate girder bridges. A detailed finite element analysis was carried out on a multi-girder bridge using available traffic data. The time histories of displacement obtained for intact and fractured scenarios show that vibration amplitude was significantly increased for fractured girder, and strain variation was recorded especially in the vicinity of fracture, conditions that can be detected with relevant sensors. Visual inspections by dedicated teams have been one of the primary tools in structural health monitoring (SHM) of bridge structures. However, such conventional methods have certain shortcomings. Manual inspections may be challenging in harsh environments and are commonly biased in nature. In the last decade, camera-equipped unmanned aerial vehicles (UAVs) have been widely used for visual inspections. However, the task of automatically extracting useful information from raw images is still challenging. In "A convolutional cost-sensitive crack localization algorithm for automated and reliable RC bridge inspection," Sajedi and Liang propose a deep learning semantic segmentation framework to automatically localize surface cracks. Due to the high imbalance of crack and background classes in images, different strategies are investigated to improve performance and reliability.

Several studies on seismic risk assessment of infrastructures have been undertaken in the past on Italian infrastructure. They mainly concerned the seismic assessment of highway and road bridges. Hence, the need to deepen the seismic assessment of railway bridges through a methodology that allows a large-scale study of the seismic vulnerability of such structures through the construction of fragility curves. The fragility curves are defined by cumulative probability distributions, which allow estimation of the probability of reaching or exceeding a given level of damage for a given severity of ground shaking. In "Large-scale vulnerability analysis of girder railway bridges," Bellotti et al. propose simplified mechanical modeling in relation to the level of the acquired knowledge. The goal of the work is to create a tool that is able to generate a numerical model, of one or more bridge/s, which is capable to derive bridge fragility curves through either the execution of numerous non-linear time-history analyses or by using a simplified procedure. Although use of non-ballasted continuous welded rail (CWR) has become inseparable portion of light-rail transit (LRT) projects due to maintainability, safety, and passenger comfort, the CWR could adversely affect the bridges supporting the LRT vehicles through rail-structure interaction (RSI). Due to temperature variations, significant axial rail stresses/deformation may develop affecting the track serviceability by increasing the probability of rail fracture. In "Nonlinear rail-structure interaction effects for multi-frame, multi-span curved Light Rail Bridges," Honarvar and Senhaji perform a nonlinear RSI analysis for five long curved bridges with a combined length of more than 2.5 mi (4.02 km) to accurately compute rail stresses, rail gap, and substructure forces. While California high-speed rail (CAHSR) project encompasses various types of infrastructure to accommodate high-speed trains (HST), a single-span concrete network tied arch bridge was proposed for the first time to support HST as a part of CAHSR Construction Packages 2-3. Such a system was utilized to accommodate a long span and to satisfy the minimum horizontal and vertical clearance requirements to the existing features on the ground. Due to compound nature of a network tied-arch bridge in combination with CAHSR seismic and track-structure interaction (TSI) analysis requirements, a comprehensive nonlinear analysis is crucial to ensure satisfactory performance of the bridge. In "Nonlinear analysis of the first concrete network tied arch bridge for California High-Speed Rail," Honarvar et al. present an overview of the nonlinear analysis required to design the bridge, including nonlinear seismic, TSI, time-dependent, and geometry analyses using a detailed 3-D finite-element model of the bridge.

The girder bridge using steel rolled H-beams is competitive and economical for short span road bridges due to low material and fabrication costs. However, the applicable span length is only 20 m to 25 m because the maximum web height is about 900 mm. To extend the span length a new steel/concrete composite bridge was developed using the steel rolled H-beam. The new bridge form has continuous-span steel H-girders which are composite with the RC slabs to resist positive bending moments at span-center. In "Experimental and FEM studies on the innovative steel-concrete hybrid girder bridge," Elmy and Nakamura provide details of experimental and modeling of the new SRC structural form. Across Europe, the need to manage roadway bridges efficiently led to the development of multiple management systems in different countries. Despite presenting similar architectural frameworks, there are relevant differences among them regarding condition assessment procedures, performance goals and others. Therefore, although existing a complete freedom of traffic between countries, this dissimilarity constitutes a divergent mechanism that has direct interference in the decision making process leading to considerable variations in the quality of roadway bridges from country to country. The need for harmonization is evident. Action TU1406, funded by COST (European Cooperation in Science and Technology), aims to institute a standardized roadway bridges condition assessment procedure as well as common quality specifications (performance goals). Such purpose requires the establishment of recommendations for the quantification of performance indicators, the definition of performance goals and a guideline for the standardization of quality control plans for bridges. By developing new approaches to quantify and assess bridge performance, as well as quality specifications to assure expected performance levels, bridge management strategies will be significantly improved and harmonized, enhancing asset management of ageing structures in Europe. In "Quality specifications for roadway bridges, standardization at a European level," Casas and Matos present the results and conclusions of the COST Action TU1406. Risk assessment and performance evaluation of individual structures subjected to seismic hazard were pioneered by the Pacific Earthquake Engineering Research (PEER) Center. A few recent studies have also explored the performance of civil infrastructure systems subjected to multiple hazards. In "Multi-hazard financial risk assessment of a bridge-roadway-levee system," Nikellis et al. conduct risk assessment of the system, through a probabilistic event-based analysis. The paper presents results of the analysis in terms of risk metrics commonly used in the field of financial engineering for portfolio optimization. A new bridge in Monmouth County of New Jersey, USA, is being constructed at a site with Class F soil classification, which requires a site-specific ground motion response spectrum analysis. Due to the highly compressible soil of about 50 to 70 feet thick in some areas, deep foundations were required and extended into the stiff clay and dense sand layers below, ending nearly 120 feet below mudline. To minimize the size of the deep foundations, seismic isolation bearings were used to reduce the seismic demand on substructures. Due to the unique nature of the isolated system involving tall piers, a 3-D time history seismic analysis was performed in addition to response spectrum analysis. In "Benefits of using isolation bearing and seismic analysis on Bridge (MA-14) with Class F Soil," Liang et al. describe the subsurface conditions, the development of a site-specific response spectrum for a 1000-year return period earthquake, isolation bearing preliminary design, and the benefits of using isolation bearings and seismic analysis.

The editor is grateful to all the authors and reviewers for their efforts in producing this volume.

Khaled M. Mahmoud, PhD, PE
Chairman of Bridge Engineering Association
www.bridgeengineer.org
Chief Bridge Engineer
BTC
www.kmbtc.com
New York City, USA

New York City, August 2019

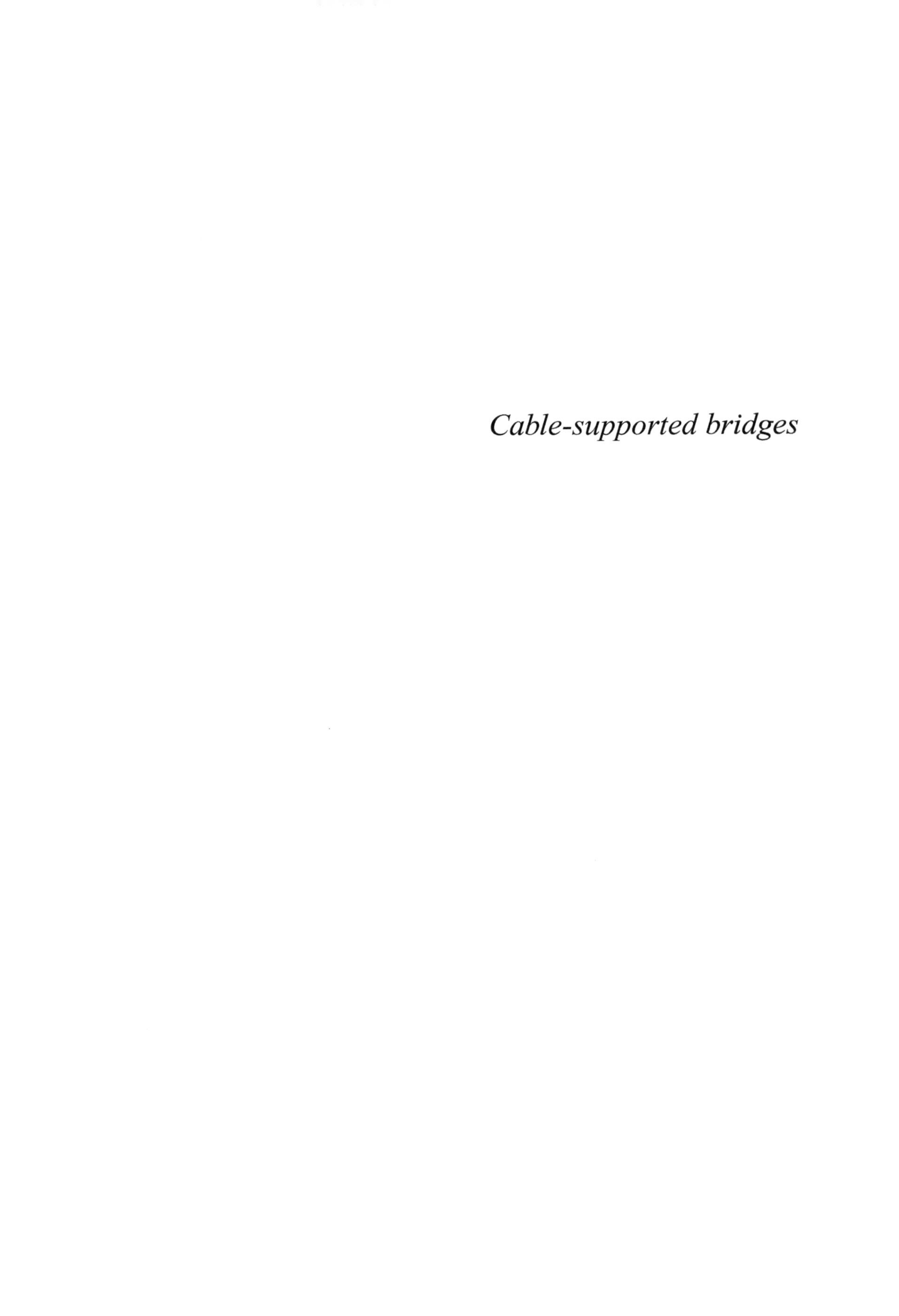

Cable-supported bridges

Chapter 1

Risk-based evaluation of main suspension cables of the Forth Road Bridge in Scotland

K. Mahmoud
BTC, New York City, USA

C. Gair & H. McDonald
Transport Scotland, Glasgow, Scotland, UK

ABSTRACT: Suspension bridges are strategic structures whose main suspension cables are composed of thousands of wires. Over time, individual wires suffer from different levels of deterioration; ranging from loss of strength and ductility, growth of cracks, to wire breaks. This eventually leads to loss of the cable load carrying capacity. Internal inspection is performed, where wire samples are extracted, and tested to evaluate the strength of the cable. Only a small percentage of wires are visible during the internal inspection, and a much lesser percentage is sampled for testing. Therefore the internal inspection of suspension bridge cables must be regarded as a risk management tool. The risk of concern is undetected serious deterioration that could lead to reduction of cable strength. The scope of inspections, the evaluation of inspection results, and subsequent decision-making are the means by which this risk is controlled and managed. This paper provides a summary of the risk-based evaluation of the ongoing main suspension cable investigation at the Forth Road Bridge in Scotland.

1 INTRODUCTION

An important element of the risk-management process is the nature and extent of the data collected pertaining to cable health. Limited or biased data, or incorrectly evaluated data, can lead to misleading evaluations of the extent of loss in the cable strength, resulting in either unnecessary corrective measures or failure to take necessary actions. Determining the nature and amount of data to collect is in part a statistical problem, as is the analytical tools by which the data is summarized and used to develop conclusions about the cable strength. Such statistical considerations must be fitted into the framework and constraints of bridge cable inspection and strength evaluation. This paper presents details of the risk-based inspection, sampling, testing and ongoing strength evaluation of the main suspension cables of the Forth Road Bridge in Edinburgh, Scotland. The Forth Road Bridge was opened to traffic, alongside the iconic Forth Rail Bridge, on September 4, 1964, with a main span of 1006 m, Figure 1. Each of the two main suspension cables contains 11,618 4.98-mm zinc-coated wires.

Following the findings of degradation during the first internal inspection in 2004, the main cables have been retrofitted with dehumidification and acoustic monitoring. A second limited internal inspection was conducted in 2008, and the third internal inspection was performed in 2012. The Forth Estuary Transport Authority (FETA) commissioned BTC to apply the BTC method to independently evaluate the cable strength utilizing the inspection and testing results from the 2012 internal inspection. In 2015, FETA was dissolved and Transport Scotland took control of the Forth Road Bridge. Subsequently, Transport Scotland appointed Amey for the maintenance of the bridge. In 2017, Amey assigned a Consulting firm to carry out the fourth internal inspection of the main cables, in accordance with the requirements of NCHRP Report 534 (Report 534 2004). In parallel to the NCHRP inspection, Transport Scotland have requested an independent

Figure 1. Forth Road Bridge, Scotland (Forth Rail Bridge in background).

investigation is carried out by BTC, based in New York, USA, using their patented BTC method for inspection, sampling and evaluating the residual strength of the Forth Road Bridge cables.

The BTC method is a patented methodology that employs a reliability-based analysis to estimate the remaining strength and service life of both, parallel wire and helical wire bridge cables (NYSDOT Report C-07-11 2011). It is included in the Federal Highway Administration (FHWA) Primer for the Inspection and Strength Evaluation of Suspension Bridge Cables (FHWA 2012). The BTC method applies to both zinc-coated and bright bridge (non-galvanized) wire. In the BTC method, wires are collected from the wedge openings, utilizing random sampling of individual wires, in each investigated panel. The randomly selected sample is tested to obtain the mechanical properties, including ultimate strength, ultimate elongation, yield strength, Young's modulus and fracture toughness. The probability of broken wires is estimated based on inspection observation of broken wires, and probability of cracked wires is estimated based on the cracks detected from fractographic examination of wire fracture surfaces. The ultimate strength of cracked wires is determined using fracture toughness criteria. All these data is utilized to assess the remaining strength of the cable in each of the investigated panels. The BTC method employs a probabilistic-based approach to assess the remaining service life of the cable by determining the rate of change of broken and cracked wires detected over a time frame, and available data from previous cable investigations, and measuring the rate of change in effective fracture toughness over the same time frame. The words reliability and probabilistic are used to describe the method of assessing the remaining strength and service life of the cable.

This paper presents a summary of the ongoing risk-based evaluation of remaining strength and residual life of bridge cables, at the Forth Road Bridge in Scotland, utilizing the BTC method.

2 INSPECTION AND STRENGTH EVALUATION OF MAIN CABLE

The current investigation of the main suspension cable at Forth Road Bridge is conducted in accordance with the two recognized methods for the evaluation of remaining strength of bridges cables:

- BTC method; employs reliability-based analysis of inspection findings and mechanical properties of sampled wires (NYSDOT Report C-07-11 2011) and (FHWA 2012), and
- Report 534 Guidelines; depends on the visual assessment of wire surface corrosion (Report 534 2004).

The two methods follow different techniques in modeling of cable degradation and strength evaluation. This offers bridge owners the advantage of well-informed decisions about cable degradation and future maintenance plans. The dual-method approach give owners confidence, as although different, the results provide upper and lower bound envelopes for the cable strength.

The BTC method has been applied alongside Report 534 at the Bronx-Whitestone Bridge, Mid-Hudson Bridge, in New York, USA, and Pierre-Laporte Bridge in Québec City, Canada. It is noted that the dual application of the two methods does not require duplication in the inspection, sampling or testing effort. BTC is currently evaluating the strength of main suspension cables of the iconic Williamsburg Bridge in New York City, utilizing the BTC method.

2.1 *Visual-based evaluation of wire degradation*

Hopwood and Havens provided the first classification of the visual corrosion for the galvanized wires in the helical strands of main suspension cables of General U. S. Grant Bridge over the Ohio River in 1984 (Hopwood and Havens 1984), as follows, Figure 2:

Stage 1: white spots on the surface of wire, indicating early stages of zinc oxidation
Stage 2: wire surface is completely covered by oxidized zinc
Stage 3: appearance of spots of brown ferrous corrosion covering (20-30%) of wire surface
Stage 4: brown ferrous corrosion covering more than 30% of wire surface

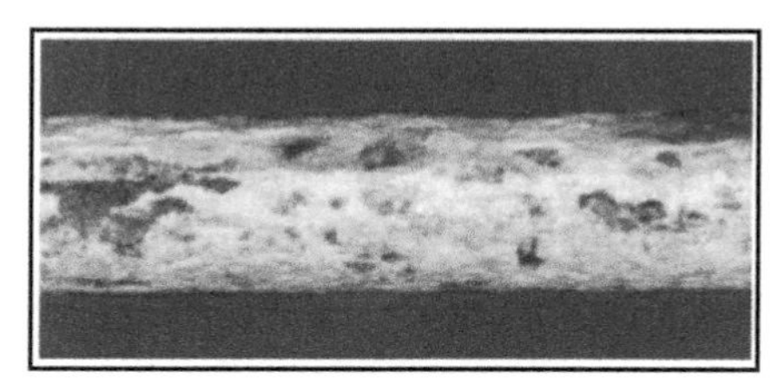

Stage 1: Spotty zinc corrosion on surface

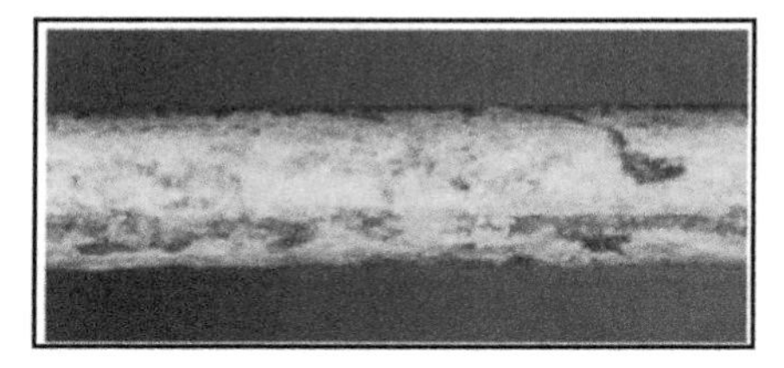

Stage 2: Zinc corrosion on entire surface

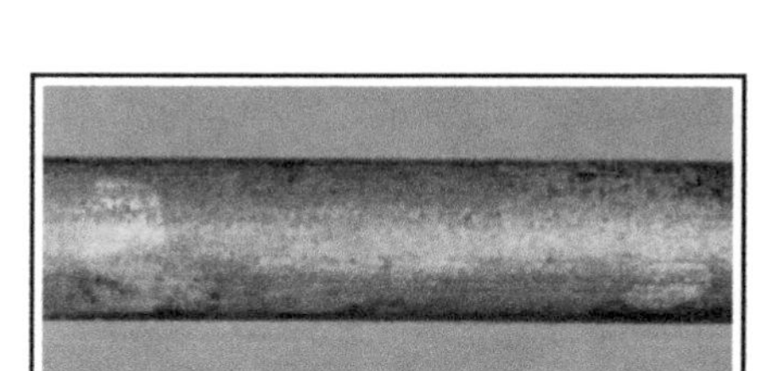

Stage 3: brown ferrous corrosion covering 20-30% of bridge wire surface

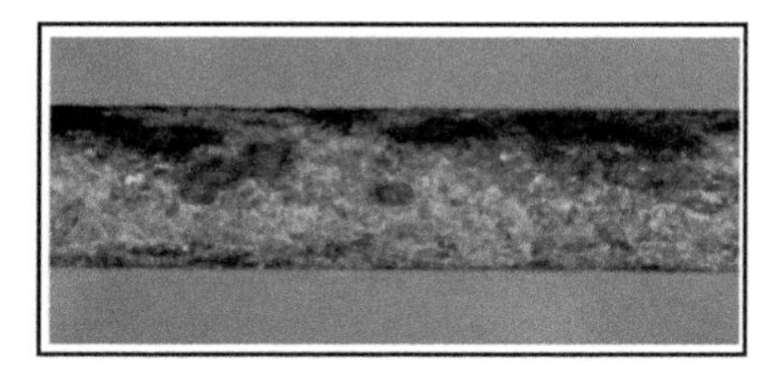

Stage 4: brown ferrous corrosion covering more than 30% of bridge wire surface

Figure 2. Stages of corrosion for deteriorated bridge cable wire.

The Williamsburg Bridge cables were investigated some 30 years ago, also utilizing a visual-based method. It was the first bridge with parallel wires whose cables conditions were inspected in-depth. The Williamsburg Bridge main cable is composed of parallel bright wires. Therefore the bridge investigation adopted a modified six (6) degrees of corrosion classification system for the metal surface of wires, ranging from Grade 0; no corrosion, almost new condition, to Grade 5; worst corrosion (Steinman 1988).

Published in 2004, Report 534 generalized the definition of Hopwood and Havens, to classify visual corrosion on surface of parallel wire cables (Report 534 2004).

With the above definition, Stages 3 and 4 represent the worst (visual) wire degradation. As such, one would expect a strong correlation between the proportions of wires classified Stage 3 and Stage 4, at different panels, and the number of observed broken wires, in the respective panels. This will be discussed later in the paper.

2.2 *Reliability-based evaluation of wire degradation*

The scope of main cable internal inspection, strength evaluation and the subsequent decision-making with regard to cable maintenance represent one of the most serious issues in the management of long span suspension bridge assets. During internal inspection, only a small percentage of wires are visible, and a much lesser percentage is sampled for testing. Therefore the internal inspection of suspension bridge cables is a risk management tool. The main risk is undetected serious deterioration that could lead to reduction of cable strength. The internal inspections, the evaluation of inspection results, and subsequent decision-making are the means by which this risk is controlled and managed.

A typical suspension bridge cable is composed of thousands of wires, and the assessment for cable strength is based on a small sample size. Therefore it is essential to employ reliability-based method, which infers the strength of the cable from a small sample of wires.

With this understanding of the limited sample from a large population of wires, the BTC method utilizes modern assessment techniques that employ reliability criteria (very similar to the Load and Resistance Factor Design "LRFD" criteria), in which the wire mechanical properties obtained from a random sample, including strength and ductility (strain) are known as probabilistic entities, from which a "probability of failure" could be estimated. If an evaluation is conducted using these criteria, it can help establish, with high level of confidence, the tempo of inspection and further evaluations in the future.

The BTC method employs random sampling to eliminate visual bias, and fracture mechanics principles to assess the strength of cracked wires. Designing the sampling plan in each investigated panel, which marks the wires to be sampled from each wedge opening, prior to the field inspection ensures that the sampling is random. The inspectors and the Contractor are strictly instructed to sample only the wires marked on the sampling plan. With the use of a random sample, the analysis evaluates the sampling error in the estimated cable strength. The method forecasts the remaining life of the bridge cable based on strength degradation and rates of growth in broken and cracked wires proportions detected over a time frame, and the rate of change in fracture toughness over same time frame (NYSDOT Report C-07-11 2011) and (FHWA 2012).

The BTC method was first utilized to evaluate the strength and residual life of the main suspension cables at the Bronx-Whitestone Bridge in New York City (Mahmoud 2013a). Afterwards, the BTC method was applied to evaluate the strength of main suspension cables at the Mid-Hudson Bridge, in New York (Mahmoud 2013b), and to provide independent evaluation of cable strength and forecast of life for main suspension cables at the Forth Road Bridge, in Scotland. The BTC method is currently being applied to evaluate the remaining strength of the suspension cables of Williamsburg Bridge in New York City and Pierre-Laporte Bridge in Québec City, Canada.

3 INTERNAL INSPECTION AND WIRE SAMPLING

During the 2004–2005 first inspection of the main suspension cables of the Forth Road Bridge, a total of ten (10) panel openings were wedged and inspected. In the 2008-second internal inspection was limited to only three panels. The scope of work included a re-inspection of the panel, which was deemed to have the worst deterioration during the first inspection of 2004–2005, and its two adjacent panels.

During the fourth internal inspection of the main suspension cables in 2018, wire samples were collected from nine panel openings. The wire samples were tested for mechanical properties, fatigue and hydrogen content. This data, along with inspection results, form the inputs for the ongoing

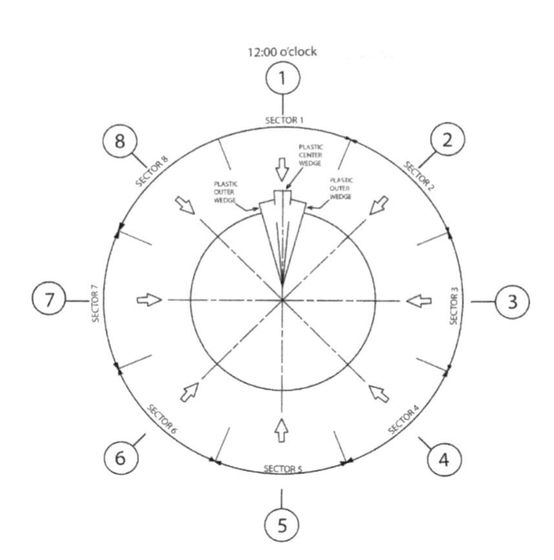

Figure 3. 8-wedge pattern for cable inspection.

Figure 4. Groove wedge at 12:00 position.

evaluation of remaining strength and forecast of life for the main suspension cables of the Forth Road Bridge.

3.1 *Cable wedging and internal inspection*

The main suspension cable is wedged in each of the nine panels, along an 8-wedge pattern, Figure 3. Wedges were driven along each wedge line to perform the inspection of wire condition along the wedge opening, Figure 4. The goal of the inspection and sampling is to identify deterioration in the wires. During inspection, engineers identify broken wires; and mark them for repairs, if accessible. Wires that are observed broken deep in the wedge groove cannot be repaired due to the tight space, which does not allow for the tools necessary to splice the wire.

It is important to increase the pool of wires being observed during internal inspection. At the panel that was inspected in previous inspections, BTC designed the sampling plan to extract wire samples from the additional wedge pattern shown in Figure 5.

This additional wedge pattern locates Wedge #1, at 15° from the 12:00 o'clock position, in a clockwise direction. Given the uncertainty of wire deterioration, it is advantageous to change the orientation of the wedge. The additional wedge pattern provides information on the condition of wires along different planes, of the cable cross-section. It also offers the opportunity to examine wire breaks, if exist, and extract samples from different locations of the cable cross-section. Further, it widens the cumulative pool of wires subject to inspection, sampling and testing throughout the cable cross-section. This broadens knowledge of wires condition in the cable cross-section and mitigates the risk of missing wire deterioration.

3.2 *Wire sampling*

Due to the large number of wires in the suspension bridge cable, the assessment for cable strength is based on a small sample size. Therefore reliability-based methods are essential to infer the strength of the cable from a small sample of wires.

Due to the fact that not every wire is actually sampled, there would be a sampling error. It describes the range that the estimated cable strength is likely to fall within. In visual-based selection of wire

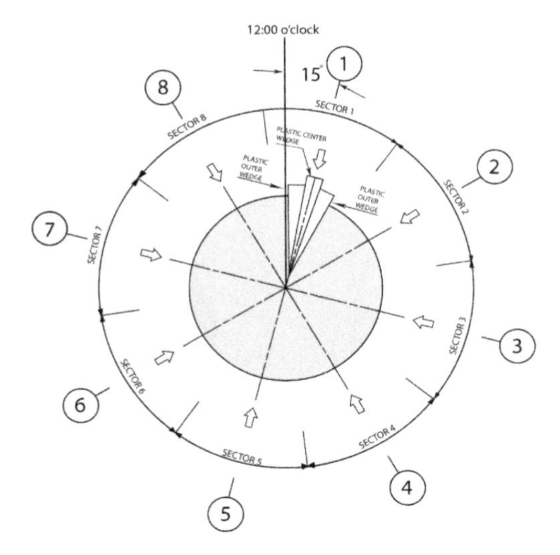

Figure 5. Additional wedging for cable inspection.

samples, there is an obvious visual bias. Further, the sampling error, which is the degree to which the sample differs from the population, remains unknown. In random sampling, it is possible to quantify the sampling error and thus provide more reliable analysis of results. Each wire in the available pool for sampling has an equal and known chance of being selected. Because random sampling is a fair way to select a wire sample, it is reasonable to infer the strength of the entire cable from the test results produced by the randomly selected wire sample. In fact, random sampling procedures do increase the probability that the randomly selected wires will be representative of the cable condition. In the design of the sampling plan, it is important to realize that sampling should be limited to provide an acceptable level of error in the estimated cable strength. It is not feasible to sample wires too deep in the wedge opening, because clearance issues would impede splicing and tensioning of the replacement wire. It is important to replace every single wire sample with a spliced new wire in its place to avoid the presence of a void between the wires. Such a void could house moisture, which is of damaging consequences to the high strength steel wire.

In the BTC Sampling Plan, the sampling frame is defined as the accessible group of wires that samples will be randomly selected from. Sampled wires constitute the sample size from which valid conclusions about the entire 11,618 wires are based. This statistical inference is performed with the aim of estimating the degraded condition of the entire cable from that found in the collected sample. By virtue of the random selection of the wire sample, the different conditions of wires would be covered in the sample, without visual bias.

The wire samples, collected from each of the nine (9) panels were subject to a regime of testing. This includes tensile testing, fatigue testing and chemical analysis.

4 EVALUATION OF CABLE STRENGTH AND METRICS FOR WIRE DEGRADATION

Test results, along with inspection findings are currently being analyzed to estimate the cable strength and forecast of cable life.

The main goal of the reliability-based analysis is to reduce the uncertainties and arrive at an estimate of the cable strength with high level of confidence. The accuracy and reliability of the estimated cable strength depend on the influential metrics of degradation listed below:

- Hydrogen embrittlement
- Broken wires

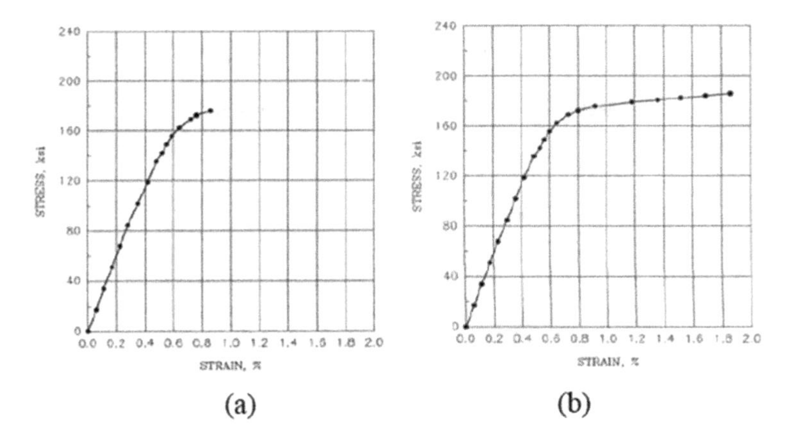

Figure 6. Stress-strain curves for tested wire specimens.

- Cracked wire
- Forecast of cable life

4.1 *Hydrogen embrittlement*

The main inputs for the assessment of remaining strength of bridge cables are the test results of wire specimens and data collected during inspection. Test results of old cable wires have shown tenuous correlation between visual-based evaluation and measured ultimate strength and ultimate elongation. In fact, the visual-based evaluation is incapable of detecting the effect of hydrogen embrittlement, a known form of degradation of cable wire. For instance, the wire shown in Figure 6(a) displays half of the elongation displayed by the wire shown in Figure 6(b), therefore the wire in Figure 6(a) is more embrittled. It is noted that the two specimens demonstrate the same yield plateau and almost the same ultimate strength. The major and significant difference displayed by the two stress-strain curves is the value of the ultimate elongation. As such, the wire could be embrittled but its ultimate strength would not be significantly affected.

The area under the stress-strain curve of a wire specimen represents the energy that it takes to break the wire. Therefore disregarding the elongation data in the analysis does not reflect the full extent of wire degradation. This is one added element of uncertainty in traditional methods of evaluation of cable strength. The BTC method includes in the analysis the ultimate strength, ultimate elongation, Young's modulus and yield strength, and therefore accounts for the embrittlement of wires in its estimation of the cable strength.

The hydrogen embrittlement causes significant reduction in the elongation capacity of the wire. The critical hydrogen concentration to cause embrittlement is 0.7 ppm (Nakamura and Suzumura 2009). With a concentration higher than 0.7 ppm, hydrogen can degrade the fracture resistance of the high strength wire material. The mechanics of this process pivots around the generation of atomic hydrogen, which causes embrittlement and eventual reduction in the fracture toughness of the wire material (Mahmoud 2003).

The attractive force, which binds various metal atoms together, is called metallic bond. In the presence of moisture, the released atomic hydrogen diffuses into the interior of the wire, at location of surface imperfection, weakens the inter-atomic bond and causes the embrittlement. Equation (1) describes the corrosion reaction that produces atomic hydrogen.

$$xFe + H_2S \underset{H2O}{\Longrightarrow} Fe_xS + 2\,[H]_{diffusible} \tag{1}$$

Bridges in rural locations may be subject to higher concentrations of corrodants than the typical surrounding environment. Hydrogen sulfide (H_2S) is produced from large amount of

Figure 7. Embrittled wire break.

vehicle-related pollution around bridges. Also, solid particulates such as sulfides generated by smokestacks, even from distant refineries and factories, may be carried airborne for very long distances. For instance, Lexington, Kentucky, has no substantial SO_2 output. Yet, a certain daily particulate sulfate concentration was relatively high (10–40 micrograms per cubic meter). Research revealed the pollution level was caused by smokestack output from power plants up to 226 miles (361 km) away (Hazarti and Peters 1981). Common atmospheric pollutants include sulfur dioxide (SO_2), chloride (Cl-), ammonia (NH_3), nitrous oxide (N_2O), and hydrogen sulfide (H_2S), (Stern 1968). Specifically, the hydrogen sulfide (H_2S) helps hydrogen remain sufficiently atomic to adsorb to the steel surface, and promote hydrogen embrittlement of bridge wire (Hopwood and Havens 1984). It has been shown that high strength cold drawn carbon steel wire is susceptible to cracking in hydrogen sulfide (H_2S) solution at room temperature at stress level less than 15% of the ultimate strength (Townsend 1972).

The hydrogen embrittlement mechanism is confirmed by evidence of hydrogen concentration testing. According to testing of wire specimens from the cables of several suspension bridges, the average hydrogen concentration measured values that are much higher than the 0.7 ppm critical threshold to cause embrittlement.

The following factors, which all exist in the bridge cable environment, are required to cause hydrogen embrittlement:

- Source of hydrogen
- Susceptible material; i.e., high strength steel
- High tensile load

The sources of hydrogen are defined as follows:

- Trapped Hydrogen: During manufacturing process, freshly drawn wires are dipped in a bath of molten zinc to provide zinc coating. Hydrogen gets trapped, with random levels of concentration in the wire.
- Diffusible Hydrogen: During the bridge service, free hydrogen generates, in the presence of moisture, as byproduct of the corrosion reaction, described in Equation (1).

Wires that are found broken during inspection display embrittled patterns, where the hydrogen assists the growth of a preexisting crack to critical depth, at which the wire breaks, Figure 7. It is clear that the broken wire shown in Figure 7, which was observed in an over 80 years old suspension cable, has no section loss, or necking, and that the wire displays unmistakable embrittled pattern. The break is driven by the growth of a crack on the lower plane of the fracture surface. In the BTC method, estimation of cable strength, wire embrittlement is quantified through the use of measurements of ultimate elongation, and evaluation of fracture toughness.

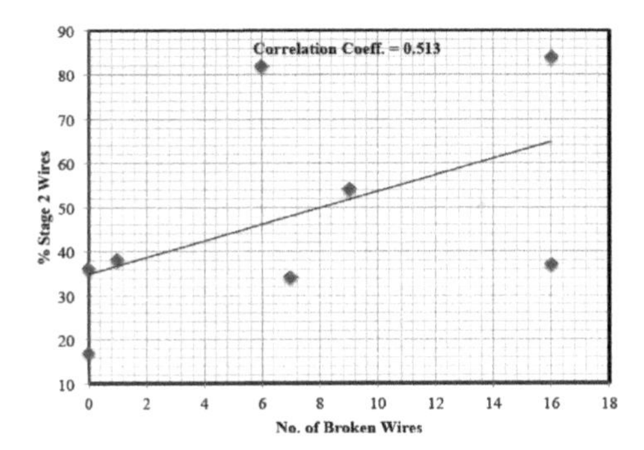

Figure 8. Correlation between number of broken wires and % stage 2 wires.

4.2 *Broken wires*

The BTC method evaluates the proportion of broken wires in each of the investigated panels as a probability. The outer wires at the surface of the cables are fully accessible for inspection, and the number of broken wires in the outer ring is observed and identified. The probability of broken wires, in the interior rings of the cable, is assessed based on the observed broken wires, as a fraction of the total observed interior wires. This approach is consistent with the random pattern of wire breaks observed during internal inspection of main suspension cables.

Analysis of the findings of broken wires observed during the third internal inspection, of the main cables at the Forth Road, reveals an inverse correlation between the number of broken wires and proportions of Stages 2, 3 and 4 in different investigated panels. The inverse correlation is defined as a contrary relationship between two variables such that they move in opposite directions. According to Report 534 Guidelines, it is expected to have less broken wires in the presence of higher percentage of Stage 2 wires and more broken wires in the presence of higher percentage of Stage 3 and 4 wires. However, the number of broken wires, observed during the third internal cable inspection in 2012, demonstrates a contrary correlation to the above expectation. For instance, Figure 8 shows a correlation coefficient of 0.513 between number of broken wires and percentage of stage 2 wires. This implies that a higher proportion of stage 2 corresponds to a larger number of broken wires, which is contrary to the definition in Report 534.

Figures 9 and 10 show correlation coefficients of – 0.495 and – 0.373, between broken wires and proportions of Stage 3 and Stage 4 respectively. These correlation coefficients suggest that the presence of higher proportions of either Stage 3 or Stage 4 wires correspond to lower number of broken wires.

This inverse correlation is inconsistent with the visual-based definition of the stages of corrosion, which assign the least damage to Stage 1 wires and the worst damage to Stage 4 wires. This observation echoes similar inverse correlation between the stages of corrosion and broken wires, observed at two bridges in the United States; the Bronx-Whitestone Bridge (Mahmoud 2013a) and Mid-Hudson Bridge (Mahmoud 2013b).

4.3 *Cracked wires*

Cracking is the major driver of wire degradation. Therefore accurate assessment of both; the proportion and ultimate strength of cracked wires are of great impact on the reliability and confidence in the estimated cable strength. There are two important factors that affect the contribution of the cracked wires in the estimated cable strength, as follows.

4.3.1 *Cracked wire proportion*

In the BTC method, the cracked wire proportion is determined, based on fractographic examination of all of the fracture surfaces of tested samples, as the ratio of number of wire samples, which contain

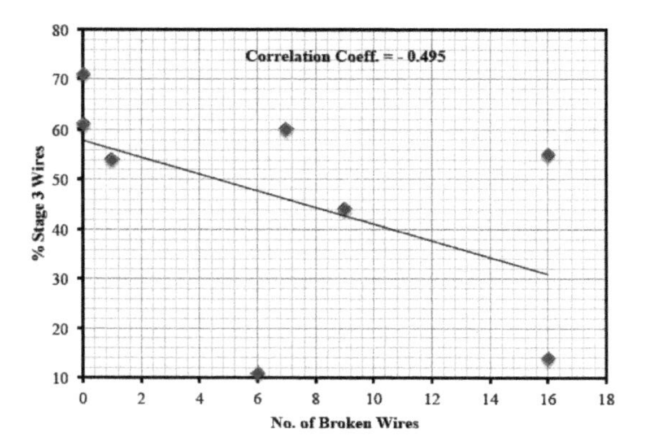

Figure 9. Correlation between number of broken wires and % stage 3 wires.

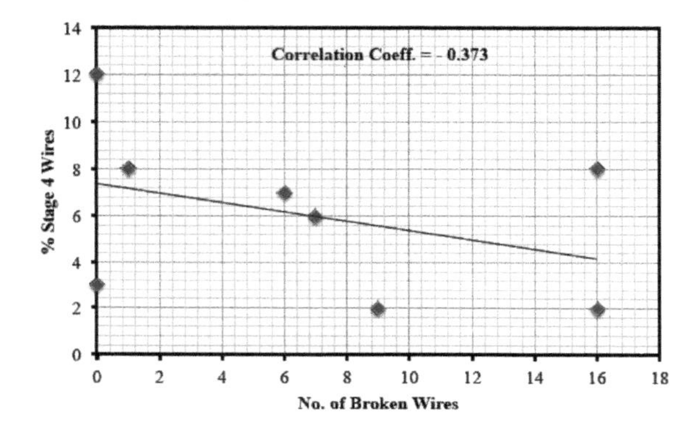

Figure 10. Correlation between number of broken wires and % stage 4 wires.

preexisting cracks, to the total number of tested wire samples. This cracking proportion, in each investigated panel, is treated as a probabilistic quantity.

4.3.2 *Strength of cracked wires*

The BTC method determines the strength of cracked wires utilizing fracture-based analysis. The cracking typically develops on the inside of the cast of the wire, which is subject to tensile stress when the wire cast is straightened out. The average value for the ultimate capacity of a cracked wire, σ_{cracked}, is determined from the following relationship (Mahmoud 2007):

$$\sigma_{cracked} = \frac{K_c}{Y(\frac{a}{D})\sqrt{\pi a_c}} \tag{2}$$

where, K_c, is the fracture toughness of the wire determined from test data, a_c is the critical crack depth measured in specimens with preexisting cracks, see Figure 11, and $Y(\frac{a}{D})$ is a crack geometry factor.

A group of wire specimens from different panels were tested under fatigue loading to identify preexisting cracks and to determine the fracture toughness of the wire (Mahmoud 2007). Figure 12 shows a wire specimen with a preexisting crack that grew under fatigue during the test.

Figure 13 helps the understanding of the strong effect of cracking on the estimated cable strength. The wire specimen shown in Figure 13(a) contains a preexisting crack in the fracture surface, which was detected following the tensile test. It is noted that the wire cross section did not suffer any

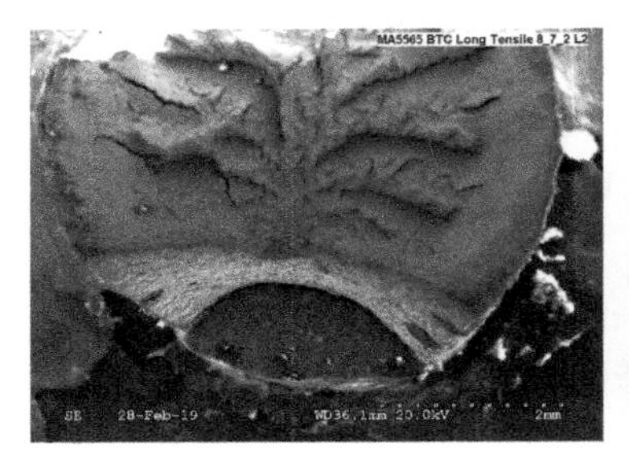
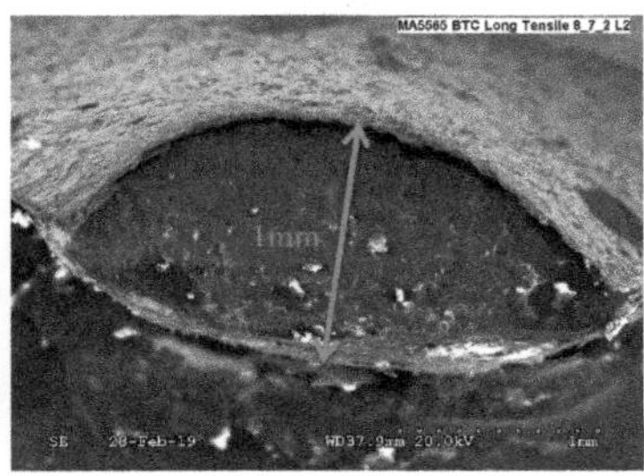

Figure 11. Preexisting crack in a tested wire specimen, Forth Road Bridge.

Figure 12. Growth of a preexisting crack under fatigue loading, Forth Road Bridge .

(a) Fracture of cracked wire test specimen

(b) Stress-strain curveof wire test specimen

Figure 13. Cracked wire fracture and stress-strain curve.

loss, and that the fracture profile shows no signs of necking. In other words, it is clear that the wire is embrittled. This is mirrored in the stress-strain curve of the same wire specimen shown in Figure 13(b), which shows an ultimate elongation of only about 1%. Consequently, the area under the stress-strain curve is only the shaded sliver, in Figure 13(b), which represents the very small amount of energy that it took to break the cracked wire.

It is worth noting that the amount of energy under the stress-strain curve in Figure 13(b) is much lower compared with the corresponding energy of a non-cracked wire. With further crack growth, it is expected that the energy at break would be even smaller.

If the wire, shown in Figure 13, were not sampled out for testing and remained in the cable, the crack would have grown under load application, until it reaches a critical depth, at which point the wire breaks, following the same mechanism of wire break shown in Figure 7. This demonstrates the strong influence of cracks on reducing the load carrying capacity of cracked wires and the overall remaining strength of the cable.

It is important to note that only wires that contain preexisting cracks would break under service loads. When the crack grows under service load, both; the load carrying capacity, and ductility are reduced. As soon as the crack reaches a critical depth, the wire breaks. In such manner, cracking is the main driver of wire degradation.

From the above explanation, it is evident that the accurate assessment of both; the cracked wire proportion and the strength of cracked wires are essential to arrive at an accurate estimate of the remaining strength of the cable.

The BTC method is the only method that considers the effect of cracking, through fracture-based analysis and thus it provides an accurate assessment for the estimated cable strength.

4.4 *Forecast of cable life*

Reliable forecast of the cable strength must quantify degradation as a function of the wire measured mechanical properties over time. The BTC method correlates the decline of wire properties with degradation kinetics. To forecast the degraded strength of the cable, at anytime in the future, t_2, the BTC method assesses the following quantities:

- Degraded strength of intact wires, $(\sigma_u)_{t2}$
- Effective fracture toughness, $(K_c)_{t2}$
- Degraded strength of cracked wires, $(\sigma_{cracked})_{t2}$
- Proportions of broken and cracked wires at time t_2, including effect of degradation in adjacent panels.

The forecast of cable life, in the BTC method, includes the effect of proportions for broken and cracked wires in adjacent panels.

There are severe limitations in forecasting of cable life, based on the visual-based condition per Report 534 and its accompanying document (NCHRP Project 10-57 2004). According to Report 10-57 (NCHRP Project 10-57 2004), the forecast of cable life is estimated "*... provided that the period over which the cable strength deterioration is estimated does not exceed 10% of the age at the time of inspection.*" In the case of the Forth Road Bridge, the forecast of future cable strength would only be projected for a period of just over five years. This apparently falls short from providing the necessary information required for long term planning.

The forecast of cable strength degradation and prediction of service life are of critical importance to bridge owners. It is acknowledged that this task is carried out based on limited data to predict uncertainties in the future. Therefore it is necessary to account for the measurable metrics of wire degradation and the increasing rate of deterioration. The BTC method employs a prediction model, which is based on strength degradation for both intact and cracked wires and includes for an increasing rate of deterioration.

5 CONCLUSIONS

With thousands of wires in the main cable of a suspension bridge, there is a significant level of uncertainty in the evaluation of strength and forecast of life of the cable. Suspension bridges are strategic structures with a significant value to the economy and environment enabling social and business connectivity. Therefore the assessment of the residual strength and service life of their main cables is of paramount importance. Only conducting internal inspection, sampling, and testing of wire specimens from different panels along the cables can provide the inputs needed for the strength evaluation. These elements of evaluation mitigate the risk of missing deterioration and provide reliable results. This assists the bridge owner in long term planning and in the management of these strategic bridges. The paper shows that cracking is the main driver of deterioration of the cable strength. Thus the fracture-based analysis of cracked wires is required to achieve accurate assessment of residual strength. The damage resulting from the corrosion reaction is assessed more accurately by the byproduct of the corrosion reaction that releases atomic hydrogen. One other source of hydrogen is the hydrogen that gets trapped, in the wire, during the processing of the freshly drawn wire in a bath of molten zinc to provide zinc coating. At locations of micro deficiencies on wire surface, atomic hydrogen penetrates the interior of the wire, breaks the interatomic bond and causes wire embrittlement. Over time, and with the application of load, the surface crack grows, leading to strength reduction and ultimately to the wire break. It is this understanding of the corrosion process, and deterioration, that is based on measureable metrics of the effect of the corrosion reaction byproducts that is significant to the strength evaluation. The forecast of residual life in the BTC method is based on strength degradation model for intact and cracked wires, and increasing rate of deterioration.

The main cables are the most critical and vulnerable elements of a suspension bridge. Therefore it would be sensible and logical to use recognized methods that provide confidence and correlation by using a separate and independent checking mechanism/method rather than just relying on one method. This would provide owners with greater assurances when dealing with perhaps the most challenging decisions they may face when dealing with the results of an internal inspection and strength evaluation of their main cables.

REFERENCES

Hazarti, A. M. and Peters, L. K., 1981. Particulate Sulfate Lexington, Kentucky and Its Relation to Major Surrounding in SO_2 Sources, Atmospheric Environment, Vol 15, No. 9, pp 1623–1631.

Hopwood, T. and J.H. Havens, J.H., 1984. Corrosion of cable suspension bridges, Kentucky Transportation Research Program, University of Kentucky Lexington, Kentucky: Kentucky.

Mahmoud, K.M., 2003. Hydrogen embrittlement of suspension bridge cable wires, *System-based Vision for Strategic and Creative Design*, Bontempi (ed.), Swets & Zeitlinger, Lisse, ISBN 90 5809 599 1.

Mahmoud, K.M., 2007. Fracture strength for a high strength steel bridge cable wire with a surface crack, *Theoretical and Applied Fracture Mechanics, 48*.

Mahmoud, K.M., *NYSDOT Report C-07-11,* 2011. New York State Department of Transportation (NYSDOT), BTC method for evaluation of remaining strength and service life of bridge cables, *NYSDOT, cosponsored by FHWA and New York State Bridge Authority*.

Mahmoud, K.M., 2013a. Application of the BTC method for evaluation of remaining strength of main suspension cables at the Bronx-Whitestone Bridge. The 8th International Cable Supported Bridge Operators' Conference, Edinburgh, Scotland 3–5 June.

Mahmoud, K.M., 2013b. Evaluation of remaining strength of Mid-Hudson Bridge main suspension cables using the BTC Method. The 8th International Cable Supported Bridge Operators' Conference, Edinburgh, Scotland 3–5 June.

Mayrbaurl, R. & Camo, S., 2004. NCHRP Report 534 – Guidelines for inspection and strength evaluation of suspension bridge parallel wire cables, Transportation Research Board, Washington DC, USA.

Nakamura, S., and Suzumura, K., 2003. Hydrogen embrittlement and corrosion fatigue of corroded bridge wires, *Journal of Constructional Steel Research*, Volume 65, Issue 2, 2009.

NCHRP Project 10-57, Structural safety evaluation of suspension bridge parallel wire cables, Final Report, 2004.

Steinman, 1988. Williamsburg bridge cable investigation program: Final Report. New York State Department of Transportation & New York City Department of Transportation.

Stern, A. C., 1968. Air Pollution-Volume I Air Pollution and Its Effects, Academic Press New York, NY, pp. 625–645.

Townsend, H., 1972. Hydrogen Sulfide Stress Corrosion Cracking of High Strength Steel Wire, Corrosion, Vol. 28, No. 2, pp. 39–46.

U.S. Federal Highway Administration, 2012. Primer for the inspection and strength evaluation of suspension bridge cables, FHWA, Washington, DC, FHWA Report No. FHWA-IF-11-045.

Chapter 2

New cable stayed bridge across Storstrømmen in Denmark

B. MacAulay & E. Stoklund Larsen
Danish Road Directorate, Hedehusene, Denmark

ABSTRACT: A 4 km long sea-crossing structure will be rising out of Storstrømmen in Denmark over the coming 3 years. The new structure will be one of the few bridges, in the world, carrying a two-track high speed railway and 2 lane road, as well as a combined footway & cycle path, on one single cross section. Comprising 44 viaduct spans and 2 navigational spans. The main bridge will be a cable stay structure with a 100 m tall pylon. The Design & Build Contract was awarded in February 2018, the detailed design is now well on the way, with construction almost about to start. The paper gives details of the overall project, outline the development of the architectural concept for the cross section, and the development of the Definition Bridge Design and the Design Basis for this mega structure. A process which started in 2013 and has continued right up to the issue of the Final Tender documents in May 2017, with significant changes being implemented throughout the competitive dialogue phase with the prequalified D&B consortia's, as a result of the dialogue. The new bridge will be an outstanding landmark, replacing what was once, and for 30 years, Europe's longest combined road and railway bridge.

1 INTRODUCTION

In Denmark, the new 4 km long Storstrøm Bridge between Masnedø and Falster will replace the existing bridge from 1937, which is in poor condition and does not have the capacity to carry the increased railway freight traffic resulting from the future opening of the Fehmarn Belt Connection. Also, the existing bridge is the only section with a single railway track, between Stockholm and Germany.

1.1 *Future-proof*

In the autumn of 2011, it became clear that the existing Storstrøm Bridge would not be able to withstand the increased freight traffic, as the main steel structure was found to have severe fatigue cracks. A feasibility study was therefore carried out to form the basis for a decision on the future of the bridge. The feasibility study analysed how the road and railway traffic could be maintained in the socioeconomically most cost-effective way, through a total of five possible scenarios – including a renovation of the existing Storstrøm Bridge. However, as the structure is in very poor condition, which would have required substantial maintenance and rehabilitation construction works to future proof the bridge, if it was to remain.

The study found that, due to the age and limited capacity of the existing bridge, a new Storstrøm Bridge would be the optimum solution. The Storstrøm Bridge is of great regional importance and is also an important part of the Trans European Network railway corridor. And in March 2013, it was decided to initiate the work of constructing a new combined road and railway bridge, with a cycle and pedestrian path, across Storstrømmen, Figure 1.

The existing Storstrøm Bridge will be demolished when the new bridge has been constructed and is open to traffic.

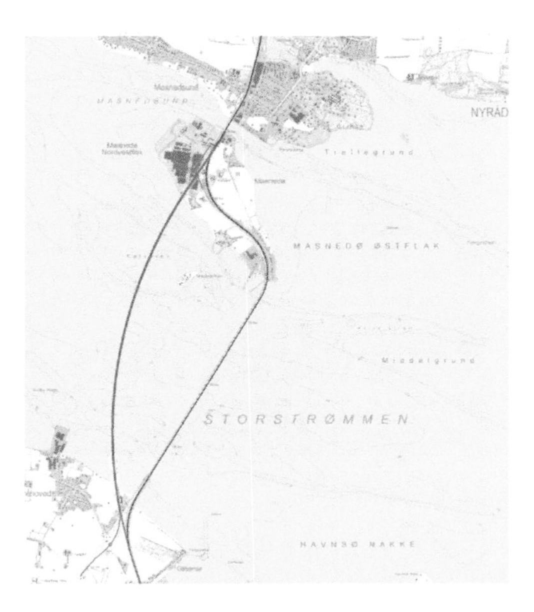

Figure 1. New Bridge alignment.

2 ARCHITECTURAL CONCEPT AND DEFINITION DESIGN

The 3.84 km long, new Storstrøm Bridge, comprise 44 viaduct spans of 80m each and 2 navigational spans of 160 m in length each, will be one of the few bridges, in the world to carry a two-track high-speed railway, a two-lane road, and a combined footway and cycle path, on a single cross section, 24m wide.

The two-span central cable-stayed bridge is 320 m long and supported by a 102 m-high tower and has a navigation clearance of 26 m. The tower is located centrally and has 20 stays on each side, ranging in length from 43 m (58 strands) to 174 m (107 strands)

The deck structure is a monocellular box girder, with a trapezoidal cross-section, internal truss diaphragm double-deck top slab. The entire structure – from foundation caissons, piers, deck box girder and to the tower – is a concrete structure.

The development of the architectural concept, and as a result the Definition Design, was a process, which had its off set in the feasibility study carried out in 2012. The feasibility study had concluded a new crossing was required for 2 railway tracks, a two-lane road and a combined cycle and pedestrian path. The solution presented by the end of the feasibility study was two parallel bridges, one carrying the railway, and one carrying the road and cycle- and pedestrian path, Figure 2.

During the course of the first 6–12 months, the project team investigated several other solutions, trying to combine the bridge deck on one single structure. This would improve the emergency access to the railway, minimise the impact on the seabed and also optimise the substructure.

A double-decker bridge deck, similar to the Øresund Bridge, was considered, but was disregarded due to the limited area available at each end of the main crossing for raising the road alignment, Figure 3. The next step in the architectural development of the cross-section, was two parallel bridge decks – concrete box girders – supported on a single substructure, Figure 4. This however, still did not solve the issue of how to provide access for emergency vehicles in case of accidents on the railway.

The next consideration was how to combine the two separate bridge deck, into a single cross section. Comprising a two-track railway, a two-lane road and a cycle and pedestrian path.

Figure 2. Feasibility Study, two parallel structures.

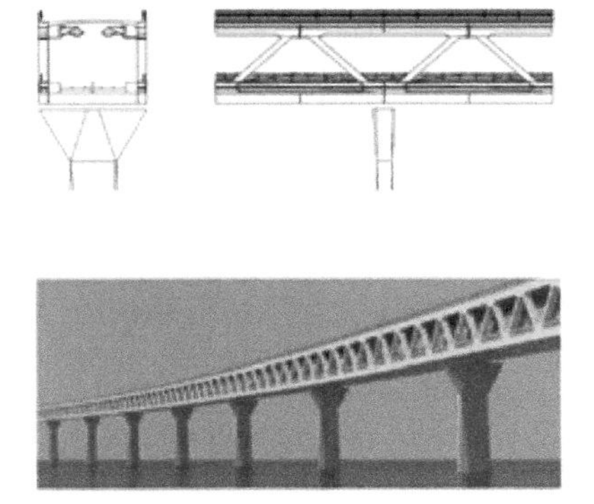

Figure 3. A single double-decker structure, steel truss.

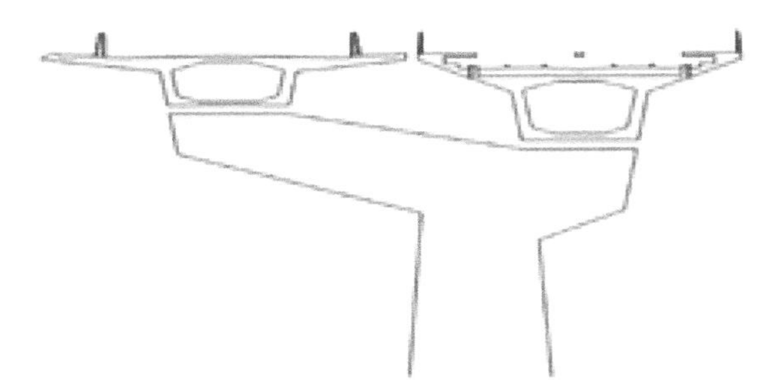

Figure 4. A single substructure with two separate rate bridge decks, concrete box girder.

The first concept had the two railway tracks, almost fully on the concrete box girder. With only a small off-set, Figure 5. The section of the bridge deck carrying the two-lane road on a cantilever, supported on either struts or concrete cross beams. Options with the cycle and pedestrian path placed on the outside of the cantilever, and below the bridge deck, were considered. A cable stay structure was decided for the navigational spans, to eliminate haunches – both for architectural reasons and to ensure the navigational clearance over the full width of the navigational spans, Figure 6. This introduced the requirement to place a pylon in the cross section.

The conclusion was a single bridge deck, with the railway slightly more off-set from the centre of the box girder, space for the pylon between the railway and the road and a kink in the bridge deck. The kink in the bridge deck is a direct result of geometric constraints (height of derailment barrier, horizontal emergency walkway and height of parapet plinth above road surface).

Many other architectural aspects have been considered, such as pier shapes, allowable geometric variations and ratios – all are detailed in the Definitions Design. The Definition Design is a separate document in the tender documents, giving all the geometric parameters the Design & Build Contractor has to comply with. These are given as maximum-, minimum-, variable- and in

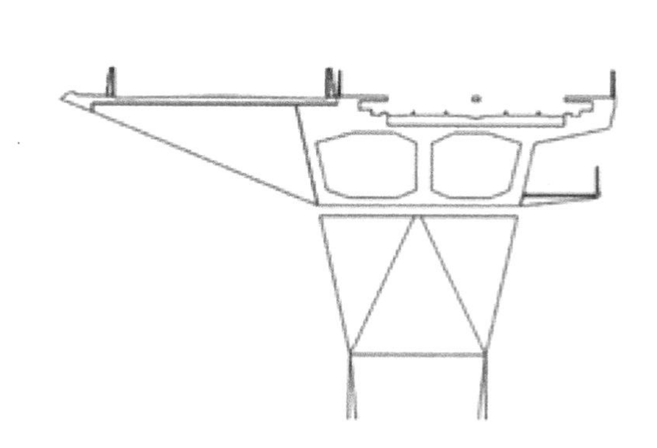

Figure 5. A single bridge deck, concrete box girder. The road section of the bridge deck on a cantilever. Cycle- and pedestrian path placed below the bridge deck.

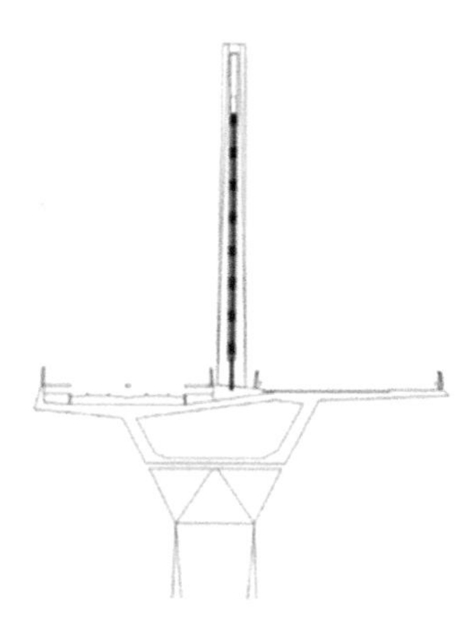

Figure 6. A single bridge deck, concrete box girder. Cable Stay Solution.

Figure 7. Rendering of the new cross-section

a few cases fixed values, as well as ratios. The concept should ensure that any given combination of the allowable values will result in a beautiful design.

Looking back there seems to have been a natural development, where the end result should have been obvious from the beginning, an aesthetic design and an optimized cross-section. Figure 7 shows a rendering of the cross-section of the new bridge.

3 PROJECT SPECIFIC DESIGN BASIS

It was decided early on to write a project specific Design Basis, as was previously done for the Great Belt Bridge and the Øresund Bridge in Denmark. This was done to compile the design requirements that are not explicitly stated in the Eurocodes, design requirements, which originate from the Danish annexes to the Eurocode, requirements from Danish Design Guides (Vejregler), as well as project specific requirements (DBS 01 2017).

Table 4-1 *50 year ice thickness at the New Storstrøm Bridge in year 2022, 2082 and 2142.*

Scenario	2022 (Start working life)	2082 (Mid working life)	2142 (End working life)
Extrapolation of historical data (1870-2014)	55cm	52cm	48cm
RCP2.6	55cm	49cm	44cm
RCP8.5	50cm	34cm	21cm

Figure 8. Table from Design Criteria Background Note ["Future ice thickness" Doc No.93200-COW-NOT-4-DBS-00280. COWI 2017].

The Design Basis furthermore included the results of special investigations carried out for standard clause, the relevant departure from Standard, was enclosed in the Design Basis (DBS 01 2017). Also, through the development of the Design Basis, several specific design issues were investigated. Some of the exceptional loads, which were studied were Ship Impact and Ice Loads, as well as railway loads and temperature loads. As compliance with these load requirements all are associated with significant costs, these were carefully considered and questioned prior to incorporating in the Design Basis.

The Design Basis developed for the New Storstrøm Bridge, has already been tested through the competitive dialogue – but has also proved to be an excellent document to clarify the requirements.

3.1 *Ice Loads*

The thickness of the ice to be taken into account, has a significant impact on the design, and should be considered in detail for the individual project. The first estimate took the observed temperature increase between 1900 and 2014 into account, and the ice thickness at the location of the New Storstrøm Bridge, with a 50 year return period, was assessed to be 56 cm for the year 2015. Based on the Met Ocean Report for Storstrømmen.

However, the future temperature changes are expected to decrease the ice thickness at Storstrømmen during the service life of the bridge. A study was carried out in order to estimate the future ice thickness at Storstrømmen and investigate the design ice thickness from a reliability perspective. Three scenarios were investigated 1) based on historical data collected from 1870–014 with no climate change expected, 2) minimum – and 3) maximum expected change in global average surface temperatures, respectively, Figure 8.

The conservative approach would be to base the design on the minimum annual reliability and not include the favourable effects of future climate changes after 2022, but this comes at a significant cost. Application of the accumulated reliability approach requires some level of future risk mitigation to avoid reduction in human safety. However, since an ice cover of 34–52 cm does not come unwarned, there will be plenty of time to take appropriate action if necessary.

In the event of occurrence of the design ice load, if necessary train speed reductions can be enforced or preventative measures can be put into place.

Thus, it was decided to reduce the design ice thickness to 49 cm in the Design Basis, for this specific bridge, Figure 9.

3.2 *A new National Annex, DS/EN 1991-1-7 DK NA:2013*

The new Storstrøm Bridge is a railway bridge, which means it will have to obtain an APIS (authorisation for placing into service), in accordance with the European Commission Implementing

Table 6-8 *Ice properties corresponding to a return period of 50 years*

Ice Property	
Characteristic compressive strength, r_c [MPa]	1.90
Characteristic flexural strength, r_b [MPa]	0.50
Ice thickness, e [m] [1]	0.49

1) Ice thickness 0.49 m instead of 0.57 m based on specific Met-ocean report for Storstrømmen, ref. /DD9/ (document ID: 93200-COW-REP-3-HYD-00042) and dispensation application ID1133.

Figure 9. DBS 01 Table 6-8 Ice Properties [Danish Road Directorate. Tender Document 93200.001 May 2017].

Regulation (EU) No. 402/2013 of 30 April 2013 on the common safety method for risk evaluation. This includes a systematic identification of all hazards of relevance during the operation of the railway, including events like ship impact.

In accordance with EU 402/2013, the risk acceptability of each hazard shall be evaluated by using the three risk acceptance principles (in the following order of preference) until an appropriate principle is found for the individual hazard being evaluated:

1. The application of Codes of Practice
2. Comparison with similar systems (ref. system)
3. Explicit risk estimation

It was considered very unclear how the Contractor's Designer, would be able to document, to an acceptable level, how the risk acceptability for hazards such as ship impact were appropriate. As similar reference systems were hard to identify, and it was considered that an explicit risk estimation approach could be challenging and time consuming to reach agreement on during the Detailed Design phase, the preferred approach would be to apply a Codes of Practice.

However, it is not clearly defined in Eurocodes, if the general requirements for structural safety which are applicable to normally occurring loads, also applies to accidental loads in case no level of acceptable risk is given in national annex, ref EN 1991-1-7 §3.2 (1). The reason behind this could partly be the very rare occurrence of these accidental loads, and partly also the very high costs associated with the compliance of general requirements to structural safety which are applied to other loads.

As the Danish Road Directorate and Banedanmark (Railnet Denmark), already had done some work in relation to structural safety levels for existing bridges with regard to ship impact, it seemed a good opportunity to develop common guidelines for new bridges, as well as existing bridges. A task force was established, to develop a new National Annex.

The term accidental loads, is for the Danish National Annex to be understood as rare occurrences, such as ship collisions, vehicle impact and other accidents which might affect a bridge's structural safety, including explosions. However, fire and seismic events are not included in the scope as they are covered by separate codes.

The methodology adopted in the new National Annex, for the risk acceptance criteria is described, as a choice between two possible methods of risk evaluations:

Method I – is the use of the same requirements of structural safety levels as used for normally occurring loads, applied to accidental loads. Applying probability of structural failure of 10^{-7} per for CC3 structures.

Method II – is developed using a cost-benefit approach, where the acceptable risk level is seen in relation to the balance between the public gain by the risk reducing measures and the costs associated with this. The public gain is measured as the reduction of probability for structural collapse, combined with the consequences of a collapse primarily in terms of user safety and cost

Figure 10. Rendering of Cable Stayed Bridge & Navigational spans.

of disruption and repair of the new Storstrøm Bridge, Figure 10. This principle is also known as a variant of the ALARP principle (As Low As Reasonably Practicable), where the risk is reduced to the lowest reasonable level seen in relation to the associated cost implications.

Application of Method II is subject to limits on the risk of loss of human lives. Represented by minimum requirements of loss of human lives of 10^{-6} and 10^{-5}, for railway and road bridges respectively, as annual probabilities

It should also be noted that the application of Method II can necessitate implementation of mitigation measures or monitoring plans throughout the structures service life. The onerous will be on the owner of the structure to monitor and actively follow up on the effectiveness of the implemented systems.

In general, it is expected that using Method I will lead to a more expensive design, than Method II, which is considered reasonable, as by using Method II a more detailed analysis is carried out. An analysis which considers the public risks by the loss of the bridges and a detailed analysis of possible risk reducing measures.

In general, for large structures, a zero risk level, is impracticable and in nearly all cases it is necessary to accept a certain level of risk.

In January 2016 a new Danish National Annex, DS/EN 1991-1-7 DK NA:2013, was published, together with a background note on risk acceptance criteria. This was a direct result of the work done for the Design Basis for the new Storstrøm Bridge.

4 TENDER PROCESS

The tender process began in September 2014, where an information meeting was held in Copenhagen. Following this a technical dialogue was carried out, prior to the prequalification of 5 international consortia's in 2015.

The Danish Road Directorate decided to carry out a Competitive Dialogue process for the Tender process.

This meant that a 'draft' version of the complete tender material was published on the 14th July 2016, and an orienteering meeting and site visit was held in September, prior to the first round of dialogue meetings.

In all, three rounds of dialogue meetings were held over the course of 6 months. The first dialogue meeting was a two-day meeting, the second and one-day meeting and the final round was only a half-day.

The overall purpose of the dialogue meetings was to obtain the Tenderer's assessments and observations of the Tender material in order to;

- Be sure that the project remained within budget
- Focus on "cost drivers" to avoid expensive solutions
- Optimize the tender documents
- Achieve a common understanding of the tender documents and the contract conditions

Following the first round of dialogue meetings, the tender material was updated, to include the changes, which were a result of the meetings. However, the final Tender documents were only published at the end of May 2017 after the third round. These included significant changes, which were a direct result of the dialogue process. Following this year long process, the tenderers submitted their bids in July 2017.

5 CONCLUSIONS

The work carried out over the past 6 years, has proven that for these Mega Structures, it is important to carefully consider the design requirements. Compliance with some of the requirements can be associated with very significant costs. Hence it is important to take an informed decision on if the requirements are realistic, practical and applicable to the structure in question, prior to incorporating them in the Design Basis.

Also, very valuable work and investigations are often carried out behind the scenes in these large infrastructure projects, which leads to new knowledge or new design approaches. These should not be 'forgotten' upon completion of the project. We should carry the results forward and make sure it is used more widely and implemented in codes of practise.

The Design & Build Contract was signed on the 26th February 2018 and the design of the bridge is well underway. The Contractor has chosen to construct the bridge from mainly large precast concrete elements, including large foundation caissons, smaller pier segments and full span girders. The coming three years will bring interesting times, and by the end of 2022, the new Storstrøm Bridge will rise out of the sea.

ACKNOWLEDGEMENTS

The authors would like to acknowledge the efforts of experts, engineers and staff from many organizations who have been involved in the development of the design and specifications for the Storstrøms Bridge, including the new National Annex for Accidental Loads.

We finally thank our own team at the Danish Road Directorate for their support, commitment and dedication to achieve the common goal of this project.

REFERENCES

93200-COW-REP-3-HYD-00042 Met-ocean report for Storstrømmen. (COWI, 2015)

93200-COW-NOT-4-DBS-00280 STORSTRØMSBROEN, Future Ice Thickness, Design Criteria Background Note (COWI, 2017)

DBS 01 – Design Basis – Bridge. Storstrømsbroen. Entreprise 93200.001. (Danish Road Directorate. Tender Document 93200.001 – May 2017)

DS/EN 1991-1-7 DK NA:2013 national annex, including Background note – Accidental Loads for new Bridges. Risk Acceptance Criteria. (January 2016).

European Commission Implementing Regulation (EU) No. 402/2013 on the common safety method for risk evaluation and assessment of 30 April 2013.

Chapter 3

Hanger cable fatigue life assessment of a major suspension bridge

S. Durukan
Accord Bridge Engineering, Istanbul, Turkey

S. Soyöz
Bogazici University, Istanbul, Turkey

ABSTRACT: The Bosphorus Bridge is an iconic suspension bridge located in the heart of Istanbul, Turkey serving as an extremely important link of the city transportation network with an average daily traffic of around 190,000 vehicle/day. Recently, the bridge has undergone a major retrofit work which also included replacement of the entire suspender system by changing the inclined orientation of the hangers to vertical. The study aims to evaluate the fatigue performance of the newly installed hanger cables under traffic loads. Three-dimensional finite element model of the bridge is established and the model is verified by using the ambient vibration measurements taken from the bridge. A realistic traffic time history loading scenario is generated. The resulting stress time histories of the hanger cables are obtained from the analytical model. Fatigue analysis and life prediction of hanger cables under variable amplitude stress history due to traffic loading is discussed.

1 INTRODUCTION

The Bosphorus Bridge is an iconic suspension bridge with a main span of 1074 m, Figure 1. Constructed in 1973, the bridge is located in the heart of Istanbul, Turkey. It is the first suspension bridge built in the country, which serves as a vital link of the city transportation network connecting Europe and Asia over the Bosphorus Strait with an average daily traffic around 190,000 vehicle/day.

The design by Freeman Fox & Partners is one of the first examples of a suspension bridge with a streamlined box girder deck instead of a stiffening truss which provides significant contribution to aerodynamic performance and at the same time employs structural efficiency and aesthetic beauty by reducing the girder depth and hence self-weight greatly. Another very characteristic feature of the bridge was the inverted-V shape orientation of the hangers in the original design. It is understood that this inclined layout of the hangers is selected in order to limit the longitudinal displacements and benefit from the inherent damping of the cables (Flint 1992). However, this design has a major drawback which is the hanger experiencing higher stress amplitudes under traffic loads. Homberg criticized the use of this orientation by noting that inclined hangers can have an early fatigue

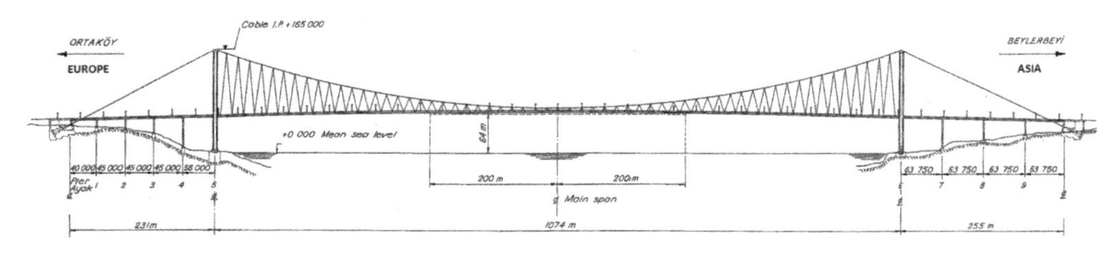

Figure 1. Bosphorus bridge elevation view – original hanger arrangement.

problem due to the high amplitude stress reversals (Homberg 1982). Severn Bridge constructed in 1966 and Humber Bridge constructed in 1982 have very similar designs to Bosphorus Bridge and share the same inclined hanger system. In both of these bridges, hangers are replaced short time after service due to fatigue cracking of wires.

In 2004, a hanger plate failed at the Bosphorus Bridge due to by fatigue. This increased the need for examining the bridge closely and after an extensive inspection period, Turkish General Directorate of Highways decided to rehabilitate the bridge. The retrofit included the replacement of hangers with vertical ones. The construction works finished and bridge is under service with the new vertical hangers since 2016.

The background information given above brings us to the topic of this study. In any cable supported bridge, the cables are susceptible to fatigue failure due to cyclic traffic loads and wind induced vibrations in combination with the corrosive environment. In this study, the hanger cables will be the primary focus since they are one of the most vulnerable elements of the structure. Cable wire breaks are very common in suspension bridge hangers and usually replacement of hangers is an inevitable activity within the service life. Since suspension bridges are naturally the most important link in a transportation network like the Bosphorus Bridge being a vital crossing for the Istanbul city, bridge operators need to be aware of this situation and plan the inspection and retrofit campaigns accordingly.

In this view, fatigue performance assessment of the newly installed hangers on the bridge is intended by simulating the traffic flow. Having suffered from the same problem in the past, this study aims to provide an idea about the fatigue performance of the new hangers in use and provide an insight about the future plans for inspection and cable replacement activities.

2 FE MODELING

2.1 *Modeling approach*

In order to capture the realistic structural behavior of the bridge and perform a condition assessment for fatigue on hangers, a precise computer model is generated. The FE model of the bridge is developed using the commercial CSI Bridge Software Version 20 based on design drawings, previously published papers and site observations. In Figure 2 below, the elevation view of the generated global 3D model of the bridge is presented. The same model is used for static and dynamic analysis.

The side spans are not modeled because they are not suspended but rest on vertical piers and are also completely separated from the main span deck. Hence, modeling side span is not likely to contribute to the structural behavior of the main span. It may affect the tower dynamic behavior due to added mass but this lies beyond the scope of this study.

Towers are modeled using frame elements. Steel box cross section is defined in the analysis software including the longitudinal bulb stiffeners. This was done by generating a CAD drawing of the section based on the as-built drawings and then importing the CAD drawing as a section file in CSI Bridge.

A spine model approach is preferred for deck modeling consisting of longitudinal frame elements representing the box girder and transverse weightless rigid link elements for load transfer to cable connections. Since the mass and stiffness of the deck has crucial importance for the behavior of the bridge, the deck is modeled precisely using the as-built drawings of the bridge. A CAD drawing

Figure 2. FE model elevation view – vertical hangers.

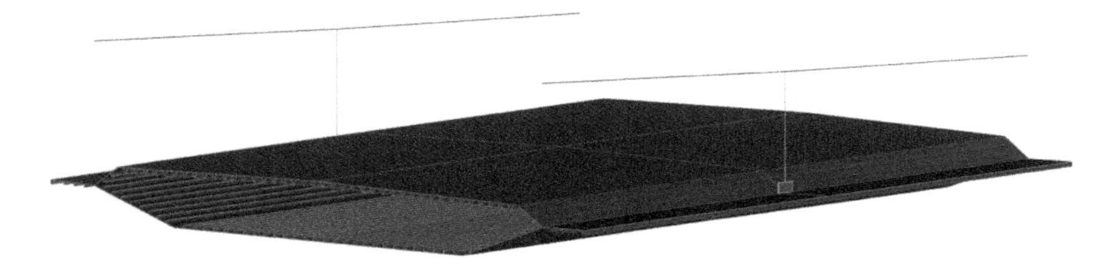

Figure 3. View of modeled deck section and hanger connection detail.

is prepared and imported to the CSI Bridge software with exact plate thicknesses and including longitudinal bulb stiffeners and V-ribs as the deck section. Hanger gusset plates are also modeled separately using frame elements with the intention of correctly defining the hanger lengths. In Figure 3, an assembly consisting of deck, hanger, main cable and gusset plate elements is shown.

The dead load on the bridge needs to be carefully determined for a suspension bridge model. This is important not just for getting the correct static results but also for obtaining the right stiffness for the modal and live load analysis that will be performed afterwards. The dead load of the structural components is calculated automatically by the analysis software using defined sections and material unit weights. The additional loads of pavement, auxiliary members (hand ropes, barriers, services etc.) are also taken into account according to the as-built drawings and the published design paper (Brown & Parsons 1976). The effect of these additional loads is included in the analysis by applying a modification factor to the deck section unit weight. The total distributed dead load for the suspended deck calculated as 110 kN/m.

Hangers and main cable elements are modeled using the cable object, which is a nonlinear element to capture tension stiffening and large deflection effects. The cable element is defined as a tension only member with no flexural stiffness. There are 30 hangers of different length in each quarter of the bridge which makes in total 120 hangers. Actually, at each hanger location, a pair of hanger cables exists on the bridge but they are modeled as one element having the total cross section area for convenience. This is a very reasonable assumption since the two cables share the same connection member at both ends and they are very closely spaced so that the loads will be shared equally. Main cable nodes are taken from the as-built drawings of the bridge which means they are defined at the deformed state of the bridge under dead load. This approach will require incorporation of initial tension at those elements in order to have zero deflection under dead load when the analysis is performed.

The young modulus used in the analysis will also have a significant effect on the force – displacement behavior especially when cables are considered. For the main cable, a Young's modulus of 193 GPa is used whereas for hangers, a modulus of 165 GPa is taken according to industry standards for corresponding cable types.

2.2 *Initial tension and shape finding*

Unlike any other structure, precisely modeling the geometry, section properties and boundary conditions is not sufficient to establish a robust structural analysis model for a cable supported bridge. If one performs a static analysis at this step, the deformation of the bridge will be unrealistically high due to very low stiffness. In our case, the midspan vertical deflection of the deck went beyond 9 meters which is not reasonable.

In a suspension bridge, the main source of stiffness is the tension in the cables. Therefore, before any analysis, this tension has to be present in the cables at the dead load state of the structure. When dealing with an existing bridge, the initial geometry of the cables is not known. The bridge is modeled based on the completed state which is actual the deformed shape under dead and superimposed dead load. Therefore, the cable tensions need to be entered so that when a static

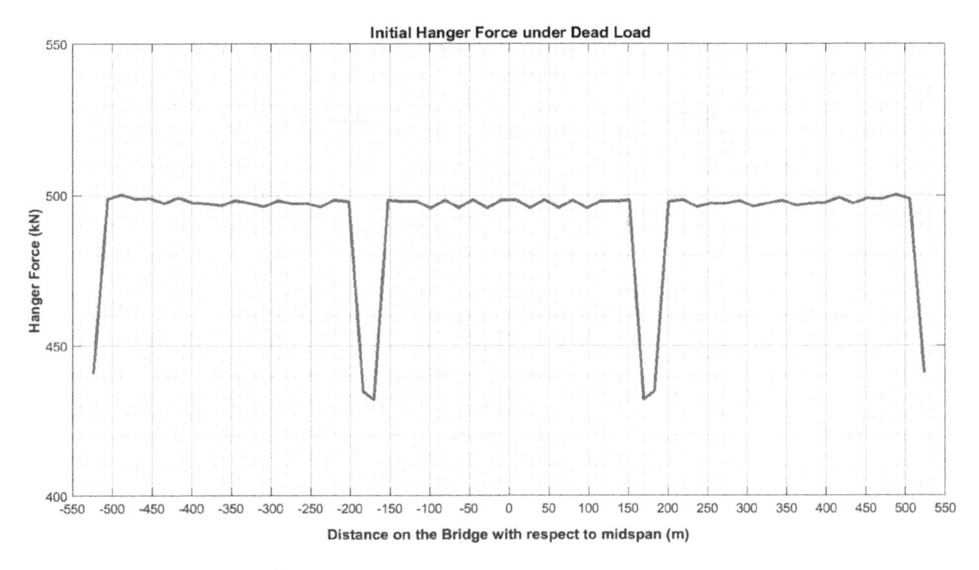

Figure 4. Hanger tension at dead load state.

analysis is performed under dead loads, the structure should be at equilibrium with no deflection meaning preserving the final geometry (Ren et al. 2004). Such process is also called shape finding.

Shape finding can be done by performing a nonlinear static analysis and using the resulting cable tensions as a pre-strain input for cables and then repeating the analysis again. This trial and error process lasts until a minimal deflection of the deck is obtained. In this study, a feature of the analysis software called CSI Load Optimizer is utilized. A key point while defining the target criteria is the rocker links (vertical deck support frames located at towers) which have to be in compression under dead load state of the bridge. If a shape finding run is performed to achieve for zero deflection of the deck only, the rocker links will take no load. Therefore, a separate criterion needs to defined as a minimal axial compression value for these members. Based on previous experience and engineering judgement, a value of 300 kN is adopted in this study. Once the target displacement criteria are defined at the selected nodes, which in this case are the deck nodes at hanger locations, the program adjusts the cable tensions automatically under the dead load case. The deck deflections obtained after this analysis are around 3 cm, which can be considered as negligible.

The dead load tension values are obtained in this way for each hanger cable (one of each pair at one hanger location) and are presented in Figure 4 below. As can be seen from the figure, the tension values are around 500 kN for each hanger. The first hanger has a lower tension due to the load sharing of the rigid vertical support provided by the tower and the distance to the tower is shorter than the regular hanger spacing of 17.9 m. Also, hanger numbers 20 and 21 have lower tension because at that location, the hanger spacing is 13 meters.

2.3 *Modal analysis*

After static equilibrium is reached under dead loads and initial strains are defined at cable elements, the model has the correct stiffness and is ready for further analysis. Modal analysis needs to be performed using this stiffness as initial state since otherwise the cables would be unstable and insignificant mode shapes and frequencies could be obtained. In Figure 5 below, the main lateral and vertical mode shape and frequencies of the deck are shown.

The frequencies obtained are in agreement with the results of field vibration measurements performed on the bridge after the vertical hangers are installed (Soyoz et al. 2017).

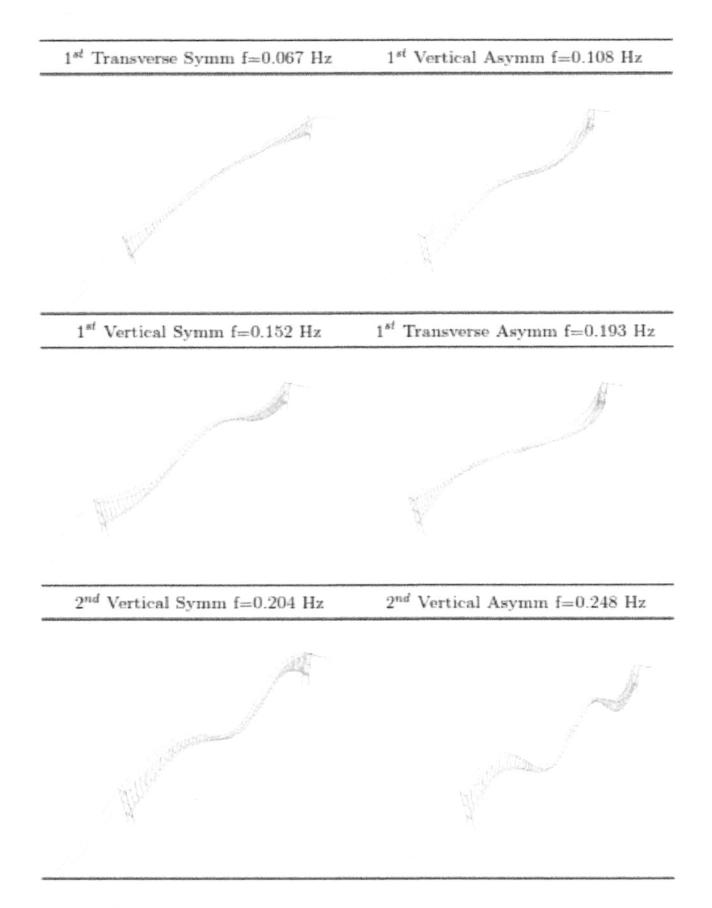

Figure 5. Mode shape and frequencies.

3 TRAFFIC SIMULATION

Since we are trying to perform a condition assessment, realistic traffic loading has to be used instead of code-based load models. WIM systems installed on bridges provide extremely valuable information about vehicle speed, arrival time, number of vehicles and weight of vehicles. Such information will ease the process of developing a realistic traffic flow model for performing analytical condition assessments. However, such system is not present on the Bosphorus Bridge. Under this circumstance, a traffic model is obtained based on the available data.

Currently, daily traffic flow over the bridge is around 190,000 vehicle/day. When the second bridge on Bosphorus, namely Fatih Sultan Mehmet Bridge, was completed in 1989, the local authority (KGM) decided to ban truck passage on the Bosphorus Bridge as a preventive measure. Therefore, only passenger cars, public transportation buses and coaches are using the bridge. In Figure 6, the deck cross section and lane configuration is shown.

The daily flow of these buses is obtained from the municipality database. The typical articulated bus using the bridge and corresponding axle loads are presented in Figure 7. This vehicle is expected to create the major part of stress ranges with the total axle load of 365 kN. A convoy is generated based on the percentage of buses using the bridge over total traffic flow. The mean speed of the flow is taken as 40 km/hr in the analysis.

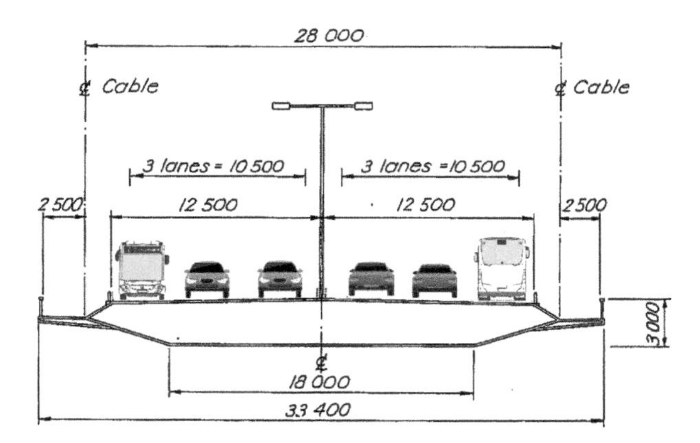

Figure 6. Deck cross section and lane configuration.

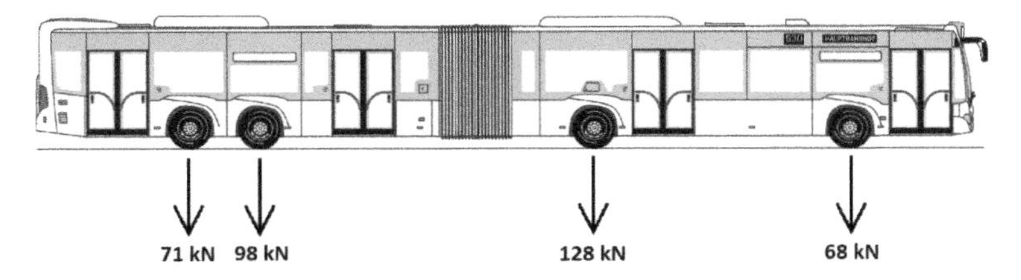

Figure 7. Articulated bus axle loads.

Apart from this, standard 20 t coaches and 1.5 t passenger cars are included in the convoy. By analyzing the daily trip numbers provided by municipality, it is found that around 10% of the total traffic flow consists of buses in the outer lanes. The artificially generated convoy is then marched over the bridge and 10 min of traffic flow is generated in the analysis software. The resulting stress time histories of the hangers are recorded for fatigue analysis.

4 FATIGUE ANALYSIS

A fatigue assessment is aimed at hanger cables under stress fluctuations caused by traffic loading of 10 minutes. Based on the traffic model established, a time history analysis is performed to directly obtain stress time histories at hangers. Time step used in the analysis is selected as 0.2 second. Many of the past studies used influence lines to perform this analysis but, in this study, we benefited from the analysis software capability of creating a time history live load case automatically based on the convoy defined and results are obtained by direct integration. This way, dynamic effects due to vehicular loads will be taken into account as well. However, in a long span suspension bridge like this, significant dynamic amplification is not expected because the bridge has vertical frequency around 0.1 Hz as shown in Section 3 whereas vehicle frequencies generally are in the order of 2–5 Hz.

Making use of a Rainflow counting algorithm, stress reversal ranges and corresponding cycle counts are extracted from the time history results. An example of a stress time history and corresponding rainflow counts are presented in Figure 8 for hanger number 1. Hangers are numbered from 1 to 30 where 1 is the longest hanger located next to the tower and 30 is the shortest hanger at midspan.

A temporary monitoring campaign was held on hanger cables and it provided the acceleration time history data for six months (data taken from the retrofit Contractor). Using the taut string

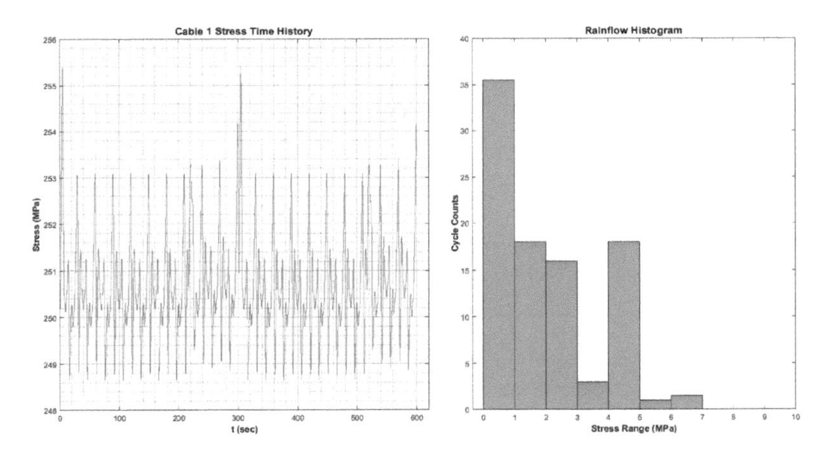

Figure 8. Simulated stress time history and cycle counts for longest hanger.

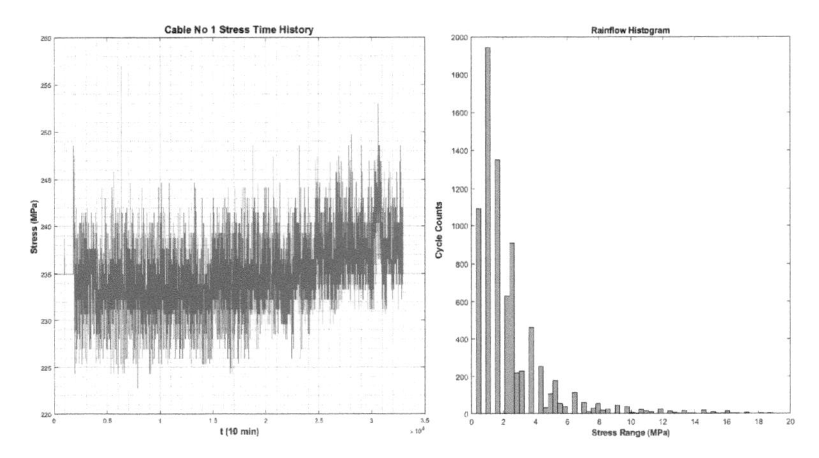

Figure 9. Measured stress time history and cycle counts for longest hanger.

theory, the tension in the cable is obtained based on the natural frequency and, the stress signal of the cable is artificially obtained. This data is valuable in the context of verifying the analysis results. As can be seen from Figure 9 below, the mean and fluctuating stress levels are in good agreement with the simulation results. Both results show the stress reversals are below 10 MPa and the mean stress on cable is around 235–250 MPa. However, acceleration data can not be used for direct fatigue analysis because the calculation of frequency has to be done over a time period. In this case, the monitoring system was reporting frequencies for every 10 min window.

After obtaining the stress cycles for every hanger, Miner's cumulative fatigue damage model is used to evaluate damage level. This method assumes that every stress cycle consumes a fraction of the fatigue life of a structure. This relation is shown in the equation below. Here, n_i is the number of cycles accumulated at stress S_i and N_i is the required number of cycles to fatigue failure.

$$D = \sum_{i=1}^{k} \frac{n_i}{N_i} \tag{1}$$

To calculate N_i, a fatigue life model needs to be used. In this study, we will focus only on axial fatigue of the cable and ignore bending effects. This is a reasonable approach since the hanger pin

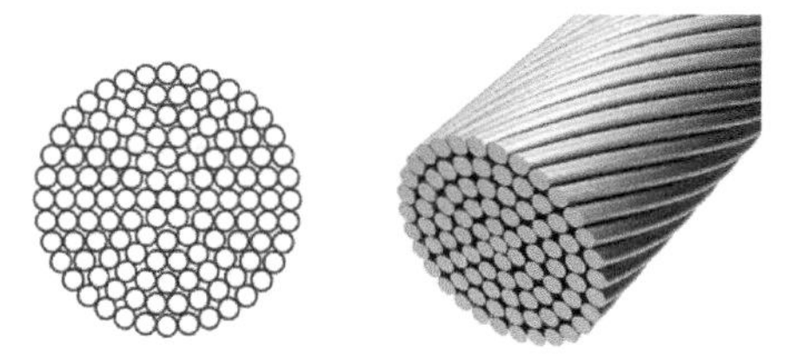

Figure 10. Open spiral strand view (taken from redaelli catalogue).

bushings on the bridge are equipped with spherical bearings which will help to prevent bending effects at cable terminations. In axial fatigue mechanism, fretting between the wires becomes significant in structural strands for fatigue life therefore structural configuration (number of wires, lay angle, shape of wires) of the strand matters. Corrosion of wires also triggers early fatigue problems and wire breaks in strands. The cable type for the new hangers installed on the bridge is 55 mm diameter open spiral strand which is made up of 127 helically laid individual hot dip galvanized high strength round wires. A sample view of the strand is given in , below.

The fatigue life model for strands is not analytically defined yet therefore it can only be determined through laboratory testing. At this point, we will benefit from bilinear S-N curves drawn in the light of laboratory testing. The S-N curve proposed by Nakamura will be used in the analysis. This curve is based on many tests performed on open spiral strands with different diameters. The S-N curve is represented by Equation 2 below (Nakamura 1987). The failure criteria of the S-N curves may differ from study to study and naturally has a significant effect on the calculated fatigue life. In this case, the failure criterion is first wire breakage in the strand. It is meaningful to note that this behavior model considers axial fretting fatigue but not the effect of corrosion. Unlike normal steel structures, there is no cut-off limit for strands that we can talk about below which no fatigue takes place (Cluni et al. 2007).

$$\log N = 3.0 - 3\log\Delta s \tag{2}$$

The term Δs in Equation 2 represents the stress range divided by the ultimate strength of the strand. The mean stress on the strand is another aspect which has an influence on fatigue damage and is taken into account by modifying the stress range with the Goodman relation given in Equation 3 (Alani & Raoof 1996).

$$\Delta s = \frac{\Delta\sigma/\sigma_u}{1 - \frac{\sigma_m}{\sigma_u}} \tag{3}$$

Here, the ultimate tensile strength, σ_u is taken as 1550 MPa and σ_m is the mean stress which varies between 240–280 MPa depending on the hanger location. The term $\Delta\sigma$ represents the stress range obtained from the stress time history record. The average value of the stress reversals is observed as 13 MPa and the peak value is 33 MPa.

The cycles and corresponding stress ranges obtained in the analysis of 10-minute traffic flow is used to calculate fatigue damage by using S-N curve equation and Miner's cumulative damage theory as described above. The calculated 10-minute damage is then extrapolated to get fatigue life, T in years with the equation below.

$$T = \frac{1}{D \times 6 \times 24 \times 365} \tag{4}$$

The calculated fatigue life of the hangers varies between 200 years to 10000 years as shown in Figure 11. The first observation made from examining the distribution of fatigue life among

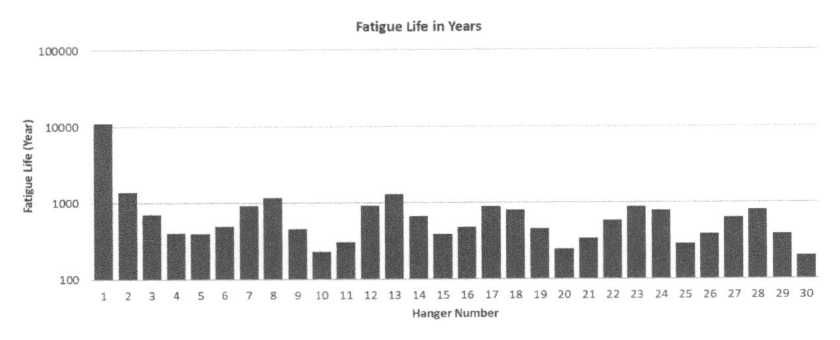

Figure 11. Fatigue life calculated for hangers based on traffic simulation.

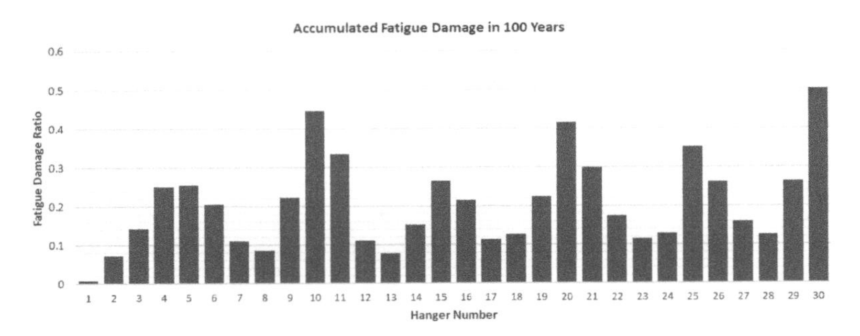

Figure 12. Fatigue damage ratio for hangers accumulated in 100 years of traffic simulation.

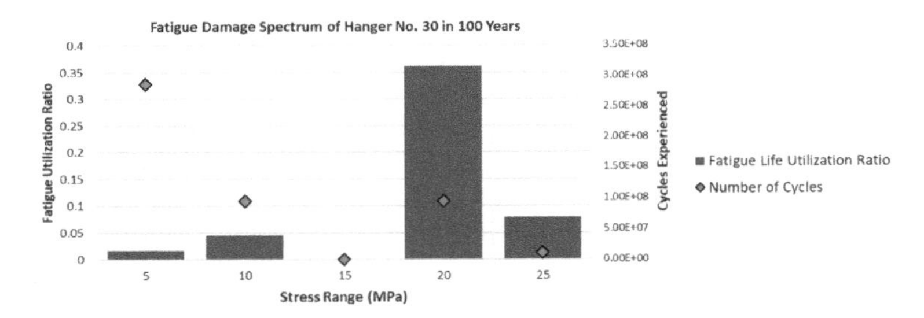

Figure 13. Damage spectrum of hanger no 30 and corresponding cycle counts.

hangers is that the hanger next to the tower has a much longer life than other hangers. This can be explained with the contribution of the rocker link support near this cable which can take up a large portion of the live load effects.

In Figure 12, the accumulated fatigue damage on the cables in a potential 100-year service life is shown. It is another way of interpreting the outcomes. This approach obviously assumes that current modeled traffic conditions will remain constant for the entire period considered. When the fatigue damage/capacity ratio (utilization ratio) shown in the figure reaches unity, it means that failure occurs. It is seen from the figure that the highest damage is experienced by the shortest midspan hanger with a utilization ratio around 0.5. It is necessary to reemphasize that failure corresponds to the first wire breakage on the strand as defined in the S-N curve adopted.

If the damage spectrum is highlighted for a hanger, like hanger 30 as shown in Figure 13, one can see each stress range contribution to damage. As displayed above, the damage utilization ratio in 100 years for this specific cable was 0.5. From examination of Figure 13, it is clear that stress ranges less than 10 MPa has a small effect on total damage whereas stress range of 20 MPa constitutes 70% of total damage occurred which shows that heavy vehicle passage is much more critical even though it happens less frequently.

5 CONCLUSIONS

Throughout the paper, a methodology is proposed to perform a fatigue performance assessment for cables of an existing bridge based on simulated traffic flow. The traffic flow scenario is developed based on publicly available traffic flow and vehicle data. Installation of WIM systems are recommended in terms of correctly modelling traffic flow. A precise FE model is developed which has paramount importance to capture structural response when dealing with a highly nonlinear cable supported structure. Dynamic properties of the structural model should be verified with field vibration measurements.

Fatigue damage on the new vertical hangers is calculated based on 10 min long traffic time history analysis. The results show that vertical hangers do not fully utilize their fatigue life in a 100-year period. The highest fatigue life is obtained at the longest hanger near the tower and on the other hand, the shortest life of 200 years is observed at midspan hanger.

Analysis of the results also reveals that stress fluctuations are in the range can go up to 33 MPa and most of the cycles are below 10 MPa. The stress cycles below 10 MPa contributes less to fatigue damage but the cables are found to be sensitive to stress fluctuations greater than 20 MPa. This shows the effect of heavy articulated buses using the bridge and provides a valuable information when planning for public transportation vehicle for future.

Testing a strand with same diameter and construction detail could increase the reliability of the damage calculations. Also, the influence of corrosion is not taken into account in the S-N curve adopted here but studies show that it could drastically reduce the fatigue life of the cables (Wu & Jiang 2016).

Monitoring the tension on the cables and storing the data with appropriate sampling frequency could enable a direct fatigue damage assessment based on stress time histories recorded.

REFERENCES

Alani, M., & Raoof, M. 1997. Effect of mean axial load on axial fatigue life of spiral strands. *International Journal of Fatigue,* 19(1), 1–11.

Brown, W. C., & Parsons, M. F. 1976. Part I: History of design. *Proceedings of the Institution of Civil Engineers*, 58, 505–567.

Cluni, F., Gusella, V., & Ubertini, F. 2007. A parametric investigation of wind-induced cable fatigue. *Engineering Structures*, 29(11), 3094–3105.

Deng, Y., Li, A., & Feng, D. 2018. Fatigue performance investigation for hangers of suspension bridges based on site-specific vehicle loads. *Structural Health Monitoring*, 1–15.

Flint, A. R. 1992. Strengthening and Refurbishment of the Severn Crossing. *Proceedings of the Institution of Civil Engineers – Civil Engineering,* 92(2), 57–65.

Homberg, H. 1982. The case against Inclined Hangers. *New Civil Engineer*,12–14.

Kim, B. H., Park, T., Shin, H., & Yoon, T. Y. 2007. A comparative study of the tension estimation methods for cable supported bridges. *International Journal of Steel Structures*, 7, 77–84.

Nakamura, S.-I. 1986. Fatigue Analysis of Spiral Strands of Cable Supported Bridges (Ph.D. thesis). University of London.

Ren, W.-X., Blandford, G. E., & Harik, I. E. 2004. Roebling Suspension Bridge. I: Finite-Element Model and Free Vibration Response. *Journal of Bridge Engineering*, 9(2), 110–118.

Soyoz, S., Dikmen, U., Apaydin, N., Erdik, M. 2017. System identification of Bogazici suspension bridge during hanger replacement. Procedia Engineering, 199, 1026–1031.

Wu, C., & Jiang, C. 2016. Fatigue Behavior of Hangers on a Three Tower Suspension Bridge. *DEStech Transactions on Environment, Energy and Earth Science*.

Chapter 4

Optimizing main cable dehumidification systems

M.L. Bloomstine & J.F. Melén
COWI A/S, Denmark

ABSTRACT: Corrosion protection of main cables by dehumidification on suspension bridges has been applied for 21 years and is currently virtually a worldwide best practice. New suspension bridges are born with a main cable dehumidification system incorporated in the design and bridge owners around the world are retrofitting their existing bridges with dehumidification systems. While dehumidification is the optimal method to protect main cables from corrosion, the design of the system should be optimized and suited for each bridge. The purpose of the optimization is to achieve a reliable and durable system that provides the necessary corrosion protection at the lowest life cycle cost possible, i.e. the lowest sum of construction, operation and maintenance costs. Maintenance "friendliness" should also be incorporated in the design, e.g. all components that require maintenance should be easily accessible and easy to repair/replace. This paper presents recommendations for design optimization and optimization of systems in service, as well as recommendations for follow-up and inspection of systems in service. The main focus of the paper is retrofit design, but all the recommendations for design optimization can also be applied to systems for new bridges.

1 INTRODUCTION

Suspension bridge main cable corrosion is a well known and serious worldwide problem and virtually all pre-dehumidification suspension bridges suffer from this problem. Many suspension bridges have a reduced factor of safety due to main cable corrosion and in the worst cases it has been necessary to strengthen the main cables or even replace them, such as the Tancarville and Aquitaine Bridges in France. In some cases, even the entire bridge has been replaced.

For roughly 120 years main cables have been protected from corrosion by the "Roebling system", developed by John Roebling for the Brooklyn Bridge in New York back in the 1880s. This system is composed of galvanization on the cable wires, lead paste on the outer surface of the bundle of wires, galvanized wrapping wire and paint on the outer surface. Over the years, numerous variations of this system with different types of paste and surface paint have been applied, but it has not been possible to prevent moisture/water intrusion, which leads to corrosion. Even if the outer surface of the main cable is well maintained and there are no outer signs of corrosion, corrosion still occurs inside the main cables. To inspect the main cables, it has been necessary to perform intrusive in-depth inspections, which require difficult access, can disturb traffic on the bridge, and are very expensive and time consuming. It is necessary to remove the wrapping wire on several panels, pound in wedges to open the cables and closely inspect and record the corrosion stage of all the visible wires. These inspections provide very limited information, as only a small percentage of the panels are inspected, and the wedging only reveals a small percentage of the wires. There is always a substantial risk that the most corroded area of the main cables is not inspected and that the calculated safety factor based on the inspection results is too optimistic.

To minimize main cable corrosion and minimize the need for in-depth main cable inspections, main cable dehumidification was developed in the late 1990s, building on many years of dehumidification experience from steel bridge components, including box girders, main cable strands in anchor chambers and anchorage boxes on cable-stayed bridges (Bloomstine et al. 1999). Since then, much main cable dehumidification experience has been gained and the systems have been further optimized.

It has been questioned, as to whether the dry air from a dehumidification system penetrates all the voids between the wires of a parallel wire cable. In the authors' opinion this is an academic question, as dehumidification has been proven to be the most effective means to minimize further corrosion. As air can more easily penetrate the voids than water, it is generally not conceivable that water will be trapped in voids when dehumidification is applied. During flow testing on the Storda Bridge in Norway (Bloomstine and Melén 2017) it was seen that the pressurized air forced trapped water out of the main cable through a small defect in the surface paint system, where water was constantly pressed out during the pressure test, illustrating the effectiveness of the injected air to penetrate voids. Should there be some small pockets of trapped water, these would have to be very well sealed to prevent the pressurized dry air from entering and drying them out. In the case of cable wires that have been oiled, it is conceivable that congealed oil could seal some voids with pockets of water. This would not however cause corrosion, as no oxygen would be available for the corrosion process, due to the sealing. A main cable dehumidification system minimizes more water from entering the main cable and will still be effective in this theoretical situation.

This paper presents the authors' recommendations for optimizing main cable dehumidification systems, based on experience from research, worldwide design of many systems, as well as continuous follow-up and analysis of numerous systems in service. The main steps of the optimization process for a retrofit system are:

1. Review of existing bridge documents
2. Preparation of preliminary layout
3. Site inspection
4. Flow test
5. Conceptual design
6. Detailed design – use of optimal details

In addition to recommendations for optimization during design, recommendations for systems in service are also presented. These include possibilities for further optimization and follow-up including inspections.

2 PRE-DESIGN PHASES

2.1 *Review of existing bridge documents*

All relevant bridge documents should be collected and reviewed. The documentation should include as-built drawings and specifications for the main cable system, including cable bands, shrouds, saddles, anchorage chambers and splayed strands. Inspection, repair and retrofit documents should also be included in the review. All the relevant information from the review should be summarized and used as the basis for planning the site inspection, preparation of the preliminary layout and preparation of the conceptual design.

2.2 *Preparation of preliminary layout*

The layout of the main cable dehumidification system is the foundation for the entire system. It defines the dry air injection and exhaust points, the flow lengths, the location of the dehumidification plant and buffer chamber (cf. section 4.3) and the schematic route of the dry air piping to the injection points. An example of a layout for a main cable dehumidification system (Great Belt Bridge, Denmark) is shown in Figure 1.

2.3 *Site inspection*

The site inspection should be planned on the basis of the document review and the preliminary layout and include all relevant components of the bridge concerning the design of the main cable dehumidification system. The purpose of the inspection is to:

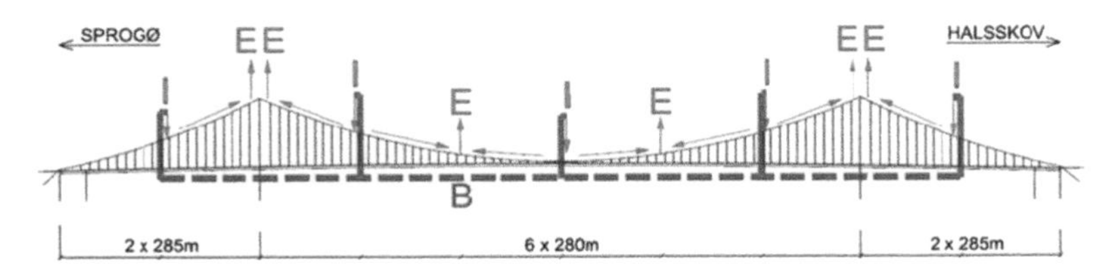

Figure 1. Layout for main cable dehumidification system, B=Buffer and plant, I = Injection, E = Exhaust.

- Confirm and/or update information from the document review
- Gain hands-on experience with the bridge and gain further relevant knowledge not obtainable from the existing documents
- Photograph and measure certain key areas, such as tower saddles
- Find/confirm a relevant location for the plant and buffer chamber – preferably part of an existing structure
- Find/confirm suitable routing for the dry air piping from the plant to the injection points
- Find any conditions that may cause an update of the preliminary layout

2.4 *Flow test*

The best practice for the longest viable flow length on main cables of parallel wires has earlier been considered to be approximately 200 m. Flow testing, as well as systems in service, have shown that the viable flow length can be as low as approximately 80 m and as high as approximately 300 m, hence a flow test is strongly recommended. The viable flow length depends on the condition of the main cable and if it has been oiled, which both can clog the small voids between the wires and increase the resistance, thereby reducing the viable flow length. Main cables with a small diameter are more sensitive to leakage than large diameter cables and may also have a shorter viable flow length.

For main cables made up of helical strands, the viable flow length is generally much longer, as there are fewer, but much larger voids and these give much less resistance. Flow tests have shown that lengths up to approximately 700 m are viable (cf. section 5.2). As suspension bridges with this type of main cable are generally relatively short, it should be viable with just one injection point at the middle of the main span, with air flowing in both directions to the respective anchorages. A flow test is however still recommended, as there could be some unknown blockage in the cables that could reduce the viable flow length. The testing also reveals leak sensitive areas, as described below.

Flow testing has previously been described in detail (Bloomstine & Melén, 2017), so it just briefly summarized here. A flow test is recommended to be carried out in connection with the design of a dehumidification system for the main cables on suspension bridges. The purpose of the flow test is twofold:

1. Establish the maximum viable flow length in order to develop the optimal layout of the system.
2. Establish where the main cable bands leak air in order to design necessary further sealing. Typical leaks are at the caulking of band joints, around the cable band bolt heads, nuts and washers, at drain holes and at other cable band details.

The test is generally carried out in the main span with an injection point near mid-span. This allows the easiest possible access and long lengths of main cables for flow testing. Further, this area is generally considered the most critical segment of the main cables with regards to corrosion and accumulation of water and oil in the case of oiled cables. At the injection point the wrapping wire is removed over a length of approximately 1–2 m, depending on the diameter of the main cable. Before cutting and removing the wrapping wire, it is secured at each end with banding. The exposed surface of the main cable is then thoroughly cleaned, removing all the paste on the surface to allow the injection air to enter the main cable.

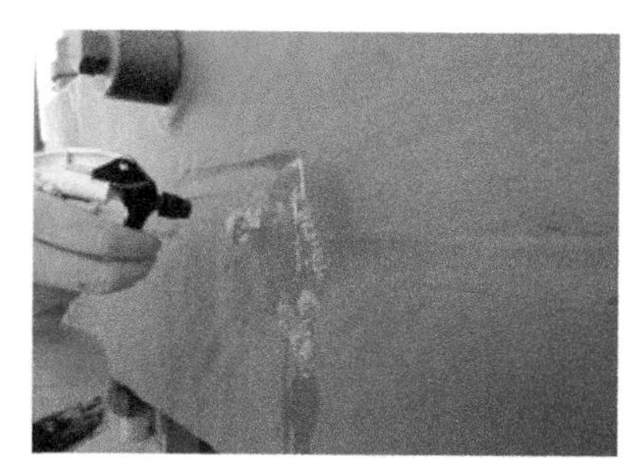

Figure 2. Great Belt Bridge, Denmark, equipment setup for flow test and leakage at longitudinal joint.

A temporary injection sleeve is then erected over the open area. The remaining equipment for the flow test is composed of a fan, a hose connecting the sleeve and the fan, instruments for measuring pressure and flow and a generator for power supply, see Figure 2. The flow testing is generally carried out in the normal operating range of pressure, dependent on the type of main cable, but generally between approximately 1000 and 3000 Pa. When the equipment is installed and tested, testing with overpressure commences. The initial activity is to discover where there are leaks at the cable bands, so that appropriate sealing can be designed in connection with a dehumidification project. Soapy water is sprayed on all possible areas of leakage and bubbles occur where there is leakage. On most suspension bridges without main cable dehumidification there is a drain hole at the lower end of the cable band. This is generally the most consistent indicator of leakage/airflow for the testing and is usually used to indicate how far the air can flow in the cable. The result will always be somewhat conservative, as the air leakage will be significantly less when the cable and bands have been sealed in connection with the dehumidification works.

3 CONCEPTUAL DESIGN

The conceptual design should be developed on the basis of the earlier completed pre-design activities, i.e. the document review, the preliminary layout, the site inspection and the flow test and these results should all be presented in a conceptual design report. If the flow test has shown different viable flow lengths than assumed in the preliminary layout, the layout my need updating. Further, the conceptual design report should include the following:

- The recommended layout
- The concepts for sealing the main cables, including panels, cable bands, saddles and shrouds
- The location of the buffer chamber and plant
- The concept for the plant
- Integration of the splay chambers (where the cable strands are exposed and anchored)
- The routing of the dry air piping from the plant to the injection points
- A preliminary estimate for the cost of construction
- A preliminary schedule for the works
- A recommendation regarding other works that could be beneficial to include in the same contract

4 DETAILED DESIGN – USE OF OPTIMAL DETAILS

4.1 *Coordination with other works*

Based on the recommendations in the conceptual design, other works that would be beneficial to include in the same project/works should be agreed upon with the bridge owner and incorporated in the project. As the costs of access and traffic regulation are substantial when working on the

Figure 3. Applying elastomeric wrap with wrapping machine and bonding the wrap with heat blankets.

main cables, all other works that can utilize the same access and/or traffic regulation should be considered. This can give substantial savings as well as reduce the accumulated impact on the traffic. As a minimum the following works should be considered:

- Surface treatment of cable bands
- Cable band bolts: tension testing, re-tensioning, replacement
- Hand ropes and stanchions: surface treatment, repair or replacement
- Electrical conduits and lighting along the main cables: repair, replacement or upgrade
- Suspender ropes: surface treatment or replacement

4.2 *Sealing of main cable panels*

The most common means of sealing the main cables is by applying an elastomeric wrap called Cableguard™ Wrap System from D. S. Brown. This material was first utilized in a main cable dehumidification project on the Little Belt Suspension Bridge in Denmark in 2003. COWI carried out research in 1999–2001 for the owner (Bloomstine & Thomsen, 2004), which led to the choice of this material.

The elastomeric wrap has a thickness of 1.1 mm and a width of 150 mm to 300 mm, dependent on the cable diameter. It is applied with slightly more than 50% overlap, so the total thickness is 2.2 mm. It is applied under tension with a special wrapping machine. The wrap is heat bonded with a special heat blanket, which bonds the two layers and shrinks the material slightly, giving an even tighter fit. The wrapping and bonding work is illustrated in Figure 3. The wrap on the Little Belt Bridge was recently inspected and after 15+ years of service the condition is still excellent, with no signs of wear or deterioration. Data from the monitoring system also confirms that no leakage has developed during this time span. Hence this is a very durable material with a lifetime that is expected to be 40 years or more.

This system requires a transitional detail where it meets the cable band at each end of a panel. In the original design, which was applied on the Little Belt Bridge and several more bridges, this detail includes a neoprene wedge that is installed over the wrap at the band. The groove in the cable bands is first filled with caulk and the wedge is then pressed up against the band. Supplemental caulking is also applied in order to fill out the area between the neoprene wedge and the cable band. After caulking, a finish strip of Cableguard™ wrap is installed on top of the neoprene wedge, see Figure 4.

Over the years, this detail has proven to be problematic and has the flowing disadvantages:

- The caulk is in a closed area and does not fully cure, as it needs air to cure.
- It is not possible to see the caulk and how well the area is filled out with caulk.
- There are often leaks and it is not possible to locate and repair these. Sometimes these leaks cause swelling up of the finishing strip, see Figure 5.
- In some cases, water gets trapped under the finishing strip.

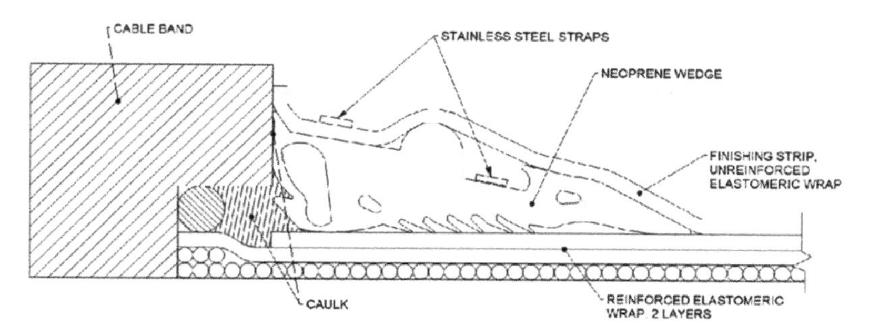

Figure 4. Original transitional detail at cable band.

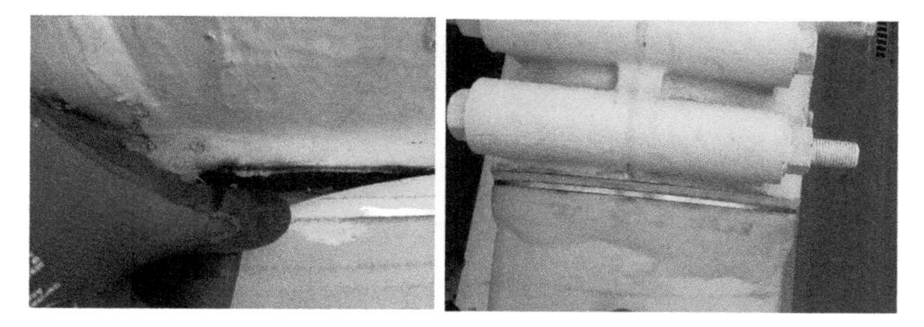

Figure 5. Corrosion on cable band at interface and bulge in finishing strip from air leak.

- Water/moisture gets trapped in the interface between the wedge and the cable band and initiates corrosion and this area is not accessible to paint repair, see Figure 5.
- It is not practical to maintain or replace the caulk at the end of its lifetime.
- It is not possible to place the normal wrap as deep into the cable band groove, as is possible with a starter strip in the optimal solution described below.

Due to these problems COWI developed an improved detail, which solves all the above-mentioned problems and eliminates the use of the neoprene wedge. Instead of using a finish strip of Cableguard™ wrap, a start/end strip is applied, which extends deep into the groove of the cable band. Normal Cableguard™ wrap is installed over the start strip, starting at the edge of the cable band. After heat bonding all three layers of the wrap, the caulk is installed in the groove with an overlap on the Cableguard™ wrap outside the groove, see Figure 6. This detail has been incorporated in the design of the dehumidification system for Hardanger Bridge, Norway in 2013 and the Great Belt Bridge, Denmark in 2015. The sealing has been 100% pressure tested on both bridges and no leaks were found. Great Belt Bridge has the longest flow lengths in the world (Nielsen and Bloomstine 2016) and Hardanger Bridge also has very long flow lengths. Despite these very long flow lengths there is very little leakage and data from the monitoring systems confirms that no leakage is developing. This detail is also applied to retrofit systems on the George Washington Bridge, NY (system under construction) and the A.L. Macdonald Bridge; Halifax Canada (system recently completed) as well two new bridges in Turkey and one new bridge in Norway.

4.3 *Buffer chamber solution*

The buffer chamber solution was first applied on the Humber Bridge, UK in 1995 (Bloomstine 2013), where there were problems with water intrusion in the tower saddles. The authority had earlier attempted to seal the saddles, but none of the attempts succeeded in keeping water out. The solution developed by COWI was to create an overpressure in the saddles by a constant flow of dry air. To achieve this, a small dehumidification plant was installed in one leg of each tower with

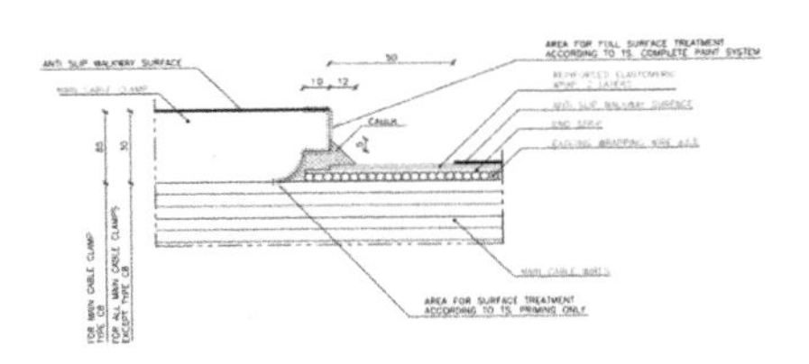

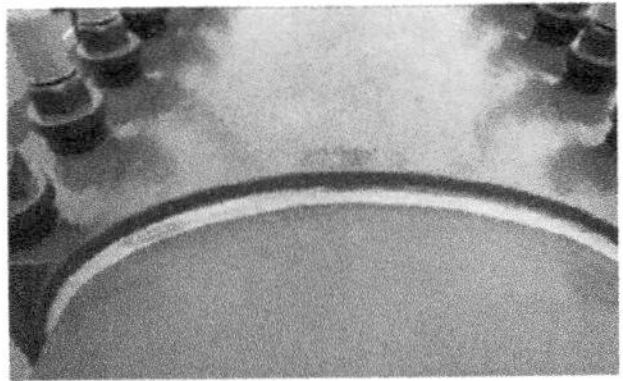

Figure 6. Optimal transitional detail, drawing and picture from Great Belt Bridge.

ducting to the respective two saddles. The dry air produced by a dehumidification unit has a relative humidity of just over 0%, which is much drier than necessary for corrosion protection. In order to conserve energy, the volume of the tower leg was utilized as a buffer chamber, where ambient air is mixed with air from the dehumidification unit to achieve a relative humidity of approximately 40%. This reduced the running time of the dehumidification unit by approximately 70–80% and saves a substantial amount of electrical energy. Further, there is also a significant reduction of wear on the dehumidification unit.

After this successful application, this solution has been successfully applied to all 17 main cable dehumidification projects prepared by COWI. The calculated ideal volume for a buffer chamber is approximately 10 times the volume of the hourly flow to the main cables, which accommodates the variation in ambient relative humidity that occur during a day-night 24-hour period. This theoretical value has been proven by practice on bridges where there has not been enough room for an ideal buffer chamber. For example, on the Högakusten Bridge in Sweden (Bloomstine and Gilliusson 2008), the Authority would not allow the volume of a tower leg to be utilized as a buffer chamber, hence only the volume of the upper cross beam of the tower was available. The volume of the cross beam is roughly 5% of the ideal volume. On the same bridge there is also a buffer chamber in the bridge box girder, where the ideal volume was available. The energy savings for the plant in the box girder are approximately 80%, whereas the energy savings for the plants in the tower cross beams are only approximately 10%. Substantial energy savings are possible for a buffer chamber with a volume slightly smaller than the ideal volume, but if the volume is significantly smaller, the savings are minimal. Placing the plant in a small room does not achieve the buffer chamber effect and does not provide any savings in electrical consumption or reduction of wear on the dehumidification unit. When the buffer chamber solution is applied with the correct volume, savings in electrical consumption in the order of magnitude of 100,000 to 200,000 kWh per year are possible, dependent on the size of the bridge and the total necessary flow of dry air.

As the dehumidification plant is placed in the buffer chamber, it is in a protected environment with dry air, which is advantageous with regards to the plant lifetime, see Figure 7. Further, the plant is also protected from other maintenance works that are regularly carried out and is only accessible by approved personnel.

The buffer chamber should also be placed in a convenient location for ease of access with regards to maintenance friendliness. This also minimizes time consumption and costs for traffic regulation and access equipment during maintenance works.

If possible, an existing bridge structure (as mentioned above for Högakusten Bridge) should be utilized for the buffer chamber in order to minimize construction costs. In the case of the Älvsborg Bridge, Sweden part of one anchor house was utilized as the buffer chamber, see Figure 7. The authority required that the deck and wall surfaces enclosed in the chamber be accessible by boom lift. The upper part of the walls was therefore designed as a steel frame covered by removeable fabric, allowing the required access.

4.4 *One plant solution contra a multiple plant solution*

In general, a dehumidification system for main cables should be designed with just one plant, which can provide sufficient dry air for the entire length of both main cables, as this is by far the most

Figure 7. Älvsborg Bridge, buffer chamber and enclosed plant.

economical solution. To provide redundancy, the plant can be designed with two dehumidification units and two fans, as these are the essential mechanical elements that are necessary for production and injection of dry air in the main cables at the target pressure. The plants are generally designed such that these units run alternately, so they are running regularly, which is essential to keep them functional. If one unit should break down, the other unit will automatically take over full time until repair work has been completed. An example of a one plant system is shown in figure 7. In this case the authority opted for one dehumidification unit and one fan, as regular maintenance is carried out.

This solution is maintenance friendly, as all the maintenance is carried out in one location and there is much less to maintain. If there are multiple plants, at least some of these will be located at areas, which are relatively difficult and/or time consuming to access. This solution can require somewhat longer air pipes, but this is a minor disadvantage when compared to the many benefits.

The one plant solution ensures a much lower life cycle cost than a multiple plant solution for the following reasons:

- One plant cost much less to construct
- One plant requires much less maintenance
- Ease of access and only one area to access saves time as well as possible traffic disruption
- One large plant consumes much less electrical energy

4.5 *Injection, exhaust and monitoring sleeves*

A basic design for all types of sleeves for dehumidification (injection, exhaust and monitoring sleeves) was developed for the dehumidification system on Little Belt Bridge in Denmark, which has been in service since 2003. The design has been somewhat refined from project to project, but the basic design is still the same, see figure 9. The main features are:

- Highly polished stainless steel – maintenance-free
- Lightweight for ergonomic installation
- Double sealing – inner closed cell compressible neoprene foam and outer durable caulk
- Flexible fit to the main cable, which can vary in diameter and is not perfectly round

There has been some discussion in the bridge design industry as to whether it is necessary to wedge the cables under the sleeves to facilitate the inflow at injection points and the outflow at exhaust points. On the Little Belt Bridge, which was the first main cable dehumidification outside Japan, some of outer helical strands were slightly spread and 2 mm thick zinc plates were inserted between these as an extra precaution (Bloomstine and Thomsen 2004). This is however, not nearly so intrusive as pounding wedges deeply into the main cables.

Since then, flow testing and many systems in service have proven that the wedging is not necessary. Abundant experience has shown that it is sufficient to thoroughly clean off the paste from the outer layers of wires (see Figure 8) in order to let the air in or out. Further, this has been proven through pressure measurement research performed in connection with the construction of the Hardanger Bridge (Bloomstine and Melén 2017). Pressure measurement at all the injection

Figure 8. Main cable cleaned and ready for injection sleeve, Storda Bridge, Norway.

points and at the corresponding adjacent cable band documented that the singular pressure loss was approximately 100 Pa when testing at high injection pressure during testing (2,500 Pa) and only 50 Pa at the final operating injection pressure (1,300 Pa). In addition to this documentation, the Honshu-Shikoku Bridge Authority in Japan has informed that none of the 17 main cable dehumidification systems in service in Japan utilize wedging, as their research and experience has also proven that wedging is not necessary.

When considering the tension in the cable wires the wedging exerts some extra tension, caused by displacing the wires. This can be especially critical for bridge wires that are in a poor condition and where the main cable capacity is already reduced. In conclusion, it is strongly recommended that the cables are not permanently wedged in connection with dehumidification. This also removes the trouble and cost of this work.

4.6 *Pressure curve adjustment*

Monitoring results of the exhaust air have shown that the humidity conditions inside the cable vary somewhat during operation. Under perfectly ideal conditions, with a 100% airtight cable and completely constant absolute water content (AWC) in the injection air, the exhaust air should have the same AWC as the injection air once the drying out process is completed. In the real world this is not the case, as the cable can never be 100% air-tight and the AWC varies somewhat as the systems are controlled by relative humidity, which gives variations according to variations in the temperature.

The air pressure inside the main cables along a given flow stretch varies from a maximum at the injection point to zero at the exhaust point, if no special measures are made at the exhaust point. The very low air pressure in the area near the exhaust points makes these areas susceptible to intrusion of moisture during certain weather conditions such as high atmospheric pressure, hard rain and high-speed turbulent wind. Due to the height, exposure and locations of many suspension bridges, these types of weather conditions occur relatively often. In order to counteract this effect, a pressure curve adjustment solution was developed. By applying a damper in the exhaust on the exhaust sleeve a small air pressure can be maintained instead of falling to zero, and thereby better protect these sensitive areas from moisture intrusion.

This solution is incorporated on recent projects e.g.; George Washington Bridge (under construction) and two systems that are in service – The Hardanger Bridge, since 2013 and the Great Belt Bridge, since 2015. During the commissioning of these systems the exhaust dampers were adjusted and locked in the correct position (determined during commissioning/adjustment of the system) and ensure an overpressure of approximately 100–150 Pa, thereby ensuring better protection of areas near the exhaust points. An exhaust sleeve is shown in Figure 9. The damper is placed in the pipe in the lower part of the instrument box.

Figure 9. Exhaust sleeve with damper (Hardanger Bridge), bolt for adjustment of damper – see arrow.

4.7 *Monitoring sleeves*

Thorough monitoring of the dehumidification system is essential to document its effectiveness. This is quite straightforward for all monitoring locations except at the exhaust flow in the anchor chambers. At these locations it is possible to measure the relative humidity and the temperature, but it is not generally possible to measure the flow, as the exhausting air continues to flow between the wires or strands and out into the very large volume of the anchor chamber and air pressure falls to zero. It is possible to insert a temperature and humidity sensor up between the strands where they splay out to monitor here. There is however a risk that the exhaust air will get mixed with the air from the splay chamber, such that humidity monitoring will be inaccurate.

To improve the quality of the exhaust monitoring a monitoring sleeve has been developed, which is placed on the main cable a short distance from the anchor chamber. The monitoring sleeve is designed in much the same way as the injection and exhaust sleeves, cf. section 4.6. The monitoring sleeve includes sensors for relative humidity, temperature and pressure. During the commissioning of the dehumidification system the relation between flow and pressure can be accurately established and the monitoring system will then be able to indirectly measure the flow close to the anchor chambers. This solution provides better monitoring of this area and has been successfully applied to several projects, including the Great Belt Bridge.

5 OPTIMIZATION OF SYSTEMS IN SERVICE

5.1 *Follow-up – Inspection and data analysis*

Analysis of the data from the monitoring system as well as inspection of the dehumidification plant(s) and the sealing system of the main cables should be carried out on a regular basis. Ideally these activities should be carried out at the same time, as leakage in the sealing system that can be found by inspection may have influence on the analysis. The inspection of the plant(s) should also include control and if necessary calibration or replacement of the sensors, as this too can affect the analysis. Performing these three activities together provides a complete evaluation of the dehumidification system and its current condition. The analysis can also be used a part of the basis for considering further optimization of the system.

5.2 *Further optimization – Case story Little Belt Bridge*

Main cable dehumidification systems have been installed on many suspension bridges since 1998. As the early projects did not have the wealth of experience that is currently available, it is possible that some of these could be optimized. One such bridge is the Little Belt Bridge, where the first system outside of Japan was installed in 2003. This was the first bridge with main cables of helical strands to be dehumidified and there was no best practice for a viable flow length at the time of

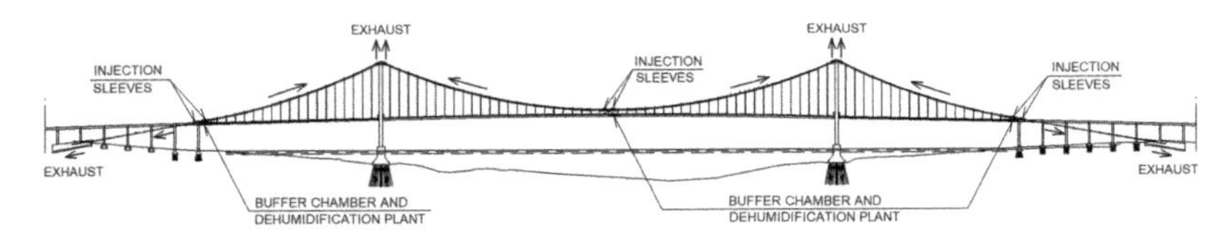

Figure 10. Original layout of system on Little Belt Bridge.

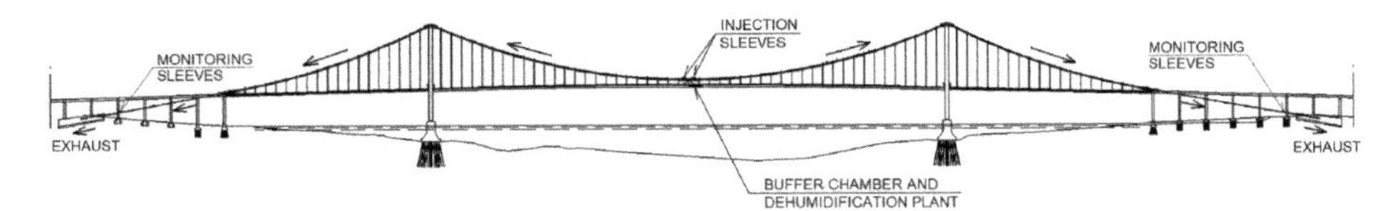

Figure 11. Optimized layout of system on Little Belt Bridge.

design. A flow test was carried out (Bloomstine and Thomsen 2004), but due to constrictions on the bridge, it was only possible to test over a length of approximately 170 m. Therefore, the system was designed with shorter lengths than is currently known to be possible.

In 2017 on-site testing was done to explore the possibility of much longer lengths. As shown in Figure 10, the original system had injection points at the middle of the main span and at the side towers, dividing each main cable into 6 flow lengths between approximately 160 and 300 m. During the testing the injection points at the side towers were closed off for one main cable, such that air would flow from the injection point at midspan and all the way to both anchorages, with a maximum length of approximately 710 m. The testing was successfully completed, and the system was updated at the end of 2017 according to the layout in Figure 11.

The optimized system has just one dehumidification plant now, which in itself is an optimal solution. The plant is located inside the box girder at midspan and utilizes a length of the box girder as a buffer chamber, another optimal solution. From midspan the dry air flows to the anchor chambers at both ends of the bridge, where it exhausts. To improve monitoring of the exhaust air, new monitoring sleeves were installed approximately 25 m from each anchor chamber. The optimized system has reduced the electrical consumption of the system by approximately 50% and there is only one plant to maintain now. The two plants adjacent to the side towers have been shut down and the components of these plants can be used to maintain the remaining plant.

The optimized system has been in service over a year now and is generally performing well. An analysis of the data from the first year of service indicates that the cables are being sufficiently protected from corrosion, but that the sealing of the main cables should be improved to ensure an even more effective system. Hence, the sealing is planned to be upgraded according to the current knowledge of optimization. This was the first bridge where Cableguard Wrap was applied in connection with dehumidification and the methods for heat bonding the wrap have improved much since then. The following optimization of the sealing is planned:

- All the splices in the wrap will be treated by heat gun and roller as is best practice today. There is currently leakage at many of the splices, roughly 30–40%.
- The transitional detail at the cable bands will be updated to the current optimal detail, as described in section 4.2. The bands, finishing strip and neoprene wedge will all be removed, then the caulk in the groove of the bands will be repaired where necessary.

When these improvements have been carried out the system will perform even more economically with a lower running time and electrical consumption. Main cable systems that are in service on other suspension bridges should also be analyzed and if further optimization is possible, it should be carried out. As main cables of helical strands have a much lower resistance than main cables of parallel wires and suspension bridges with helical strands are also generally shorter, it should be feasible with just one injection point at midspan for this type of bridge. The Little Belt Bridge has to the best of the Authors' knowledge the longest main cables of helical strands in the world.

6 OPTIMIZATION THROUGH INTEGRATED SYSTEMS

Integrated dehumidification systems are systems that include the main cables as well as one or more other elements, an optimization that goes beyond optimizing the system for the main cables alone. The following examples illustrate some of the possibilities for further optimization through integration.

The Angus L. Macdonald Bridge, Canada has recently been through a major rehabilitation including dehumidification of the main cables. The dehumidification system is an integrated system, with one plant protecting the entire length of both main cables, as well as the exposed strands in the splay chambers of both anchorages. The layout of the dehumidification system is shown in Figure 12.

There is one dehumidification plant placed in the Halifax anchorage, where the two splay chambers together provide an ideally sized buffer chamber, without the need for a new structure. As the exposed strands here are in the buffer chamber, they are fully protected from corrosion. Dry air flows through piping to the center of the main span, where it is injected in both main cables. The dry air flow towards both anchorages, where it exhausts. In the Halifax anchorage the return dry air is recycled, which reduces the need to produce dry air, giving extra savings in electrical consumption as well as less wear on the plant. The flow of dry air from the main cables into the Dartmouth anchorage is sufficient to protect the exposed strands in the splay chambers. The splay chambers have been sealed and equipped with exhaust pipes, where the flow, temperature and humidity are monitored to document the effectiveness.

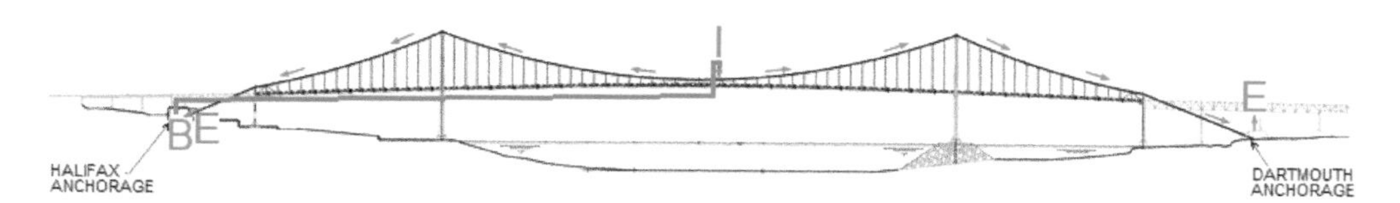

Figure 12. Layout for Macdonald Bridge, I = Injection, E = Exhaust, B = Buffer chamber and plant.

Figure 13. Great Belt Bridge, exhaust pipes from the saddle lead dry air to the structure beneath.

The Great Belt Bridge is another good example of an integrated system. The layout for this system is shown in figure 1. There is one plant in the bridge box girder that provides protection of the box girder in the main span, the entire length of both main cables and the interior of the supporting structures for the tower saddles, where the exhaust air from the main cables flows through these structures, see Figure 13.

It is highly recommended that the possibility for developing an integrated system is explored and developed during the conceptual design phase, as an integrated system provides a very high degree of optimization.

7 CONCLUSIONS

As described in this paper, there are many ways to optimize a main cable dehumidification system, starting already in the earliest design phases. An optimal layout for the system is one of the first steps and it should include integration of other elements, which is usually possible. A flow test is the best method for ensuring the longest viable flow length, which also helps minimize construction and maintenance costs. Pre-design activities should be summarized and presented in a Conceptual Design Report, including a preliminary budget and schedule for the works. This provides the bridge operator with all the necessary information for budget planning and scheduling and presents an easily understandable overview of the system and other recommended works. Further, it provides a complete and reliable basis for the detailed design.

This paper also describes many optimal details and methods, which have been developed since the year 1995. These details include sealing of the main cables and cable bands, the buffer chamber solution, the one plant system, sleeves for injection, exhaust and monitoring, pressure curve adjustment and integrating other elements.

Recommendations for follow-up on existing main cable dehumidification systems and further optimization are also presented. As the early projects did not have the wealth of experience that is currently available, optimization of a number of existing systems should be considered and carried out, where further optimization is possible. Some of these possibilities are illustrated by an example from Little Belt Bridge in Denmark, where a system was installed in 2003. In general, if the recommendations in this paper have not been followed, further optimization is most likely possible.

It is strongly recommended that the guidelines in this paper for developing an optimal main cable dehumidification system are followed. These methods and optimal details together ensure an optimal, reliable and durable system with the lowest possible life cycle cost (the lowest sum of construction, operation and maintenance costs).

REFERENCES

Bloomstine, M.L. & Melén, J.F., 2017. Main Cable Dehumidification – Flow Testing and Other Innovations, *2017 NYC Bridge Conference*, New York, USA

Nielsen, K.A. & Bloomstine, M.L., 2016. The Storebælt East Bridge Main Cable Dehumidification, World's Largest Retrofit and Latest Design Optimization. *8th International Cable Supported Bridge Operators' Conference*, Halifax, Canada

Bloomstine, M.L., Latest Developments in suspension Bridge Main Cable Dehumidification, 2013. *NYC Bridge Conference*, New York, USA

Bloomstine, M.L. & Gilliusson, S., 2008. The Högakusten Suspension Bridge – Corrosion Protection of Main Cables and Maintenance of Major Components, *6th International Cable Supported Bridge Operators' Conference*, Takamatsu, Japan

Bloomstine, M.L. & Thomsen J.V., 2004. Little Belt Suspension Bridge – Corrosion Protection of the Main Cables and Maintenance of Major Components, *4th International Cable Supported Bridge Operators' Conference*, Copenhagen, Denmark

Bloomstine, M. L., Rubin, F. & Veje, E., 1999. Corrosion Protection by Means of Dehumidification *IABSE Symposium, Structures of the Future – The Search for Quality*, Rio de Janeiro, Brazil

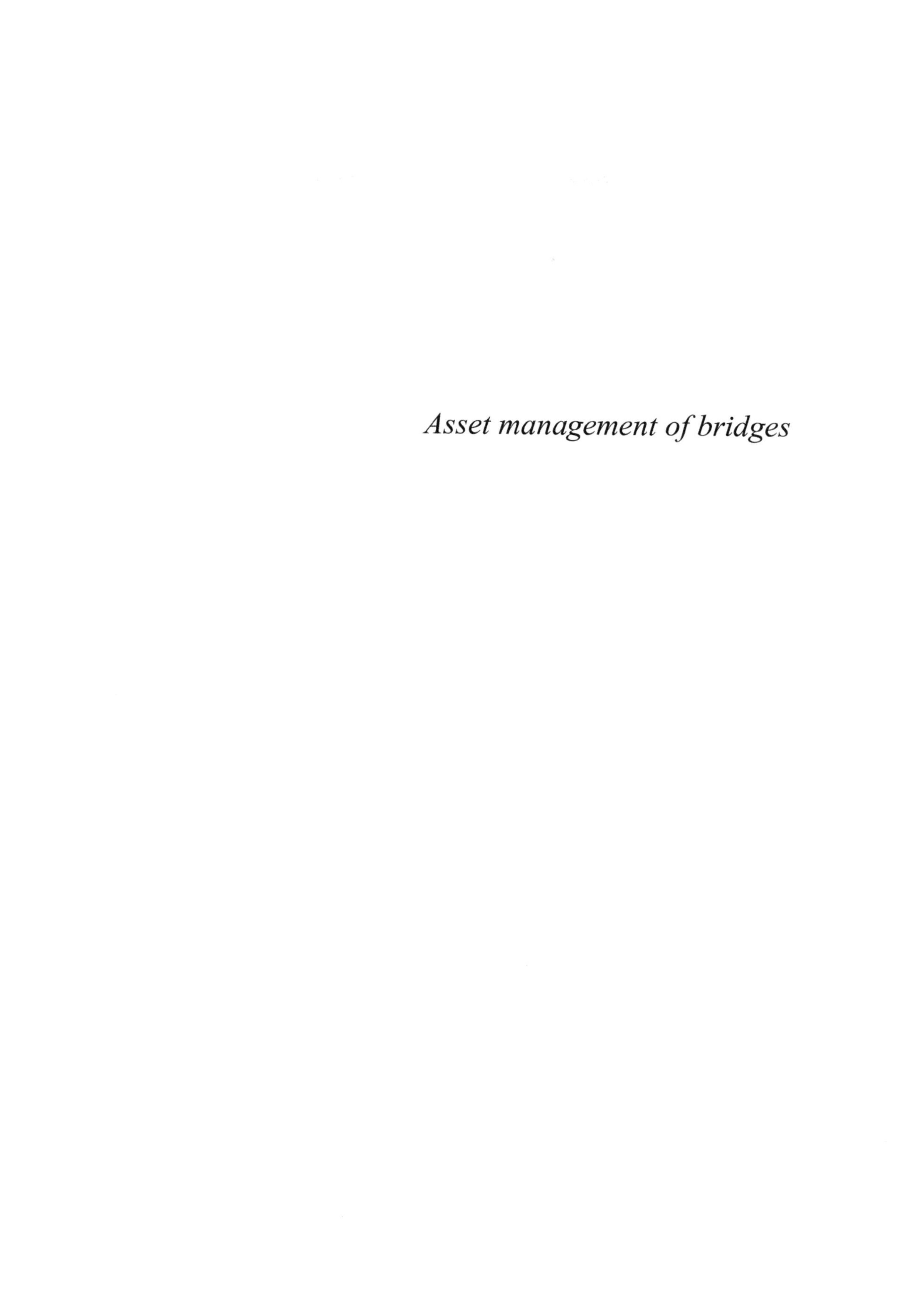

Asset management of bridges

Chapter 5

Asset management of Honshu-Shikoku Bridges based on preventive maintenance

R. Uchino & N. Toyama
Honshu-Shikoku Bridge Expressway, Kobe, Japan

ABSTRACT: The Honshu-Shikoku Bridge Expressway (HSBE) has 17 long-span bridges located over straits where environment is very harsh. In order to keep these bridges in sound state for more than 200 years, the Honshu-Shikoku Bridge Expressway Company Limited (the company) is conducting maintenance based on the concept of the preventive maintenance. This paper describes the concept of asset management and each maintenance element of the HSBE.

1 INTRODUCTION

The HSBE consists of three routes (total length of 172.9 km) between Honshu and Shikoku (Figure 1). It unifies the region of Seto Inland Sea and plays an important role as a part of the arterial high-standard highway network. The feature of the HSBE can be described as maintenance of 17 long-span bridges (Table 1) located over straits where environment is very harsh. Figure 2 shows signature bridges of the HSBE. The Akashi Kaikyo Bridge is the longest suspension bridge in the world. The Seto-Ohashi Bridges are one of the longest highway-railway combined bridge link. The Kurushima Kaikyo Bridges are the unprecedented three consecutive suspension bridges.

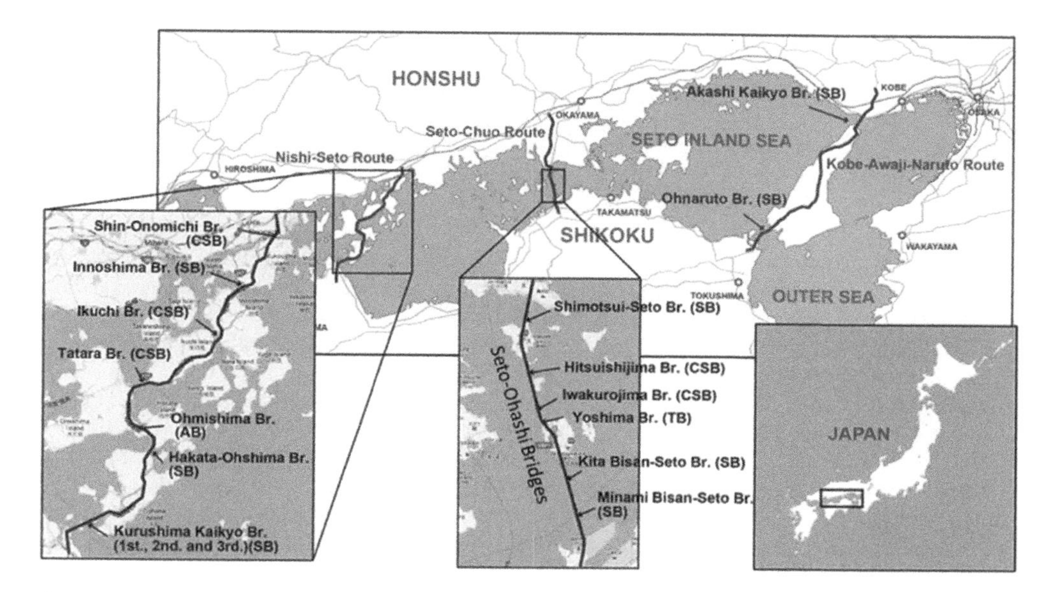

Figure 1. Honshu-Shikoku Bridges (SB: suspension bridge, CSB: cable-stayed bridge, TB: truss bridge, AB: arch bridge)

Table 1. Long-span bridges of HSBE.

Bridge	Completion (year)	Type	Length (m)	Height of Tower (m)
Kobe-Awaji-Naruto Expressway				
Akashi Kaikyo	1998	Suspension	3,911	297
Ohnaruto	1985	Suspension	1,629	144
Seto-Chuo Expressway (Seto-Ohashi Bridges)				
Shimotsui-Seto	1988	Suspension	1,400	149
Hitsuishijima	1988	Cable-stayed	790	152
Iwakurojima	1988	Cable-stayed	790	161
Yoshima	1988	Truss	850	–
Kita Bisan-Seto	1988	Suspension	1,538	184
Minami Bisan-Seto	1988	Suspension	1,648	194
Nishi-Seto Expressway				
Shin-Onomichi	1999	Cable-stayed	546	77
Innoshima	1984	Suspension	1,270	146
Ikuchi	1991	Cable-stayed	790	127
Tatara	1999	Cable-stayed	1,480	226
Ohmishima	1979	Arch	328	–
Ohshima	1988	Suspension	840	97
Kurushima Kaikyo 1st.	1999	Suspension	960	149
Kurushima Kaikyo 2nd.	1999	Suspension	1,515	184
Kurushima Kaikyo 3rd.	1999	Suspension	1,575	184

Figure 2. Signature bridges of HSBE (left: Akashi-Kaikyo Bridge, center: Seto-Ohashi Bridges, right: Kurushima Kaikyo Bridges)

In 1979, the first portion including Ohmishima Bridge, a steel arch bridge, completed and opened to traffic. Then the Seto-Chuo Expressway that includes the Seto-Ohashi Bridges and the Kobe-Awaji-Naruto Expressway that has the Akashi Kaikyo Bridge opened to traffic in 1988, and 1998 respectively. And the completion of the Nishi-Seto Expressway brought the HSBE from the construction era to the maintenance era.

In the business policy of the company, it is stated that the company strive for appropriate management of the bridges aiming for long-term service life for more than 200 years (Honshu-Shikoku Bridge Expressway Co., Ltd. 2018). Currently, the ages of the HSBE's long-span bridges range between 20 and 40 years old and they are considered to be relatively young bridges; however, some deteriorations in the bridge members are observed through the inspection and investigation activities. The company is now trying to avoid large-scale repair or replacement and to minimize life cycle cost (LCC) according to the idea of preventive maintenance strategy. In order to achieve proper maintenance based on the strategy, the concept of asset management is introduced to the maintenance process utilizing inspection data and repair history.

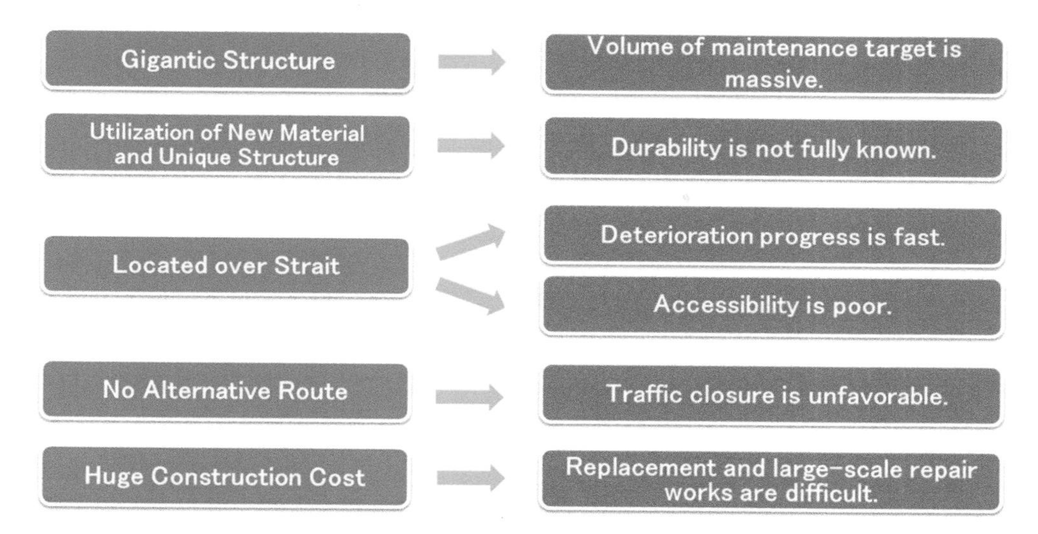

Figure 3. Features of HSBE's long-span bridges in maintenance.

The asset management is implemented in the following sequence; inspection of the bridge members according to a maintenance plan, evaluation of the bridge condition, prediction of the deterioration and countermeasure taken at an optimum time. Then, the management process is appraised and the feedback is reflected in the plan for the next period. During the above process, the inspection and repair histories are accumulated in an integrated system and utilized as a database for more efficient management.

2 FEATURES OF HSBE'S LONG-SPAN BRIDGES IN MAINTANACE

The company is conducting maintenance based on the following features of the HSBE's long-span bridges (Figure 3).

1) Volume of the maintenance target is massive. Signature example is 4,000,000 m^2 of the external surface of the steel structures that needs to be recoated multiple times during the life time of the structures.
2) For new materials or special structures utilized in the bridges, its maintenance experience is scarce and durability is not fully known.
3) Since they are located over straits, they are considered to be in very harsh corrosive environment such that they are exposed to strong wind that brings chloride particles from the sea. And, since the superstructures are in high location over straits, they are not easily accessible and construction of scaffoldings required for inspection or repair works becomes extensive work.
4) Since there are no alternative routes, repair works under traffic closure are not favorable.
5) Since construction cost is huge, replacement and large-scale repair works are difficult.

3 INTRODUCTION OF ASSET MANAGEMENT CONCEPT

A systematic maintenance management based on a preventive maintenance concept is being conducted aiming for bridges with service life of more than 200 years. In the preventive maintenance, immediate actions are taken before deterioration gets serious (Hanai et al. 2014).

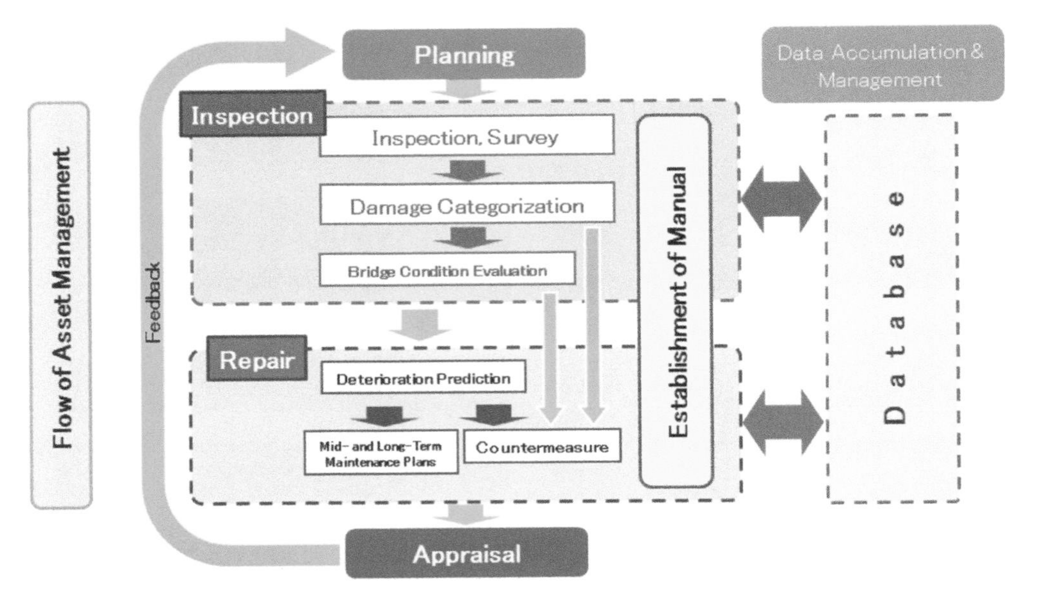

Figure 4. Asset Management Procedure.

Figure 5. Long-term recoating plan until 2100.

It is important to take into account the LCC when implementing the maintenance management. The company have established a maintenance framework centered on around the Maintenance Department of the company to promote an asset management concept in cooperation with subsidiary companies. This framework is expected to enable the company to minimize the LCC and to implement an efficient maintenance management.

A flow of the asset management being promoted by the company is shown Figure 4. It is being implemented in the following sequence; inspection of the bridge members according to a maintenance plan, evaluation of the bridge condition, prediction of the deterioration and countermeasure taken at an optimum time. Then, the consequence is appraised and the feedback is incorporated into the plan. In order to implement the flow, the inspection and repair histories are accumulated in an integrated system and utilized as a database for more efficient maintenance management.

In order to maintain the long-span bridges for more than 200 years, the company drew up long-term maintenance plan until the year of 2100 based on the concept of asset management. Figure 5

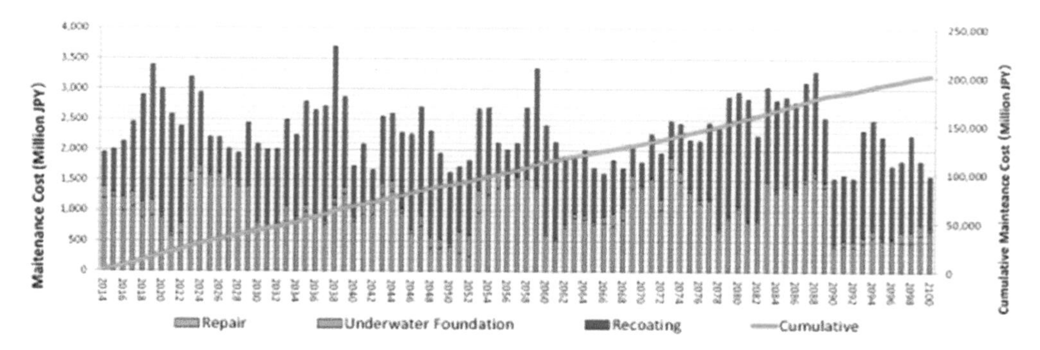

Figure 6. Maintenance cost until 2100.

is a long-term recoating plan until 2100. Red bar indicates girder, blue bar indicates tower, and yellow bar indicates cable. This plan is based on the data as of March, 2013. Accordingly, the plan will be updated as new maintenance data is accumulated.

Figure 6 shows a maintenance cost until 2100. It is said that cumulative maintenance costs of some old suspension bridges in the USA have increased exponentially and exceeded the initial construction costs due to large-scale repair works against serious damages, such as breakage of cable wire, after 70 years of its service life. On the other hand, HSBE's cumulative maintenance cost shows steady increase and remains low level even after 70 years. The company is following out the preventive maintenance concept in order to avoid the maintenance cost from increasing exponentially and to minimize the LCC.

4 INTRODUCTION OF INNOVATIVE TECHNOLOGY

Innovative technologies were introduced for long-span bridges maintenance to reduce the LCC. This chapter describes "Coating for Steel Structure", "Cable Dehumidification System for Suspension Bridge", "Non-destructive inspection method of Suspender Ropes", "Non-destructive evaluation based on Infrared Thermography Measurement", "Anti-Deterioration Measure for Concrete", and "Anticorrosion System for underwater steel structure" as representative innovative maintenance technologies.

4.1 *Coating for steel structure*

Since the total external surface of the steel members of the HSBE's long-span bridges is about 4,000,000 m^2, maintenance of coating is a major task. The long-span bridges were built across straits where corrosive environment is severe. Therefore, the heavy duty coating system is used to prevent corrosion. The coating system consists of rust-preventive thick inorganic zinc-rich paint as a primary layer and weather resistant fluorine resin paint as a top layer. Before the Akashi Kaikyo Bridge, polyurethane resin paint was used for top layer. Coating specification for external surface of steel structures is shown in Figure 7.

Recoating policy restricts recoating layers to deteriorated top and middle coats, keeping the inorganic zinc-rich paint in a sound state (Figure 8). This is because the recoating of the entire coat requires on-site blasting that is disadvantageous in terms of quality control, cost, and environment. In early recoating, fluorine resin paint was used as top coat. After 10 years of research, highly durable fluorine resin paint that has higher durability than existing top coat was developed and has been utilized for recoating since 2010.

In order to grasp the optimum recoating cycle, i.e., the timing when the top and middle coats disappears, consumption rate of coating must be understood with satisfactory accuracy. Since visual observation is not sufficient for this purpose, fixed observation points were set up on the existing

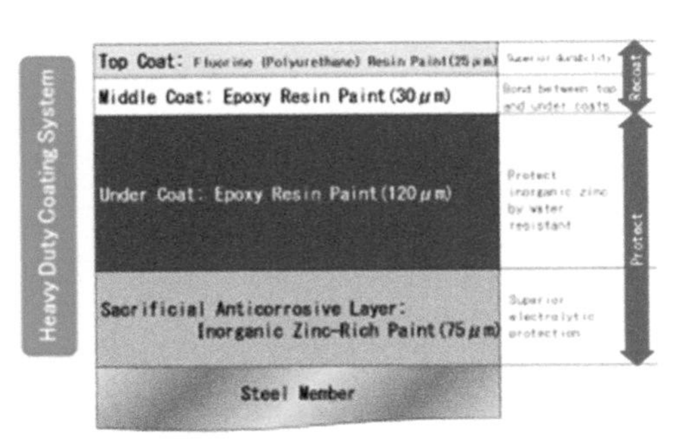

Figure 7. Heavy-Duty Coating.

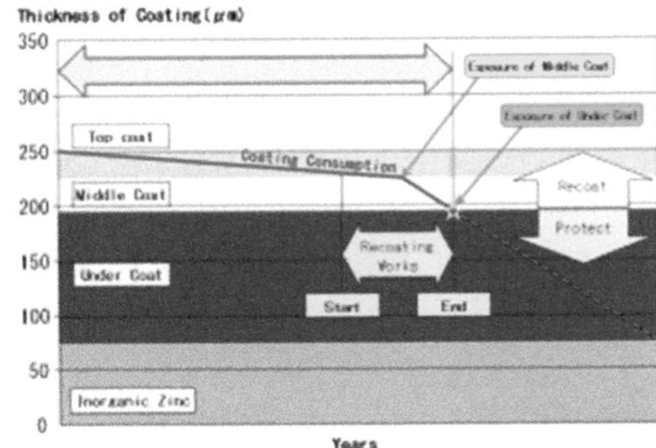

Figure 8. Consumption of Coating.

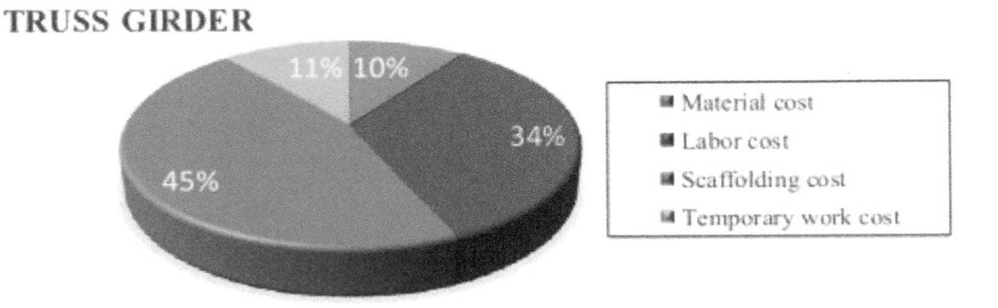

Figure 9. Breakdown of repainting cost at Seto-Ohashi Bridges.

bridges. At these points, thickness, glossiness, and adhesion of coating are tracked periodically. From the past results, measured consumption rates were found to be smaller than those initially estimated. Since the prolongation of recoating cycle significantly reduces LCC, the company is trying to improve the estimation accuracy by increasing the number of observation points.

In the current recoating process, different materials are used for top and middle coats and therefore it requires two painting processes and drying time between the applications of two coats. If the function of these two layers can be replaced by one layer coating, recoating cost can be reduced significantly by the reduction of the work period.

Figure 9 shows the breakdown of repainting cost at the Seto-Ohashi Bridges. Whereas the material cost of paint is only 10%, the majority of the repainting cost consist of labor and scaffolding cost. In other words, even if the newly developed paint was more expensive than the existing paint, increased cost of paint material has an insignificant effect to total repainting cost. Furthermore, if the repainting cycle was extended by applying new paint, the scaffolding cost, which account for the majority of total cost, becomes small in the long run. And the LCC becomes small consequently.

The company is working on the development of the paint with simplified application process (Figure 10). In order to verify the durability of simplified application process paint, exposure tests have been conducted.

4.2 *Cable dehumidification system for suspension bridge*

The main cable is one of the most important members of the suspension bridge. Therefore, protecting the cable from corrosion is an important task in maintenance. The conventional anti-corrosion measure is shown in Figure 11, which aims to prevent water from penetrating into the cable from the surface.

Between 1989 and 1993, investigations of the cables of the existing suspension bridges in Japan, which had been in service for 4 to 6 years, were conducted. In the investigations, it was found that the

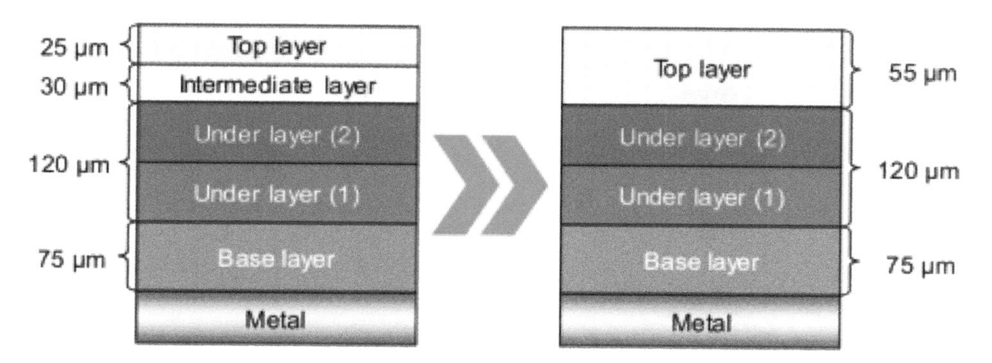

Figure 10. Simplified application process paint.

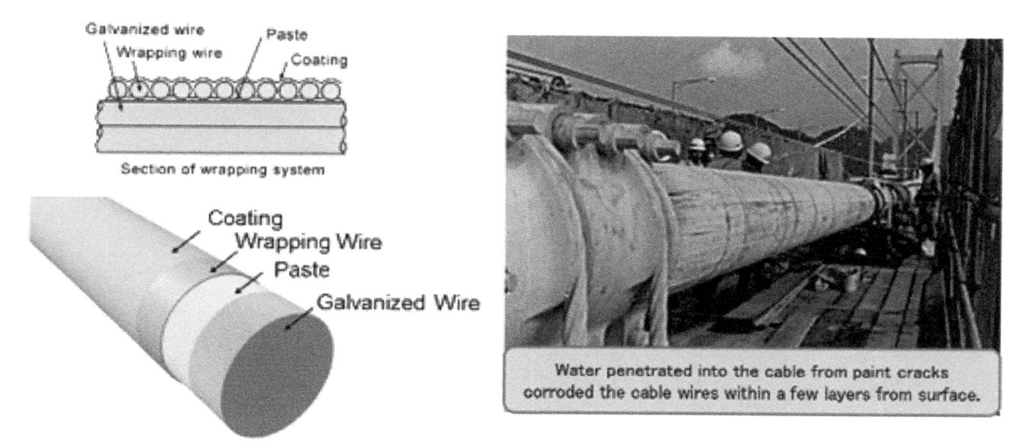

Figure 11. Conventional anti-corrosion measure.

Figure 12. Unwrapped main cable of Innoshima Bridge.

deteriorated paste retained water and that induced corrosion of the cable wires within a few layers from surface (Figure 12). Following the investigation, the company initiated the development of a new anti-corrosion measure that will improve corrosive environment inside the cable in addition to preventing water penetration (Kusuhara et al. 2015).

Cable dehumidification system is to dehumidify inside of the cable by injecting dry air into the cable (Figure 13). By laboratory tests, it was confirmed that if the relative humidity inside the cable is lower than 60%, corrosion of the cable wire will not occur. The system was developed in parallel with the construction of the Akashi Kaikyo and Kurushima Kaikyo Bridges. For these new bridges, new wrapping system was used for air and water tightness. These wrapping system were installed in the all of the HSBE's suspension bridges.

4.3 *Non-destructive inspection method of suspender ropes*

Out of HSBE's 10 suspension bridges, 7 use center fit wire rope core (CFRC) ropes for suspender ropes as shown in a left photo of Figure 14. Although each wire is galvanized and external surface of the rope is coated with polyurethane paint, corrosion was occasionally observed in these ropes as shown in a right photo of Figure 14. Intrusion of water induces corrosion in this member. In order to evaluate remaining strength of the corroded suspender ropes, some ropes were removed and tested. From the test, if the reduction of the cross section reaches 20%, the remaining strength becomes half of that of the intact ropes.

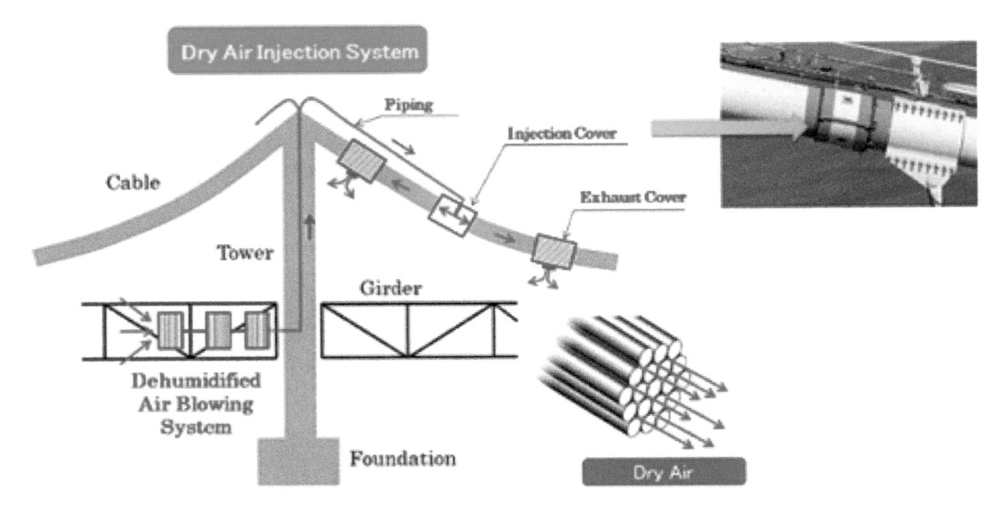

Figure 13. Cable dehumidification system.

Figure 14. CFRC suspender ropes (left: general part, right: anchoring part).

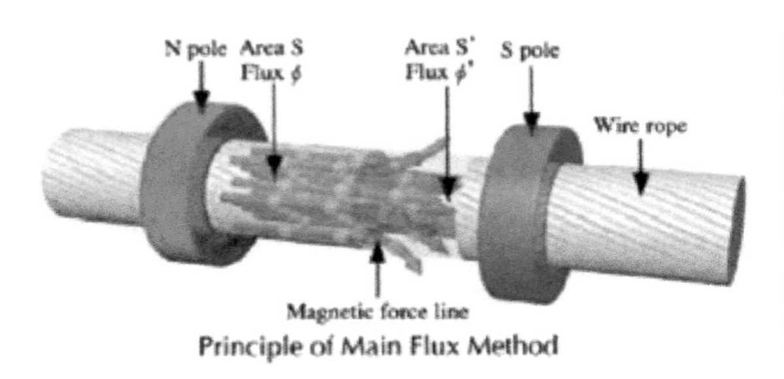

Figure 15. Inspection by main flux method.

Figure 16. Inspection by main flux method.

It is difficult to evaluate internal corrosion state of CFRC ropes by visual inspection. The company developed a non-destructive inspection method using the electromagnetic method, main flux method (Takeuchi et al. 2013). When a rope is strongly magnetized in longitudinal direction, magnetic flux is generated. Since the cross sectional area is proportional to the magnetic flux, remaining cross section can be calculated from the measured data. Principle and application of main flux method is shown in Figure 15 and Figure 16, respectively.

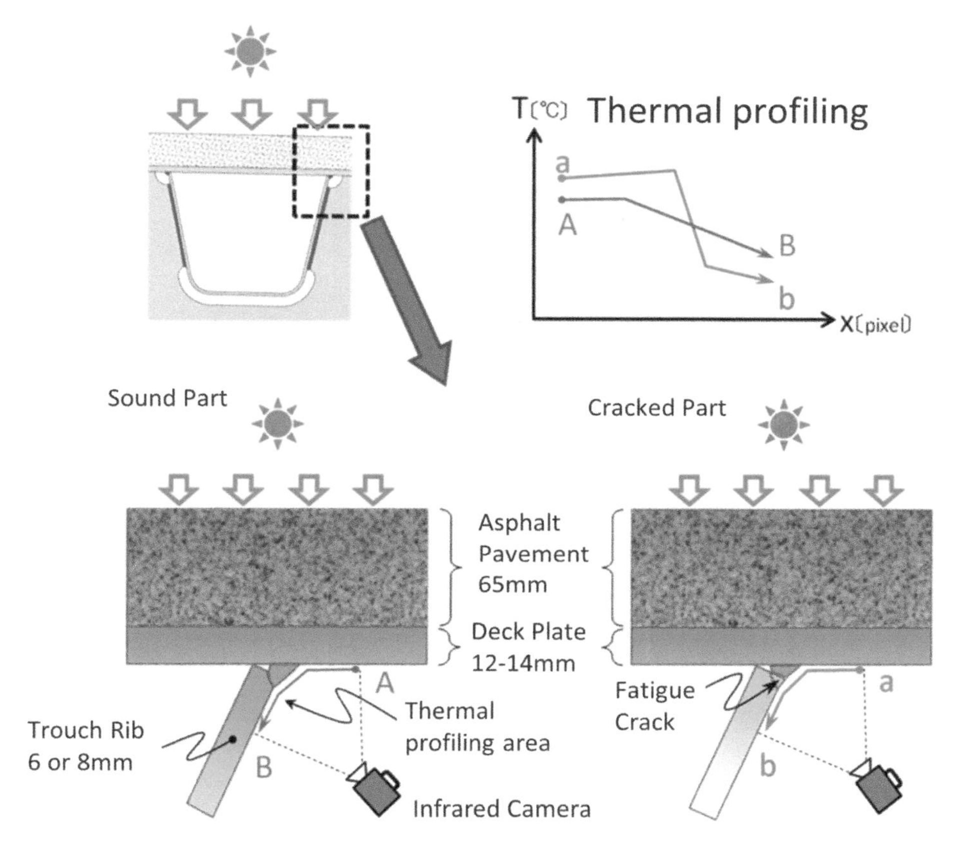

Figure 17. Basic principle of infrared thermography measurement.

4.4 *Non-destructive evaluation based on infrared thermography measurement*

Some of aging steel bridges may suffer fatigue damages because of heavy traffic loading, and periodic inspections, structural integrity assessments and repair works are required during the life cycle of the bridges.

HSBE collaborated with Kobe University and University of Shiga Prefecture, and developed non-destructive evaluation techniques employing infrared thermography enabling us to inspect the fatigue cracks in the welded connections of steel bridges remotely, effectively and efficiently, and to evaluate the structural integrity of the bridges based on the result of infrared thermography measurement (Sakagami et al. 2015, Izumi et al. 2015, Okumura et al. 2017).

Basic principle of infrared thermography measurement based on temperature gap detection for weld-bead-penetrant-type fatigue crack is shown in Figure 17. Asphalt pavement on the surface of steel deck is heated by sun shine. Heat conduction occurs from the pavement to trough rib through deck plate and this makes temperature gradient in the trough rib (U-rib) as shown in the figure. The temperature change is continuous when there is no crack in weld bead, on the other hand temperature gap is observed between deck plate and trough rib due to thermal insulation effect of the crack in weld bead. The temperature gap can be detected by the continuous temperature monitoring along the weld bead using traveling infrared thermography equipment.

In order to apply this inspection technique to practical use, it was necessary to develop move efficient detection device and flaw detection algorithm that is free from human error. Therefore we are working on the improvement of automatic flaw detection method, effective detection device and analysis device.

Figure 18. Camera device.

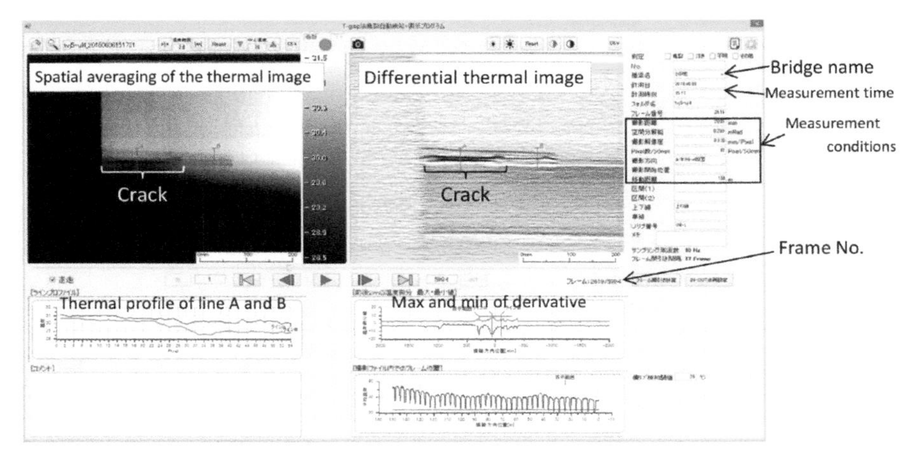

Figure 19. Browsing system's screen.

The developed crack detection system can automatically detect cracks through the infrared camera (Figure 18) and analysis device. It can reduce the evaluation time dramatically and improve the reliability in comparison with the conventional visual check of the thermal image. In addition, to support evaluation, it can produce thermal images. Furthermore, it can judge "float and dirt of paint". It can distinguish peeling of paint and dirt on paint, and the data is synchronized with positional information.

The browsing system (Figure 19) enables to check the result of the detected position quickly by the jump function. Moreover, the crack length can be estimated by spatial averaging of the thermal image and differential thermal image. The flow of the crack detection system is shown in Figure 20.

4.5 *Anti-deterioration measure for concrete*

HSBE's long-span bridges are under severe salty environment because those are located over straits. Penetrated chloride ion may corrode steel members inside the concrete, then expansion pressure of corroded steel members may damage the concrete, and finally, cracking, flaking and spalling may occur. In order to prevent intrusion of chloride ion, coating is applied to surfaces of some concrete structures.

A non-destructive testing (NDT), which measures the extent of deterioration, such as chloride ion content, carbonation depth, and other check items, is periodically conducted in order to identify the necessity of countermeasures (Hanai et al. 2012). Also, future chloride ion contents, which

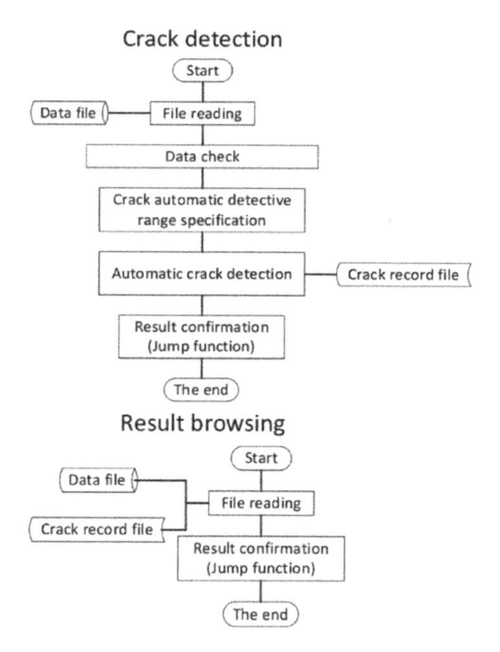

Figure 20. Flow of the crack detection system.

Figure 21. Concrete coating of the 2nd Kurushima Kaikyo Bridge.

penetrate into the concrete with time at the depth of rebar in from several decades to 100 years are predicted and are compared those with a threshold of corrosion limit. The company comes to a decision to application of the coating for concrete structures in order to prevent further intrusion of chloride ion if generation of rust of the rebar is predicted.

Figure 21 shows a case of the 2nd Kurushima Kaikyo Bridge; the top of the tower foundation was coated. Figure 22 shows a case of the Ohnaruto Bridge; the entire side surface of the anchorage was coated because salty environment is very severe. Figure 23 shows a case of Bisan-Seto Bridges; lower portion of the side surface of the anchorage was coated because salty environment is not as severe as in the Ohnaruto Bridge

4.6 *Anticorrosion system for underwater steel structures*

For the most of the underwater foundations of the HSBE's long-span bridges, steel caisson foundations were selected. Although the role of the steel caissons was initially considered as formworks

Figure 22. Concrete coating for anchorage of Ohnaruto Bridge.

Figure 23. Anchorage of Bisan-Seto Bridges.

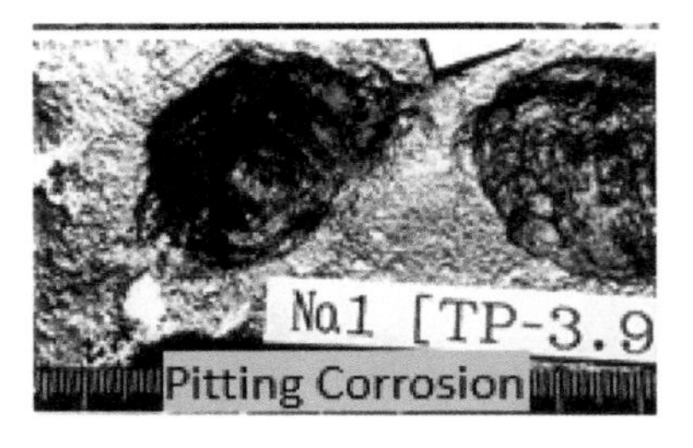

Figure 24. Pitting corrosion on igure steel caisson.

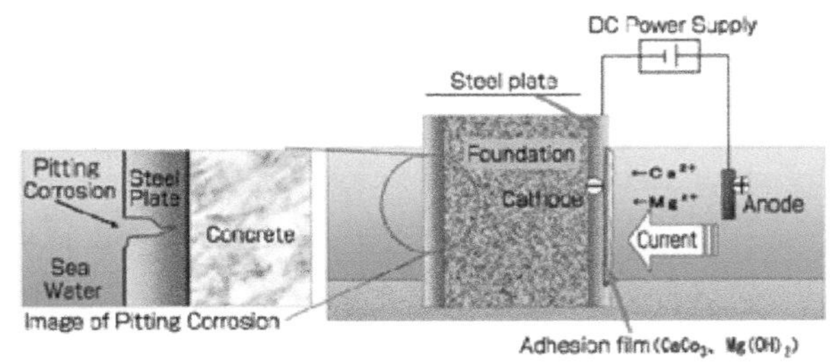

Figure 25. Concept of electrodeposition method.

for underwater concrete which was cast in the seawater at the construction stage, it is advantageous to protect them in order to keep the foundations in sound condition for more than 200 years.

Ten years after the completion of the Seto-Ohashi Bridges, corrosion survey of the underwater foundation was conducted. Progress of corrosion was the most severe in the tidal zone and the reduction of the thickness of the steel plate of 10 mm thick was up to 3 mm. On the other hand, while the corrosion rate in underwater zone is smaller, local pitting corrosions at which whole thickness of the steel plate was lost were observed (Figure 24). If this kind of pitting corrosions is left unrepaired, seawater may corrode steel members inside the concrete, expansion pressure of corroded steel members may damage the concrete, and finally, integrity of foundations may be lost. In order to avoid this kind of risk, the company decided to protect the steel caissons.

For the steel caissons of the HSBE's long-span bridges, except for those of the Tatara Bridge and the Kurushima Kaikyo Bridges in which cathodic protection was applied at the construction stage, long-term anticorrosion measures were not taken. As a result of comparison of various existing methods, electrodeposition (ED) method (Figure 25), in which calcium and magnesium contents in the seawater are deposited on the surface of the steel caissons and form protective layer, was tested in the actual structures because the method does not require diver work and therefore is considered as safe and economical.

On the other hand, since the splash zone and tidal zone are not always in the seawater, the ED method is not applicable. For these zones, an anticorrosion measure with coating system which is applicable to wet surface is selected. In order to long-lasting stable coating, rust on the surface of the steel caissons must be removed completely by blasting. Furthermore, at the boundary of each

Figure 26. Cathodic protection of the Kurushima Kaikyo Bridge.

	Akashi		Hitsuishi		Iwakuro			Kita/Minami Bisan						Tatara		Kurushima		
	2P	3P	2P	3P	2P	3P	4P	2P	3P	4A	5P	6P	7A	2P	3P	3P	4A	5P
Depth (m)	-50	-47	-25	-20	-10	-20	-10	-4	-4	-4	-25	-40	-40					
(base) (m)	-60	-57	-28	-25	-15	-23	-14	-10	-10	-10	-32	-50	-50	-33	-13	-40	-30	-29
Status	C	C	A	A	B	B	B	B	B	B	C	C	C	A	A	A	A	A

ED : electrodeposition
: cathodic protection
CC : cathodic protection + coating
: coating
S : under study
BF : back-filling

Anti-corrosion system was applied since construction stage

Status A : Completed B : Under construction C : Under study

Figure 27. Anticorrosion measures for steel caissons of HSBE's long-span bridges (Kusuhara et al. 2013).

method, neither ED film nor coating may be applied and it becomes weak point. Therefore, for this boundary, cathodic protection method was applied with aluminum alloy anodes.

For shallow foundations, since the ED facility may be excessive and uneconomical, cathodic protection method was selected for under water zone. And for deep foundations, safe and economical method has been studied because the application is considered not to be economical by the current ED facility.

In the Kurushima Kaikyo Bridge, countermeasures were applied for underwater foundations at the construction stage. In addition to coating, cathodic protection was adopted as an anti-corrosion measure (Figure 26). In the cathodic protection, aluminum alloy anodes were attached on the surface of the caissons as sacrificial material. Several surveys have been carried out to identify the condition of the steel caissons and the consumption rate of the aluminum alloy anodes after its completion. The results showed that the steel caissons were in sound condition and the consumption rate of aluminum alloy anodes was smaller than estimated at the design phase. The effectiveness of the anti-corrosion measure was confirmed.

Anticorrosion measures for the steel caissons are summarized in Figure 27.

5 CONCLUDING REMARKS

Deterioration of the certain components of the HSBE's long-span bridges has already started. However, it requires long time to evaluate new maintenance technology or high-precision deterioration prediction method. In this situation, it is important to continue the development of the maintenance technology that will enhance durability and reduce the LCC in order to execute secure maintenance.

The company is promoting the maintenance aiming to minimize the LCC by the reliable inspection, accumulation of data, and analysis and evaluation of that database and introduces innovative technologies to reduce the LCC.

REFERENCES

Hanai, T., Sakai, K., and Kagoike, T. Maintenance Management and Condition of Concrete Structure of Honshu-Shikoku Bridge Expressway, *Proceedings of the 28th US-Japan Bridge Engineering Workshop, 2012.*

Hanai, T., Kuwabara, T., Shimomura, M., and Ito, S Preventive Maintenance Strategy of the Long-Span Bridges of Honshu-Shikoku Bridge Expressway, *Proceedings of the 30th US-Japan Bridge Engineering Workshop, 2014.*

Honshu-Shikoku Bridge Expressway Co., Ltd. 2018. Disclosure Book of Honshu-Shikoku Bridge Expressway Co., Ltd., pp26–37

Izumi, Y., Sakagami, T., Nishiwaki, S., Mizokami, Y., and Kobayashi, Y. Examination about detectability of crack by the measurement using the infrared thermography and increasing precision of it, *JSCE Annual Meeting. 2015.*

Kusuhara, S., Sakai, K., Moriwaki, M., and Hanai, T. Development of Anticorrosion System for Underwater Steel Structures of Long-Span Bridges, *Strait Crossings, 2013.*

Kusuhara, S., Moriyama, A., Toyama, N., and Kagawa, A. Current Condition and Improvement of Dry Air Injection System for Main Cables, *Proceedings of the 8th New York City Bridge Conference, 2015.*

Okumura, A., Mizokami, Y., and Moriyama, A. Fatigue inspection for orthotropic steel deck with infrared thermography, *Proceedings of the 39th IABSE Symposium, 2017.*

Sakagami, T., Mizokami, Y., Kobayashi, Y., Izumi, Y., and Nishiwaki, S. Examination about the crack distinction method in the temperature gap measurement using the infrared thermography, *JSCE Annual Meeting, 2015.*

Takeuchi, M., Sakai, K., Moriyama, A., and Kishi, Y. Development of a New Non-Destructive Inspection Technique of Suspender Ropes of Suspension Bridges, *36th IABSE Symposium, 2013.*

Chapter 6

Implementation of a bridge management system in the Ukraine

L. Bodnar
M.P. Shulgin State Road Research Institute State Enterprise – DerzhdorNDI SE (DerzhdorNDI SE), Kyiv, Ukraine

M. Koval
Scientific-industrial enterprise "Triada" Ltd., Co, Lviv, Ukraine

ABSTRACT: In Ukraine, there are 16 149 bridges on public roads out of which 35% are surveyed and the survey data are introduced in the database. Since 2004, the works on the development of the Analytical Expert Bridge Management System of Ukraine have been carried out. The system was implemented in 2006. The purpose of the system is to issue the recommendations on the bridges operation strategy taking into account their technical state and funds for bridges repairs. This paper outlines the experience of implementing the AESUM, the functional characteristics of the system, and the ways of its further development. The attention is paid to such areas of research as expert assessment of the bridge, the degradation model of the bridges elements, the model for determining the cost of operational measures at the network level, developing the strategy to optimize the bridges operation.

1 INTRODUCTION

Bridge Management Systems were developed for making managerial decisions and effectively using the funds for the maintenance and repair of bridges in many countries. The world's first Bridge Management System was developed in the early 70's of the last century in the United States in response to a collapse of the bridge (Shesterikov et al. 2007). At present, almost all European countries, the United States and Canada have their own official Bridge Management Systems which were established about 50 years ago.

The main global objectives that have to be implemented by Bridge Management Systems are the following:

- rational and systematic approach for the organization and implementation of planning, designing, construction, maintenance, repair and replacement of bridges;
- help to make the choice of an effective alternative to achieve a serviceable state of facilities within the limits of allocated funds and to plan future costs;
- contribute to making the effective managerial decisions by administrators, technical specialists and managers of all levels of operational services.

The modern approach to the Bridge Management Systems can be formulated as follows:

- the system for evaluation of the facilities state should be as objective as possible;
- the management apparatus must make a decision on the rational allocation of the funds based on the research and prediction of the bridges state with the constant updating of information;
- the system should be based on the objective database entered in the computer of new generation with high performance.

Thus, the general scheme of the Analytical Management System (AMS) includes two the most important parameters:

- collection of information on inspection (certification) and its analysis;
- determination of the optimal strategy taking into account technical and economic parameters;

Today, the advanced Bridge Management Systems are built on a clear understanding of the objective which includes tasks of a specific purpose. A target function is formed which means a formal statement of the rule of the decision making which includes the function of priority that is compared with the optimization criterion. The objective of the facility evaluation by priority is determination of the relative urgency of the bridges maintenance. In this case, to determine the urgency of the measures and to plan funding, an approach based on the score evaluation of a given factor is used.

Available publications on the problem of bridges management (Shesterikov et al 2004.) indicate that Bridge Management Systems that are created and implemented abroad are much in common:

- there is a database on the bridges which involves many modules – the bridge passport data and survey data, evaluation of the bridges state;
- planning of works (long-term and short-term) with the definition of the work type for the bridge with the possibility of choosing the repair strategies, with the prediction of changes in the bridges state in time;
- ranking of objects for assignment of the priority work.

But there are significant differences in Bridge Management Systems.

For example, the Danbro system (Denmark) does not use mathematical methods to predict the change in the state of the elements (everything is based on expert evaluation) and does not estimate transport losses.

Only the US and Sweden systems take into account transport losses from the unsatisfactory state of the bridges. In the NATS (UK) system, traffic costs are not taken into account and optimization calculations are not performed as in some others, Table 1. Only the US and Danish Bridge Management Systems provide long-term planning.

2 OPERATION OF AESUM SOFTWARE

Since 2004, the Analytical Expert Bridge Management System (AESUM) which accumulates information on bridges on public roads of Ukraine has been developed and implemented by the State Road Agency of Ukraine (UKRAVTODOR) based at the M.P. Shulgin State Road Research Institute State Enterprise (DerzhdorNII SE) with the involvement of specialists from NTU (National Transport University) (Bodnar 2010.). Since 2006, the implementation of the system in the Regional Road Services has begun. Starting from 2013, scientific support is carried out comprehensively and systematically in all Regional Road Services of Ukraine. Analytical information from the database of the AESUM software is used at all levels of the road sector system, work on the filling of the AESUM with information from inspections and certification is carried out regularly, all changes in the classification of roads and bridges are taking into account etc.

In Ukraine, as of the beginning of 2019, there are 5 835 bridges on public roads about 70% of which (namely 4 096 bridges) were surveyed.

It is known that one of the most important components of the operation of the facilities is the repairs that maintain the proper technical state of their structures. Such repairs should be scheduled, it is necessary in advance to determine the term and volume of repair works. In order to determine technical state of the bridges, it is necessary to carry out inspections during which the defects are detected, the defects severity is fixed and their impact on the operational performance of the facility is evaluated. According to the results of the inspections, the decisions are made to carry out repair work. It is a well-known fact that the survey costs are paid off by optimizing the costs on carrying out operational measures and their timely implementation.

Table 1. Brief description of foreign Bridge Management Systems (Bodnar 2010).

Parameters of Bridge Management Systems	Countries - systems								
	USA								
	Pontts	Bridgit	Finland – SINA	Canada – OBMS	United Kingdom – NATS	France– OA-MeGA	Denmark– Danbro	Sweden – Safebrow	Germany– SIB-Bauerwerke
Storage of bridge passport data	+	+	+	+	+	+	+	+	+
Prediction of the state:									
– by expert evaluation	-	-	-	-	-	+	+	+	+
– by probabilistic evaluation	+	+	+	+	-	-	-	-	-
Consideration of different maintenance strategies	+	-	+	-	+	-	+	+	-
Assessment of transport losses	+	+	-	-	-	-	-	+	-
Determination of maintenance prioritization	+	-	-	-	-	+	+	-	+
Planning:									
– current	+	+	+	+	+	+	+	+	+
– short term	+	+	+	+	-	-	-	+	+
– long term	+	-	-	-	-	-	+	-	-

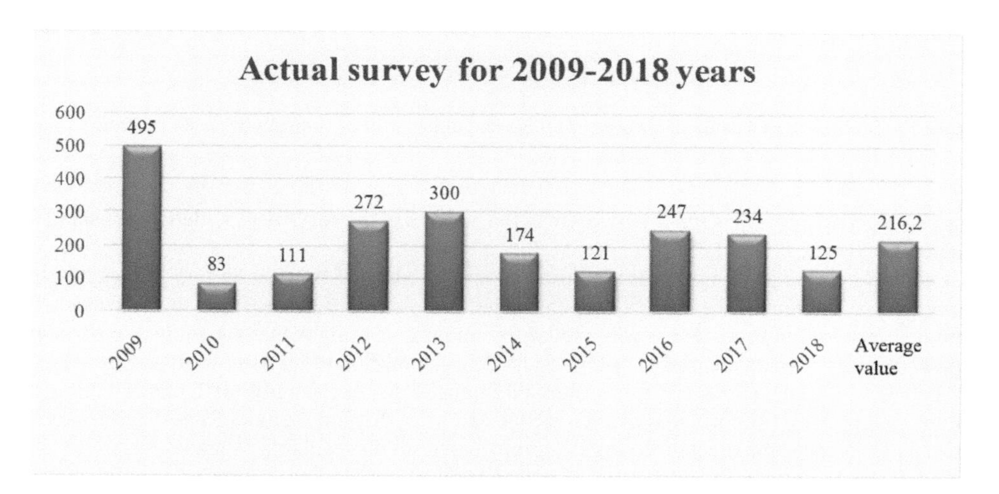

Figure 1. Retrospective analysis of actual surveys.

AESUM database contains limited information on the uninspected bridges including index, highway title the kilometer of location, year of construction, length, width, bridge scheme, and detailed information on the main elements is absent, as well as the information on the defects on the bridges and, respectively, the bridge state is not defined.

A retrospective analysis of performed bridge surveys and an analysis of the survey schedule generated in the AESUM and based on the terms of the standards (Figure 1) showed that the

Table 2. Frequency of bridges inspection according to the standards.

	Age of bridge				
	1–20	21–40	41–60	61–80	>80
Bridge	Periodicity of survey, years				
Steel, Composite	5	4	3	2	1
Steel reinforced	7	6	5	3	1

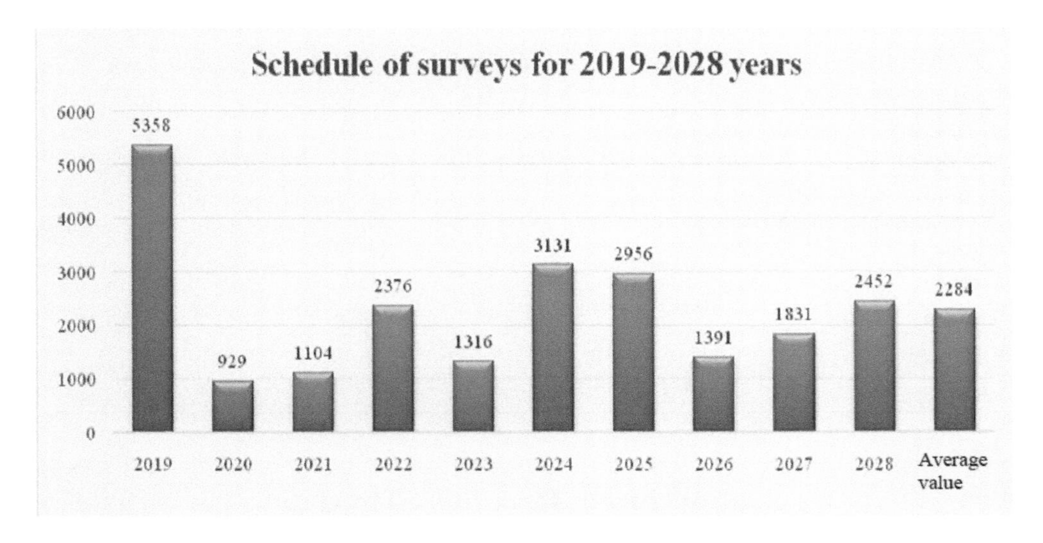

Figure 2. Prospective schedule of surveys of bridge on public roads of state importance of Ukraine.

number of actual surveys is much smaller than the number of the surveys provided by the standards (Table 2).

It should be noted that in the prospective survey schedule of bridges on public roads of state importance of Ukraine (Figure 2) it was envisaged to hold additional surveys of the bridges that were not surveyed in time; that is why the largest number of bridges that are subject to surveys is expected during the first year of the program (5358). In addition to the long-term survey schedule for 10 years, bridges survey plan is generated annually for each region of Ukraine taking into account the priority (importance) of each bridge and after that the regional Road Services locally form the final order for specialized organizations based on the available financial resources.

The AESUM software is in a state of constant development and improvement. To date, the functional of the software is proportional to the major foreign Bridge Management Systems. The AESUM database contains about 150 technical parameters for each bridge; detailed information on individual elements of the bridges such as bridge span, bearing, and foundations, as well as information on defects of the elements, their operational status and carrying capacity of the structure is stored, this information is obtained in accordance with the requirements of the state regulations (DBN V.2.3-6: 2009, DSTU-N B.V.2.3-23: 2012).

The passport contains photos of the general view of the bridge, bridge deck, main elements and defects. Drawings (or schemes) of general view and cross-section of the facilities with the main sizes are in the passport.

From the results of the surveys, defects are recorded in a defects record for each group of elements with a detailed description, distribution area and photos with the evaluation of their impact on the state of the facility. As a result of the information provided, an evaluation of the impact of the defects on the element state is carried out and the operational state for each group of elements is determined based on the worst element state of this group.

During determining the state of the elements, the possibility of expert intervention is expected which can adjust the evaluation of the operational state based on the additional load-carrying capacity calculations, safety characteristics calculations and own experience (DSTU-N B.V.2.3-23: 2012).

After that the operational state of the facility as a whole is calculated in the automatic mode. There is also the possibility for an expert to establish the operational state of the facility in accordance with the worst state of one of the defining elements of the bridge.

Quantitative indicator of an integral evaluation of the technical state of the bridge as a whole is the rating in accordance with the DSTU-N "Guidelines for the evaluation and prediction of the technical state of highway bridges.

It is determined using a scale of 100 scores and is a generalized characteristic of operational suitability for the state of all its elements. The rating is the main determining parameter for evaluation of the bridges for defining the priority of repairs.

Based on the rating, bridges are evaluated by five operational states, namely: state 1 – serviceable, state 2 – limitedly serviceable, state 3 – operable, state 4 – limitedly operable, state 5 – non-operable.

3 THEORETICAL BASIS OF EXPERT FUNCTIONS OF AESUM

3.1 *Assessment of the operational state*

The theoretical basis of the modules of the operational state assessment and forecasting the residual life of the facility is the novel model of bridge elements degradation that is formulated in the studies (DBN V.2.3-6: 2009). The model establishes a connection between the reliability and the operational time of the element. The transfer of the discrete state from one to another is described as a Poisson process with discrete states and continuous time. The model of the element degradation is calculated as follows:

$$P_t = 1 - 0{,}008333\,(\lambda t)^5 e^{-\lambda t} \qquad (1)$$

where P_t is the probability that the element will transfer to state k during $t < Tk$; λ – process parameter – failure rate, t-time, e – constant, $e = 2.718$.

Dependence (1), with a given failure rate λ, establishes the connection between the reliability of the element P_t in the i-th state and the time t passed from the beginning of the operation to the state $i = 2, \ldots 5$.

The solution of equation (1), relatively unknown t – operational time, can be obtained by a given parameter λ (the reliability of the element P_t in the i-th state is taken from the state classification table).

The failure rate λ_i is determined for the element from equation (1) as its solution by given initial conditions [8]:

- the reliability of the element in the i-th discrete state P_t obtained from the classification table of discrete states;
- time t passed from the beginning of the element's operation until the time of the classification of its discrete state.

3.2 *Forecast of the residual life of the facility*

The residual life of the element is determined from the equation of the element degradation (1), by the given reliability of the element in the i-th state and – $P_{t,i,}$ and by the parameter of the failure rate of element λ_i; the predicted time T_n which will pass from the beginning of the element operation to the state n is determined. In the case of $n = 5$, T_n will be a prediction of the residual life.

The initial data for determining the residual life is the reliability of the element $P_{t,i}$ and the time passed from the beginning of operation to the state i – t_i. These data are obtained by the engineer based on the inspections and surveys, classification of the operational state, checked calculation of load carrying capacity and safety characteristics β (DSTU-N B.V.2.3-23: 2012).

The given algorithm of the assessment and forecasting the operational state is normative and carried out in accordance with the current document DSTU-N B.V.2.3-23: 2009 "Guidelines for the assessment and forecasting the technical condition of highway bridges".

An example for implementation of this algorithm in the AESUM, a window for calculating the residual life of the bridge span is shown in the next section.

3.3 *Formalized expert evaluation (rating) of the facility*

An expert operational evaluation of the operational state is proposed for the ranking of the facilities by the need for repairs. This evaluation is determined by the scale of dimensionless coefficients of 100 scores and is calculated by the formula:

$$E = \frac{80\,(5 - \sum_{i=1}^{i=7} \alpha_i D_i)}{4} + 20 \tag{2}$$

where D_i is the number of operational state of structural elements of the facility; α_i is coefficient of the impact of the state of the i-th element on the general state of the facility (normalized weight coefficients); and $i = 1,2, \ldots, 7$ are elements of the facility according to (DSTU-N B.V.2.3-23: 2012). For example, for the bridge over the water barrier, the coefficients α_i for the bridge decking = 0,04, for the bridge spans = 0,40, for the bridge bearings = 0,25, for the foundations = 0,15, for the headwork structures = 0,05, for the channels = 0,09, for the entrances to the bridge = 0,02.

An expert operational evaluation of the operational state along with the residual life of the bridge span is the basic parameter for generating the recommendations for the strategy of operation and optimization of repair and reconstruction costs.

A quantitative indicator of expert determination of the operational state of the bridges is a formalized expert evaluation – a rating that serves as an indicator for attributing the facility to a certain operational state and the appointment of certain operational measures (Table 3).

Figure 3 presents information on the operational state of the bridges on the public roads of Ukraine that were inspected during 2006–2019 years and the information was entered into the AESUM database. The number of bridges in the worst condition (4 and 5) causes concerns.

The determined state of the facility allows assigning the necessary operational measures.

As shown in Figure 4, for each bridge element it is possible to determine the residual lifetime with the construction of the degradation curve, which allows predicting the state of the facility elements for a certain period of time in the future. Figure 4 shows the modified diagram of the distribution of bridges according to operating states, taking into account the model of degradation. So, these values have been brought up to the current year.

In this example for the element "span 7–8", in 2009 its operational state (state 4) was obtained while the operational state of the bridge as a whole is 3. The evaluated total resource of the span is 46 years. According to the predicted calculation on 2019, the current state of the span corresponds to state 5.

Taking into account the low level of financing on the bridges maintenance at the current stage of economic development of Ukraine, it is seen that it is sufficient ranking of the bridges for repairs need by tree indicators – administrative importance of the road, the operational state (or rather,

Table 3. Classification of operational state of facility elements.

Operational state	Rating	Title of the operational state	General characteristics of the state
State 1	100 – 95	Serviceable	The element meets all requirements of the project and current operation standards
State 2	94 – 80	Limitidely serviceable	The element partially does not meet the requirements of the project, but the requirements of neither the first nor the second groups of limit states are not violated.
State 3	79 – 60	Operable	The element partially does not meet the requirements of the project, but the requirements of the first group of limit states are not violated. Partial violation of the requirements of the second group of limit states is possible, if it does not restrict normal facility operation
State 4	59 – 40	Limitidely operable	Partial violation of the requirements of the first group of limit states is possible. The requirements of the second group of limit states are violated. The facility is operated in a limited mode and requires special control on the state of its elements
State 5	≤39	Non-operable	The element does not meet the requirements of the first group of limit states and it is impossible to meet them which indicates the need to stop the operation of the facility

Operational state	Quantity	%
state 1 (servisable)	97	2
state 2 (limitedly serviceable)	475	12
state 3 (operable)	2398	59
state 4 (limitedly operable)	994	25
state 5 (non-operable)	82	2
Σ	4046	100

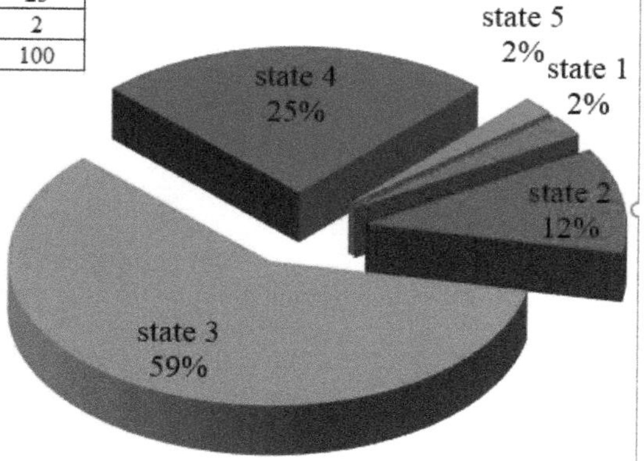

Figure 3. Distribution of bridges according to operational state on public roads of state importance of Ukraine.

the rating) and the road category. Consequently, it is assumed that in the first place repairs are needed for bridges on the roads of the highest administrative importance, highest categories with the highest traffic volume and with the worst technical state or even more precisely with the lowest rating. This is, in fact, a delay strategy, that is, a repair strategy when repairing those bridges that are already in a critical situation.

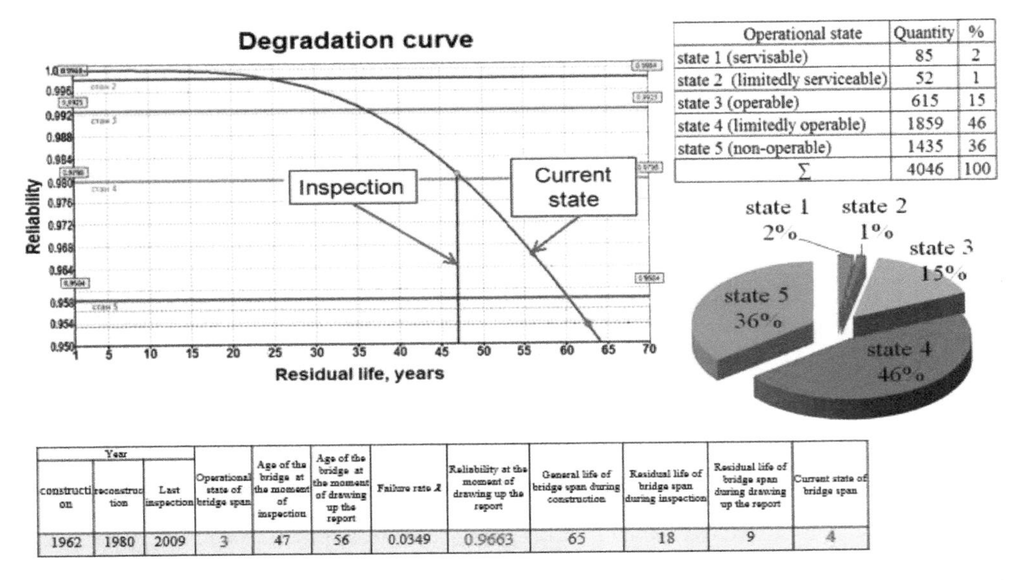

Operational state	Quantity	%
state 1 (servisable)	85	2
state 2 (limitedly serviceable)	52	1
state 3 (operable)	615	15
state 4 (limitedly operable)	1859	46
state 5 (non-operable)	1435	36
∑	4046	100

Year			Operational state of bridge span	Age of the bridge at the moment of inspection	Age of the bridge at the moment of drawing up the report	Failure rate λ	Reliability at the moment of drawing up the report	General life of bridge span during construction	Residual life of bridge span during inspection	Residual life of bridge span during drawing up the report	Current state of bridge span
construction	reconstruction	Last inspection									
1962	1980	2009	3	47	56	0.0349	0.9663	65	18	9	4

Figure 4. AESUM software mode for calculating the residual life of the bridge elements and the calculated report of the residual life and the predicted calculated current operational state.

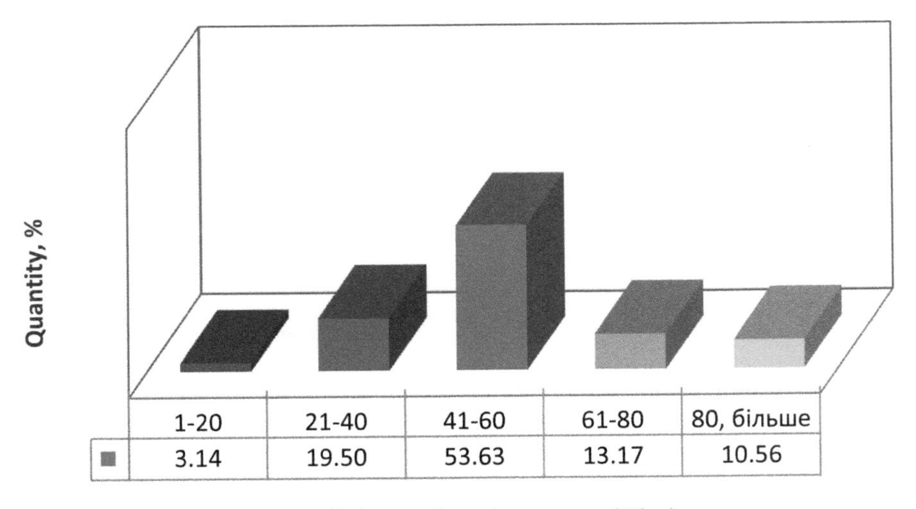

Figure 5. Ranging bridges by ages on highways of state importance of Ukraine.

An additional factor that requires increased attention to the effective operation of bridges is the fact that the middle age of the bridges is 53 years, the distribution of bridges by year is presented in Figure 5, and the project lifetime of bridges according to the standards of Ukraine is 70–100 years depending on the design (DBN V.2.3-6: 2009) but this is subject to proper maintenance.

It is needed to take into account that the repairs that were carried out at the beginning of the destruction of the bridge elements, which can be achieved by relatively small financial investments, delay the implementation of more serious overhauls or rehabilitation. Moreover, for rational operation of bridges it is needed to have a strategic long-term program. Of course, such a program undergoes annual adjustments based on the results of the actual performed work.

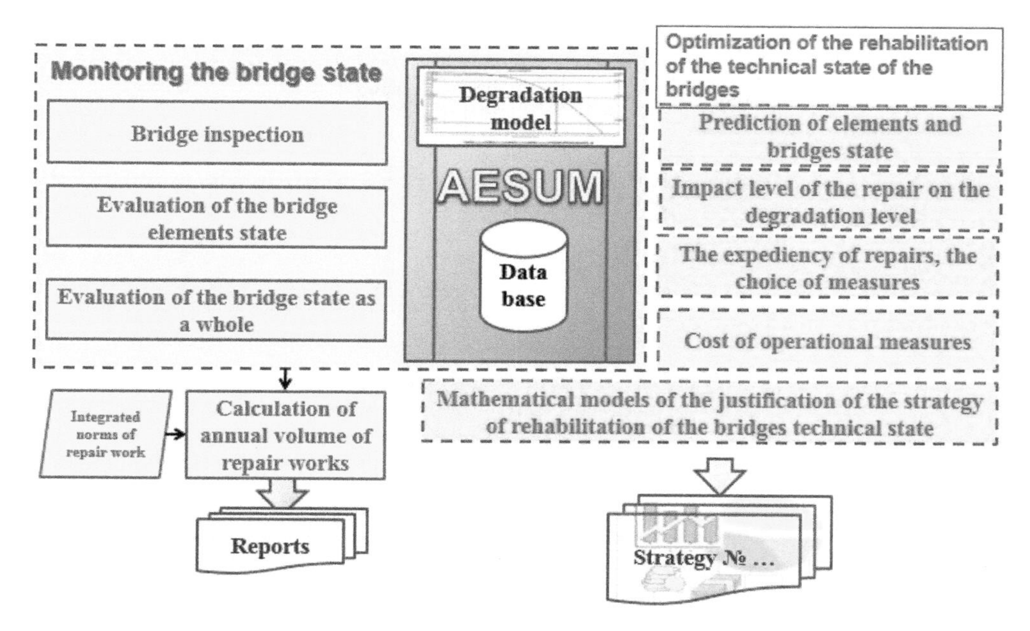

Figure 6. Basic components of the justification of the optimal bridge operational strategy.

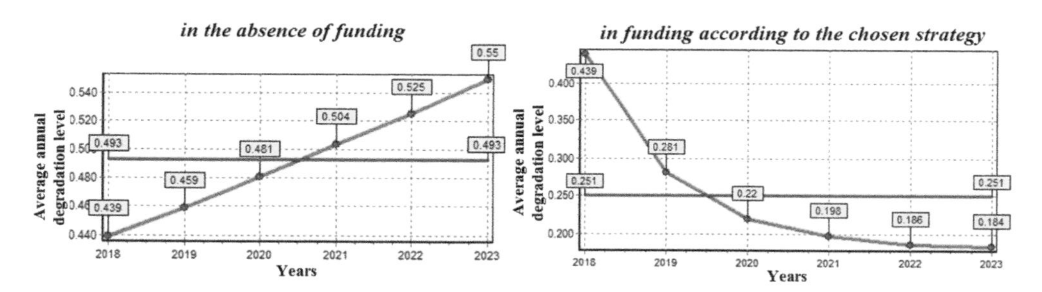

Figure 7. Change of the average annual and average weighted degradation level of the bridges by years.

4 STRATEGY OF BRIDGE OPERATION IN AESUM SOFTWARE

Research has been carried out and a module for strategic management of repairs and operational maintenance of bridges has been developed within the AESUM software. A scheme of the basic components of the justification of the optimal bridge operational strategy is presented in Figure 6.

Planning of bridges operational strategies is at the network level, that is, planning at the network level of public roads, on the network of the state, local roads, at the region level is predicted etc.

Two function of the target were proposed by which the optimal solution is adopted. Namely – minimizing the average annual level of degradation by period and average weighted by the bridges area and minimizing the cost of repairs and operation maintenance (Bodnar et al 2018). Options for constructing operational strategies with a limited budget are predicted.

Changes in the level of bridges degradation in the absence of funding and in funding according to the chosen strategy are demonstrated in Figure 7. On the graphs the red line shows the change in the average annual degradation level for a number of bridges for which a certain strategy of operation is planned. Moreover, the left graph shows a change of degradation level in the absence of funding and, accordingly, in the absence of repair work.

Year	State at the beginning of the Year	Degradation level at the beginning of the year	Work types	State at the end of the year	Degradation level at the end of the year	Estimated cost of works (Thousand UAH)	LxB, м bridge area, м2
			P-44 Sumy-Putyvl-Hlukhiv		E = 36; λ = 0.0318; Ke = 11.3;		
2018			(40040) Bridge on km 55 + 578, age - 60 years	5 (B)	0,741		11.1x7.1 78.81
2019	5	0,741	Overhaul	3	0,208	1011,9	
2020	3	0,208	Operational maintenance	3	0,221	15,8	
2021	3	0,221	Operational maintenance	3	0,234	15,8	
2022	3	0,234	Operational maintenance	3	0,247	15,8	
2023	3	0,247	Operational maintenance	3	0,260	15,8	
			T-19-04 Bilopillia-Terny-Lypova Dolyna-Hadiach				
2018			(39661) Bridge on km 0 + 782, age - 65 years	4	0,414		7.4x9.1 67.34
2019	4	0,414	Operational maintenance	4	0,423	20,2	
2020	4	0,423	Overhaul	1	0,050	720,5	
2021	1	0,050	Operational maintenance	1	0,050	6,7	
2022	1	0,050	Operational maintenance	1	0,050	6,7	
2023	1	0,050	Operational maintenance	1	0,050	6,7	
			P-44 Sumy-Putyvl-Hlukhiv		E = 39; λ = 0.0565; Ke = 6.9;		
018			(40035) Bridge on km 65 + 883, age - 29 years	5 (Б)	0,602		18.1x12 217.2
2019	5	0,602	Reconstruction	1	0,050	2823,6	
2020	1	0,050	Operational maintenance	2	0,052	21,7	
2021	2	0,052	Operational maintenance	2	0,070	32,6	
2022	2	0,070	Operational maintenance	2	0,088	32,6	
2023	2	0,088	Current minor repair	2	0,057	830,8	

Figure 8. Fragment of the work program results by each bridges for 5 years.

Midterm programs for the operation of bridges (for 5 years) are developed by the authorized persons of State Road Agency of Ukraine (Ukravtodor) based on the calculated strategies by the AESUM software. Fragment of the work program for 5 years is shown in Figure 8. For each bridge, information on the change of the operational state subject to the performance of the proposed work types for each year of the calculated period is provided.

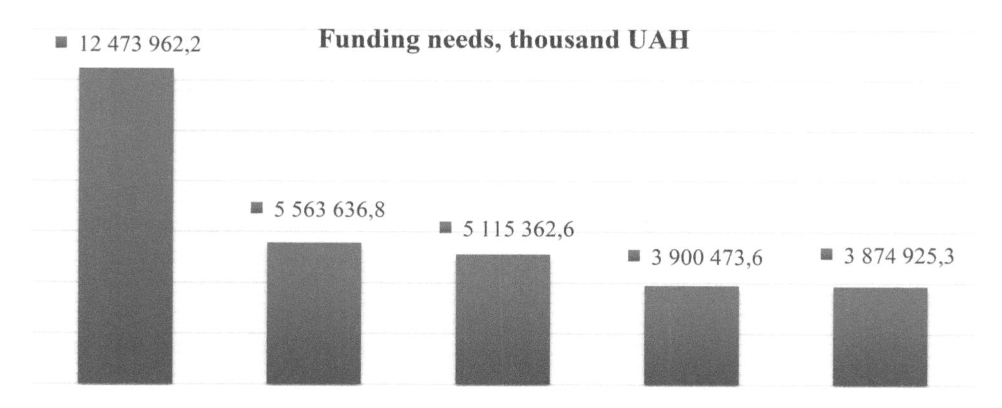

Figure 9. Graph of funding needs for the bridges operation according to a certain strategy.

The graph (Figure 9) presents the distribution of required funds by years for a certain strategy the fragment of which is presented in Figure 8.

5 CONCLUSIONS

The AESUM software of Ukraine contains all main parameters of Bridge Management Systems and solves main global objectives of the systems of this type. Work on its improvement will be continued in the following areas: update of the AESUM database with information on inspections / certificates of bridges; justification of reconstruction / new construction taking into account risks; refinement and improvement of models and methods of calculating the degradation level of bridges to increase their reliability, durability; improvement and implementation of the module for making effective decisions for strategic planning of repairs; development of the comparison mechanism of results for different strategies, models; improvement and implementation of the mechanism for determining the cost of repairs per 1 m^2 by type of repairs due to different indicators.

At present, the AESUM software does not evaluate the transport losses from the unsatisfactory state of the bridges; therefore, it is necessary to study the experience of the AMS (Analytical Management System) of the USA for the development of the corresponding module in the Ukrainian software.

REFERENCES

Bodnar L.P. 2010. AESUM software. Current state and concept of further development. "Roads and Bridges": collection of scientific works – K .: DerzhdorNDI – Issue 12. – 31–39 p.

Bodnar L.P., Kanin A. & Stepanov S. 2018. Genetic Algorithm to optimize the strategies for bridge repair works. 10 p. Access mode:https://zenodo.org/record/1485402#.XDX1nGmLmUk.

DBN V.2.3-6: 2009 Transport facilities. Bridges and culverts. Inspection and testing.

DBN V.2.3-22: 2009 Transport facilities. Bridges and culverts. Main design requirements.

DSTU-N B.V.2.3-23: 2012 Guidelines for evaluation and prediction of the technical state of highway bridges.

Shesterikov V.I. et al. 2007. Management of the bridge structures state on the federal network of highways of Russia: Overview. Inform. / V.I. Shesterikov, L.I. Gorobets, I.K. Matveev; Informavtodor.-M.-96 p.: ill.- (Highways and bridges, Issue 2). – Access mode: http://internet-law.ru/stroyka/text/56251/

Shesterikov V. & Gorobets L. 2002. Bridge operation management system in Russia. Intertraffic Asia 2002: PIARC Seminar. – Bangkok, Thailand,

Speiran K. & Reed M. Ellis. 2004. Implementation of a Dridge Management System in the Provinct of Nova Scotia: Annual conference of the Transportation Association of Canada. – Guebec,

Chapter 7

Replacement strategies of existing highway bridges in Germany

M. Schumm
Leonhardt, Andrä und Partner Beratende Ingenieure VBI AG, Stuttgart, Baden-Württemberg, Germany

ABSTRACT: This paper deals with the replacement of existing highway bridges by the example of two bridges over the river Rhine in Duisburg-Neuenkamp and in Leverkusen. Both mainly welded cable stayed bridges are under traffic since circa 1970–1965 and are constructed as welded steel girder bridges with an orthotropic bridge deck. Recent damage to the steel structure of the bridge deck reduced the service lives of both bridges significantly, therefore, the bridges needed to be replaced within a short time, while maintaining the traffic during all construction stages. The replacement strategies and the design requirements are described in the following.

1 GENERAL

1.1 *Actual situation of infrastructure and traffic flow in Germany*

Today the total length of the federal trunk highway roads in Germany is about 50,000 km being one of the densest traffic in whole Europe. Among these are approximately 40,000 bridges with a total area of 30 million Square meters. Fifty percent of these bridges were built in the years between 1960 to 1980.

At that time these bridges were mainly designed according to German Standard DIN 1072 for bridge class SLW 60 (60 ton truck). Based on the traffic given at those times, no fatigue checks were required. This assumption seemed fair enough because the massive increase, especially in heavy traffic, could not have been anticipated (see Figure 1).

1.2 *Existing Rhine bridges in Duisburg-Neuenkamp and Leverkusen*

Both existing bridges in Leverkusen and Duisburg-Neuenkamp are under traffic since built in 1965 and 1970 respectively (Figure 2). They are cable-stayed bridges and constructed as steel girders

Figure 1. Traffic on motorways – actually 2017 © FUNKE Foto Services / Ulla Michels vs. 1970 (Meyer 1965).

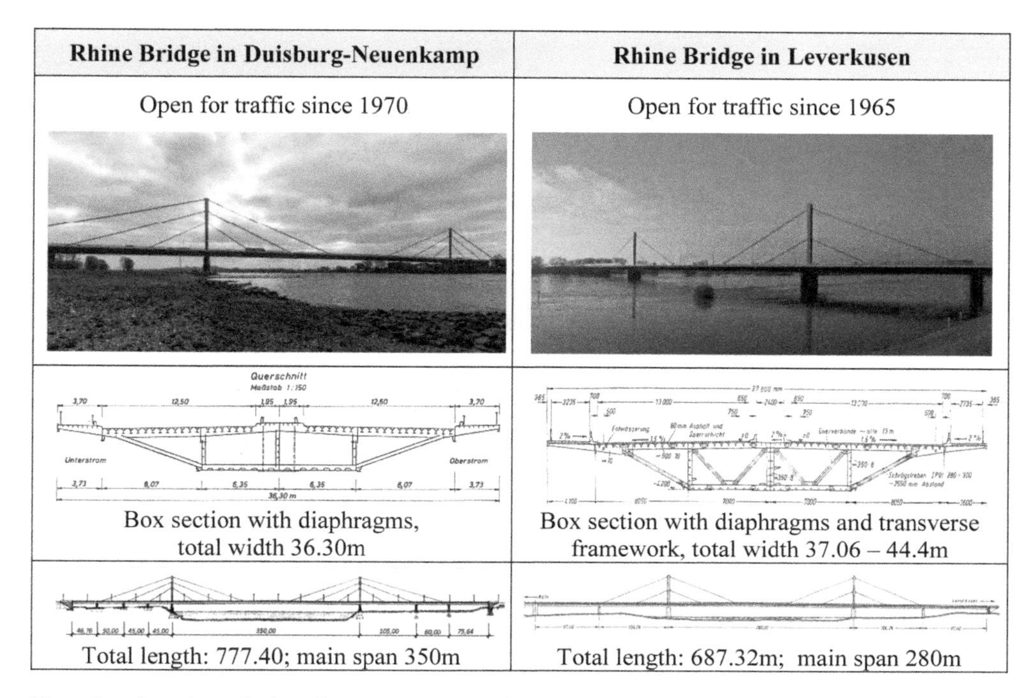

Figure 2. Overview of selected construction data of the existing bridges.

with orthotropic bridge decks. Most parts of the bridges are welded. The bridge over the Rhine near DuisburgNeuenkamp is one of the first completely welded bridges in Germany.

Currently more than 100,000 vehicles per day pass the bridges, with a portion of more than 10% heavy lorries. Due to the increasing traffic, in connection with some inadequate fatigue design details a steady increase of damages in steel bridges is noted on many steel bridges such as in Duisburg-Neuenkamp and in Leverkusen.

In the last 5 years not only the area of the orthotropic bridge deck but also the main structural elements were damaged. In Duisburg-Neuenkamp damages occurred at the connection of the bracings with the main girder web due to laminar tearing defects at the 12mm web plate to which the gusset plate was welded. The cyclic through-thickness compression stress range caused fatigue issues starting at the laminar defects of the web plate (Paschen et al. 2017). A repair at those locations has been conducted to deal with the worst damages, but to strengthen all such details makes economically no sense.

Both bridges can only stay under traffic by constant monitoring and repair works, so that they have to be replaced in the very near future. The bridge in Leverkusen is reduced to a total weight of 3.5 ton per vehicle. In Duisburg-Neuenkamp the six lane traffic configuration had to be reduced to four traffic lanes centered in the middle of the bridge. In-motion weighting units are to be installed on both bridge ends, to ensure no vehicle above 40 tons will pass the bridge.

2 BRIDGE REPLACEMENT

2.1 *General strategies*

The Primary requirement for the replacement of the existing bridges includes minimizing the time required for construction and maintaining the traffic flow during all construction stages and to arrange eight lanes on the new crossing. The new design shows separate superstructure decks for

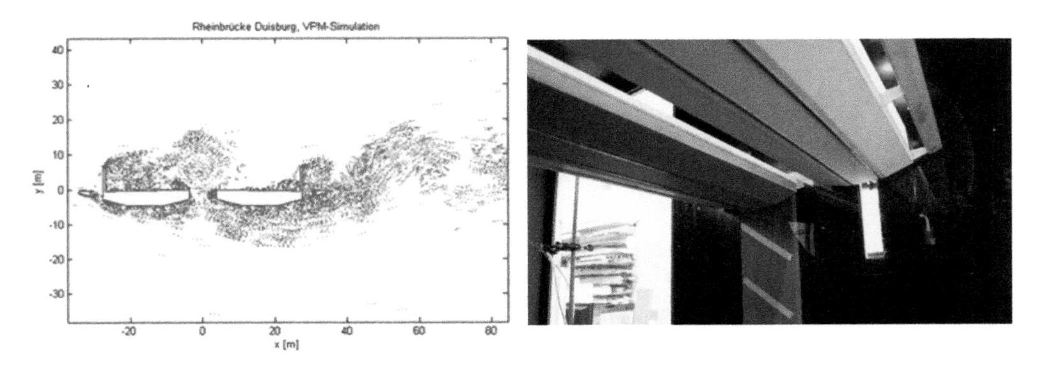

Figure 3. CFD analysis and wind tunnel test (Wacker 2018) for the bridge in Duisburg-Neuenkamp.

each carriageway with a width to allow for a temporary traffic management in both directions on one bridge deck with 6 lanes during maintaining and repairing the other bridge deck.

2.2 *Design requirements*

The design of the new bridges must be in accordance with the Eurocode Standards. Ultimate and service limit states have to be checked carefully. The design of the bridges has to fulfill all fatigue requirements. Apart from this, the following additional requirements have to be considered:

- A minimum thickness for the orthotropic steel deck plate of 16 mm in the carriageway and 12 mm in the area of the walkways has to be kept.
- The stiffeners of the orthotropic deck shall have a minimum thickness of 8 mm and the slenderness ratio shall consider class 1 or 2 section. The stiffener arrangement at the orthotropic deck is such that the span of the deck plate between the trapezoidal stiffeners does not exceed 300mm.
- The properties in through-thickness direction shall fulfill a grade Z15 as a minimum, to avoid the effect of laminar tearing in welded structures.
- The transverse stiffening of the box girder has to be very efficient. In case of using steel cross frames as bracings it has to be considered that they are weaker than solid steel diaphragms regarding section warping and that their connections with gusset plates is being seen critical with respect to fatigue stresses.
- The steel composite section at the backspan of the cable-stayed bridge shall be designed in a way that no uplift forces occur in ULS. Therefore, it is necessary to increase the corresponding thickness of the concrete slab to gain sufficient counterweight.
- For the stay cables a mono strand system is chosen. Cable exchange and sudden cable loss, considering an impact factor of minimum 1.5 need to be considered in the design.
- Inspection of all structural elements must be possible; all box sections (pylon and deck) must be accessible. For the inspection of the deck in the main span a stationary bridge inspection vehicle has to be designed, especially for the bridge and it shall be parked in the abutment.
- Ship impact forces up to maximum of 17 MN (1,700 tons) shall be considered.
- Wind tunnel tests shall be performed for the wind design (Figure 3) to confirm aeroelastic stability and to ensure vortex shedding will not result in unacceptable passenger comfort or cable vibrations.

3 BRIDGE OVER THE RIVER RHINE IN DUISBURG – NEUENKAMP

3.1 *Essential design conditions and preferred alternative*

The German highway A40 runs in east – west direction over the river Rhine and links the Dutch border with the industrial sites in the Ruhr area in Nordrhein-Westfalen (NRW). Due to the extremely

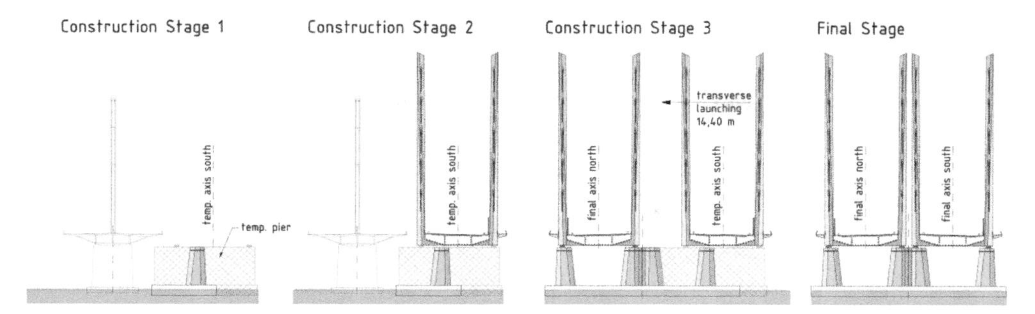

Figure 4. Principle sketch of the construction stages – Variation 3.

high density of traffic, traffic jams occur every day. The massive damages of the existing bridge deck required, especially due to the increase of heavy goods traffic on the highway A40, a reduction to only four traffic lanes on the bridge which resulted in even more traffic jams. Existing traffic data from 2016 shows around 115,500 vehicles per day with a portion of more than 15% being heavy lorries. Therefore, the development of an efficient crossing of the river Rhine and the expansion of the motorway to 8 lanes was an essential part of the federal transport infrastructure plan, executed by the German government.

The horizontal alignment of the bridge replacement has to consider the existing buildings which are situated close to the highway on both sides of the Rhine as well as the environmental effects due to the widening of the highway A40.

The following horizontal alignments for the replacement were investigated:

- Variation 1 – lateral off-set of the highway axis towards North
- Variation 2 – lateral off-set of the highway axis towards South
- Variation 3 – optimized highway axis with a temporary lateral off-set towards South and a transverse launching of the southern bridge deck in final stage

While variations 1 and 2 show significant interferences with the existing buildings and the environment, variation 3 is more favorable because the lateral off-set especially in final stage is much smaller. Therefore it is planned, that the southern bridge is first built with a temporary lateral off-set of about 14.40 m. After that, the temporary six-lane traffic is rearranged from the existing bridge to the new southern bridge and the existing bridge can be dismantled. Then the northern bridge is to be built in its final position. Once finished the temporary 6-lane traffic will be transferred to the northern bridge. Finally, the southern bridge has to be launched in transverse direction into its final position (Figure 4) and the connections to the road network is finalized. Finally the East bound carriageway is transferred back to the south bridge allowing a 8 lane arrangement on the finalized new crossing.

The most important component for this motorway section is the replacement of the existing bridge over the river Rhine in Duisburg-Neuenkamp. In preliminary studies, the following bridge types were investigated (Figure 5):

- Two double wing cable stayed bridges with two separate bridge decks
- One double wing cable stayed bridge with one single bridge deck
- Two suspension bridges (self-anchored to allow transversal shifting)
- Two asymmetric cable stayed bridges with one single bridge deck
- Two arch bridges

With respect to the redundancy of the bridge system, two double wing cable stayed bridges with separate bridge decks were mandated in the tender design.

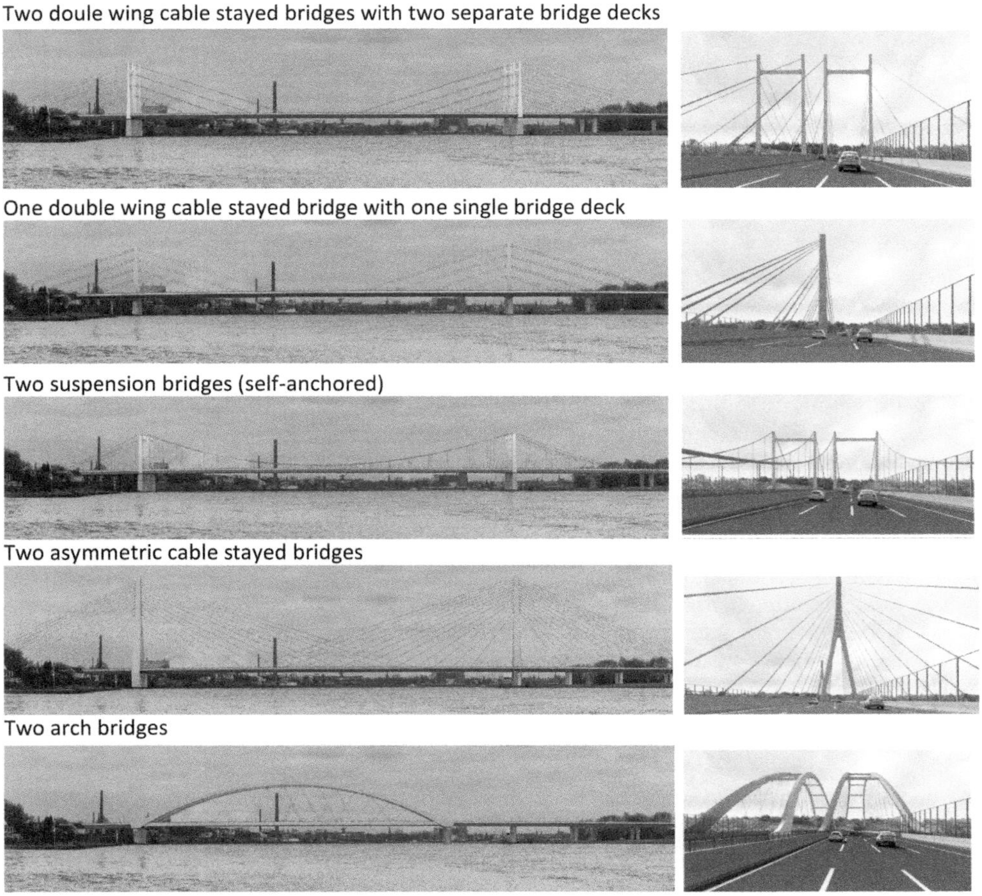

Two doule wing cable stayed bridges with two separate bridge decks

One double wing cable stayed bridge with one single bridge deck

Two suspension bridges (self-anchored)

Two asymmetric cable stayed bridges

Two arch bridges

Figure 5. Investigated bridge types during the preliminary design phase.

3.2 *Tender design*

The design of the bridge is a double-wing cable-stayed bridge with two pin pylons and separate bridge decks for each carriageway. The harp-shaped arrangement of the stays with a mono strand cable system ensures a clear setup avoiding visual clutter (Figure 6). The stiffening truss is a closed three-cell box girder with a width of 30 m and a height of 4 m in the bridge axis. The pylons are also steel box sections and are rigidly connected to the bridge deck and the pylon cross beam to allow for the required transverse launching of the southern bridge. The bearings under each pin pylon have a capacity of up to 13,000 ton. At the main span, the bridge deck is an orthotropic steel plate with diaphragms at a distance of 3.75 m. At the side spans a steel composite box section with a concrete slab of 55 cm (15 cm precast elements as formwork with an overlay of 40 cm cast in situ) acts as a counter weight and prevents uplift reactions.

To protect the environment against traffic noise, transparent acoustic walls with a total height of 6.50 m have to be installed on both sides of the bridges.

The following criteria governed selection of the final main span of the replacement:

- Guideline of the German Federal Waterways and Shipping Administration (WSV) that the main span shall not be less than 350 m to avoid restriction of the shipping

Figure 6. Visualization final bridge layout © A. Keipke, Rostock.

- The new main piers shall be shifted towards landside to avoid interference between the new foundations and the caisson foundations of the existing bridge
- Reduction of hydraulic and morphological effects
- Simple construction of the foundations outside of the river Rhine

This results in a main span of 380 m and the following individual spans for the two double-wing cable stayed bridges are thus:

48 m – 70 m – 70 m – 380 m – 60 m – 60 m – 60 m – 54 m resulting in a total length of 802 m (Figures 7–9).

Pylons:

The pylons are rigidly connected to the bridge deck. In total eight single pylons with an overall height of about 70 m above deck level are to be installed. Each pylon leg rests on spherical bearings with a capacity of about 13,000 ton. The ratio between main span and pylon height is $71/380 = 0.19 \approx 0.2$ which leads to cable angles of 21°.

For aesthetic reasons, no pylon cross girders are used. The stability in longitudinal and transverse bridge direction is only given by the stay cables and the fixation of the pylon into the deck.

The width of the pylon legs is 4.50 m at the bottom and 3.0 m at the top and in transverse bridge direction it is 3.0 m at the bottom and 2.60 m on top. The cross section is a steel box with a hexagonal geometry (6 corners) at the bottom and a pentagonal geometry (5 corners) on top (Figure 10).

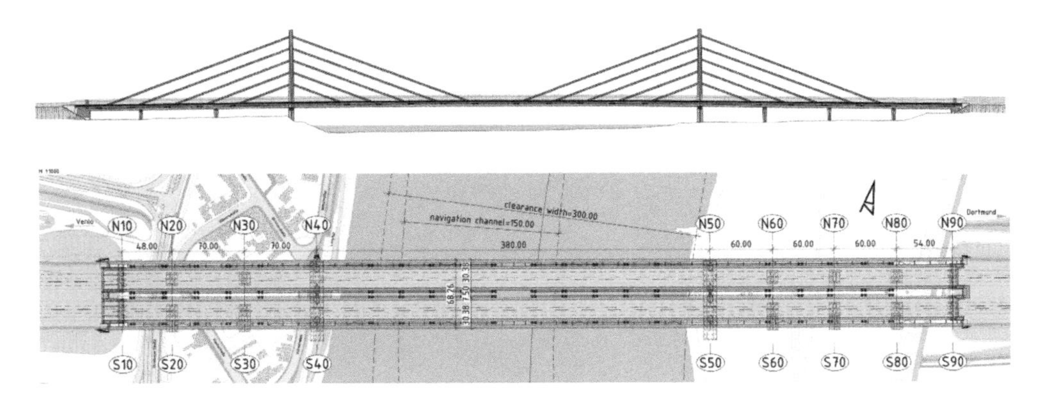

Figure 7. View from the south and plan view of the final bridges.

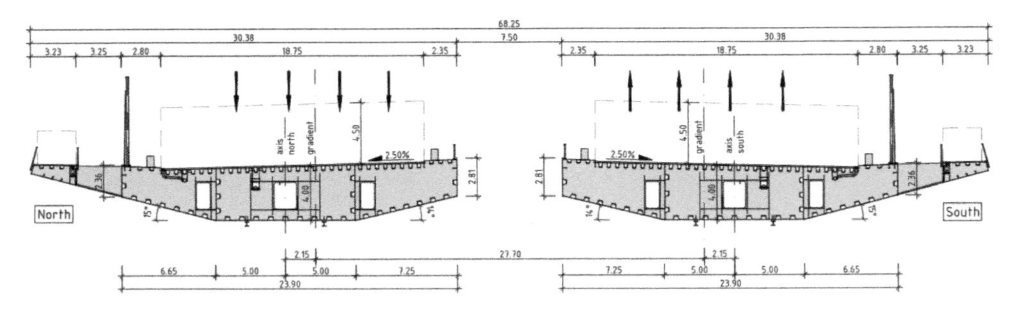

Figure 8. Cross section main span – orthotropic steel deck.

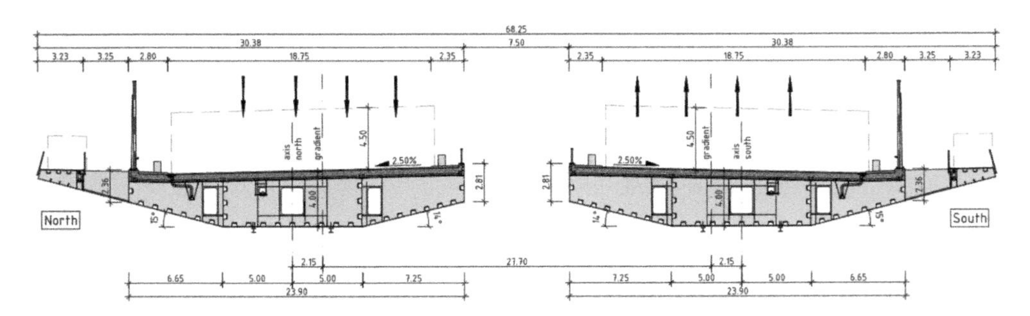

Figure 9. Cross section side spans – steel composite section.

3.3 *Construction stages and time schedule*

The steel section of the bridge deck at the side span will be incrementally launched from the assembly areas behind the abutments without use of temporary piers. The main span erection will be carried out by free cantilevering method.

According to the scheduled construction sequence, the southern bridge will be built first temporarily in lateral position to the existing bridge. After that, the traffic is transferred to the southern bridge and the demolition of the existing bridge will start. Subsequent the northern bridge will be built in its final position. After a transfer of the traffic, the southern bridge with a total weight of approximately 36,000 ton will be launched transversely into its final position (see Figure 3).

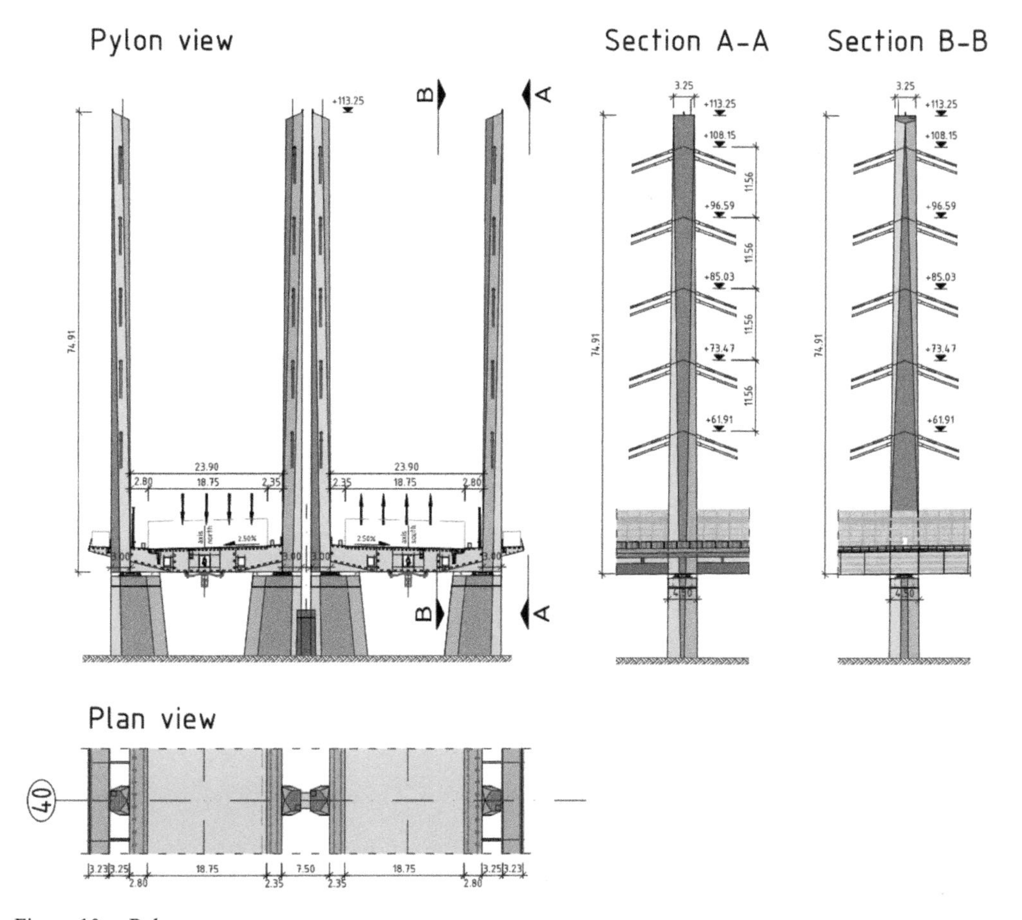

Figure 10. Pylons.

Planned time schedule for the replacement of the Rhine bridge:

Start preliminary and final design	November 2018
Submittal of tender documents	October 2019
Construction Contract award	March 2020
Construction of South Bridge and transfer of traffic from old bridge	End of 2022
Demolition of old bridge and Construction of North bridge	End of 2023
Transfer of traffic to North Bridge and lateral shift of south bridge	April 2026
Opening of finalized crossing	July 2026

4 BRIDGE OVER THE RIVER RHINE IN LEVERKUSEN

The main bridge is likewise a double-wing cable-stayed bridge with separate bridge decks for each carriageway and a main span of 280 meters with a fan-shaped arrangement of the cables (locked coil ropes) (Figure 12). The bridge has a total width of 34 m, carrying six traffic lanes each, consist of two outer steel box girders with crossbeams at a distance of 4 m. The A-shaped towers consists of steel boxes and are rigidly connected to the bridge deck. As for the bridge in Duisburg-Neuenkamp a steel composite cross section in the side spans is used here. According to the construction sequence, the northern bridge is built first in its final position, than after dismantling of the existing bridge,

Figure 11. Visualization during the competitive phase © LAP.

Figure 12. Bridge visualization – final design © Landesbetrieb Straßenbau NRW (Ritterbusch 2019).

the southern bridge can be built in its final position. Thus the alignment highway will be shifted in northern direction.

During the competitive phase, LAP proposed an alternative bridge design of two unsymmetrical cable stayed bridges set apart from each other (Figure 11). Although this design was considered to be quite good, it could not be followed up because of requirements to reduce the blocking by the stay cables to minimize the risk for the birds.

5 CONCLUSIONS

Many bridges in Germany exceed their service life mainly due a considerable increase in heavy traffic volume causing in particular fatigue issues for the steel bridges. Since strengthening and repair is extremely costly and a further widening is not possible, the decision is taken by the responsible authorities to replace such live line bridges to maintain reliability on the European Road Network.

The paper describes the design of the replacement for two major bridges and explains some specific requirements such as:

- Design according to Euro Codes with a high demand on traffic design loads and structural detailing especially with respect to fatigue resistance and material choice (requirements are stricter than for a FCM material being in accordance with AASHTO)
- Separated bridge structures for each carriageway allowing closing one of them by transferring both traffic direction on one girder if required for strengthening or exchange of elements at the other bridge

REFERENCES

Meyer, H. 1965. Die Bundesautobahn, Nördliche Umgehung Köln. Zeitschrift Bitumen 6.

Paschen, M., Hensen, W. & Hamme, M. 2017. Instandsetzungs- und Sicherungsmaßnahmen bei den Rheinbrücken Leverkusen und Duisburg-Neuenkamp – ein Zwischenbericht (Teil 1 und 2), Stahlbau 86, Heft 7 + 12.

Ritterbusch, A. 2019. Errichtung der Rheinbrücke Leverkusen und anderer Ingenieurbauwerke – Ausbau der A1 zwischen Köln-Niehl und Leverkusen-West. 19. Brückenbausymposium in Leipzig, Ausgabe 1/2 2019, Verlagsgruppe Wiederspahn.

Wacker Ingenieure 2018. Windgutachten Rheinbrücke Duisburg-Neuenkamp.

Bridge failures

Chapter 8

Two collapses of the Ontario & Western Railway's bridge at Fish's Eddy

D.F. Mazurek & K.M. Tarhini
U.S. Coast Guard Academy, New London, Connecticut, USA

ABSTRACT: Built in 1882, the New York, Ontario & Western Railway's three-span pin-connected through truss bridge over the East Branch of the Delaware River near Fish's Eddy, New York experienced two collapses in its short 15-year life. The first occurred as a result of a derailed caboose striking the end post of the north span. The second took place while a lightly loaded train was crossing the bridge, collapsing the middle span; the cause of this failure was never determined. A historical review is presented that begins by examining relevant aspects of the specifications used in the design of the structure. Details of an innovative bridge protection system developed in response to the first collapse are discussed, and related to the subsequent development of industry standards. Analysis results are given that demonstrate that the hangers were excessively loaded, and that a hanger failure may have initiated the second collapse event.

1 INTRODUCTION

The New York, Ontario & Western Railway (O&W) was formed in 1880 from the ashes of the bankrupt New York and Oswego Midland Railroad, and soon engaged in a program to renew and upgrade the right-of-way. In 1882, the Central Bridge Works of Buffalo, New York, was contracted to design a new bridge over the East Branch of the Delaware River, near Fish's Eddy, New York. The new structure was a three-span, wrought iron, pin-connected through truss, and accommodated a single track. The bridge functioned with no known incidents of significance until March 3, 1886, when the caboose of a southbound train derailed due to a broken rail and struck the end post of the first truss, collapsing the span and crushing the car, killing all four of its occupants (*Orange County Press* 1886). The railroad restored the span, and in apparent reaction to the tragedy, soon developed an innovative bridge protection system specifically designed to prevent future accidents of this nature.

Some eleven years later, the middle span of the bridge collapsed in the early morning hours of April 28, 1897 as a northbound train was crossing. Figure 1 shows the collapsed structure. The 30-car train was under orders not to exceed 4 miles per hour while traversing the bridge, as it was to immediately take a siding to allow another train to pass. The locomotive had safely traversed the span prior to the collapse, and the train itself consisted of empty coal gondolas, eleven of which fell into the river with the failed span. Because the train was traveling so slowly, it was felt that derailment was not a likely cause, and the bridge "was considered so perfectly safe in every particular that the officials express the greatest surprise that an accident of this kind should happen" (*The Middletown Daily Times* 1897a). After a thorough investigation, officials were still at a loss as to the exact cause, and speculated that the collapse might have been triggered by either the train becoming separated, with the rear portion catching up and colliding with the forward part of the train, or that a car simply derailed and struck the bridge. The Union Bridge Company (which had absorbed the Central Bridge Works) was contracted to build a new structure to replace all three spans (*The Middletown Daily Times* 1897b). The new structure continued to be a three-span through truss, albeit a much more robust design than the former bridge, and now consisted of two tandem

Figure 1. Middle span collapse of the Fish's Eddy Bridge in 1897. (DeForest Douglas Diver Railroad Photographs #1948. Division of Rare and Manuscript Collections, Cornell University Library.).

Figure 2. Fish's Eddy Bridge in 2017.

bridges instead of just the one to accommodate the two-track mainline. Though the railroad was subsequently abandoned in 1957, one of these two bridges survives today as shown in Figure 2, converted to highway use.

After reviewing various details of the Fish's Eddy Bridge and the specifications used in its design, this paper will examine the unique bridge protection system developed by the O&W in response to

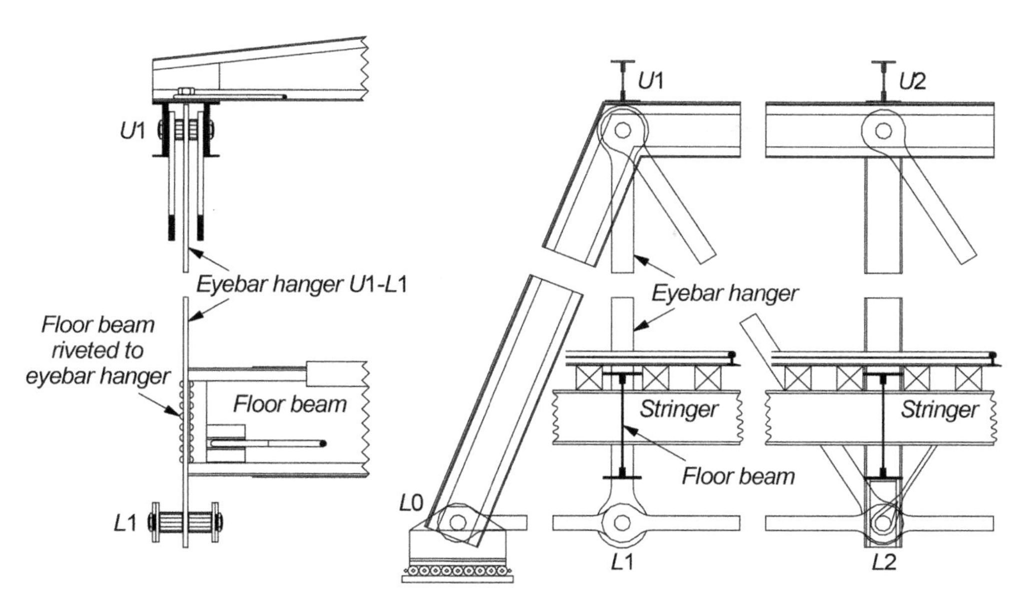

Figure 3. Truss details of the Fish's Eddy Bridge, showing a partial section (left) and elevation (right).

the 1886 collapse, and consider the possibility of a hanger failure as being the cause of the 1897 collapse.

2 BACKGROUND

2.1 *Structural aspects of the 1882 bridge design*

The bridge consisted of three identical Pratt trusses skewed at 53.5°, with each being 43.9 m long and consisting of nine panels. The compression members (upper chord, end posts, and vertical posts) essentially consisted of double channels reinforced with various configurations of cover plates, tie plates, and lacing. Tension members (lower chord, diagonals, and first-panel hangers) used eyebars with the exception of the diagonal counters, which were rod members. The skew was accommodated entirely within the first panel at the ends of each truss, resulting in a sizable difference in lower chord lengths on each side (approximately 6.7 m versus 3.0 m). Figure 3 highlights the main structural details of the first few panels. Except for the pins at all panel points, connections were generally riveted. With the exception of the hangers (such as member $U1$-$L1$), all tension members were comprised of at least two eyebars. On the other hand, each hanger consisted of just a single eyebar, which in turn supported a floor beam that was riveted directly into the eyebar's shank as shown in Figure 3. This unusual design results in a significantly reduced cross section of the hanger, potentially lowering capacity. A more common practice would have been to either connect the floor beam directly to the pin joint, or to rivet the floor beam into a hanger consisting of a more robust, built-up section. The influence of this reduced section on hanger strength will be examined in detail later in this paper. Since the riveted connection itself possessed a measure of redundancy, discussion of rivet capacity is not included.

2.2 *O&W's 1881 Specification for Iron Bridges*

Shortly following the reorganization of the Midland into the O&W, the *General Specifications for Iron Bridges* (Katte 1881) was developed, governing the design and construction of the Fish's Eddy

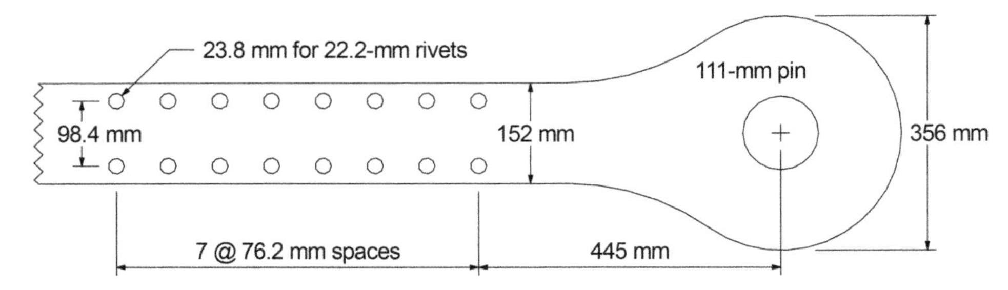

Figure 4. Hanger connection details for Fish's Eddy trusses.

truss spans in 1882. These specifications required that wrought iron be used for all parts of the superstructure, except in the case of bed plates and washers, where cast iron was permitted. As demonstrated by testing, the iron procured for the bridge was to have a minimum ultimate tensile strength of 345 MPa and an elastic limit of not less than 172 MPa. Spans over 22.9 m were to consist of pin-connected trusses (which, at 43.9 m each, the three Fish's Eddy spans were so constructed). Recognizing that hangers supporting floor beams are particularly vulnerable to impact loads, the allowable tension stress for such members "and other similar members liable to sudden loading" was limited to 41.4 MPa. Among all the types of tension members addressed by the specification, this allowable stress was the most restrictive. For comparison, the allowable tension stress for counter rods and long verticals was 55.2 MPa, while that for bottom chords and main diagonals was 69.0 MPa. The specification also required that in the case of tension members, "full allowance shall be made for reduction of section for rivet-holes, screw-threads, etc." In addition, any members that are subjected to bending "from local loadings (such as distributed floors on deck bridges)," the affected members must be proportioned to support these bending effects in combination with the primary member stresses. In comparison to the practices of others during this period, these allowable stresses are generally more conservative, especially in the case of hangers. For example, the general specification used by Clarke, Reeves & Company of Phoenixville, PA, stipulated an allowable stress of 69.0 MPa for all members subject to tension under the application of dead and live loads (Vose 1881).

By the O&W specification, "all the connections and details of the several parts of the structure shall be of such strength that, upon testing, rupture shall occur in the body of the members rather than in any of their details or connections." Figure 4 shows the hanger connection details for the Fish's Eddy trusses; since it is apparent that the net section at the riveted connection is considerably less than the gross section in the body of the eyebar, this rupture requirement was obviously not satisfied. (On the other hand, conventional practice today is to use yield of the gross section and rupture of the net section as appropriate limit states; based on this it is possible for such a connection to comply with a requirement of yielding on the gross section before rupturing on the net section.) Regarding allowable stresses, the specification required that pins be sized such that the shearing, bearing, and bending stresses not exceed 51.7 MPa, 82.7 MPa, and 103 MPa, respectively. Riveted connections were to be designed such that the shearing and bearing stresses not exceed 41.4 MPa and 82.7 MPa, respectively.

The O&W specification also stipulated a number of fabrication requirements pertinent to hangers. The eyebars were to be free from flaws, full thickness through the necks, and perfectly straight before boring, with the pin holes bored in the center of the head and on the centerline of the bar. The pins were to be straight and smooth, fit the pin holes within 0.508 mm, and be of diameter not less than two-thirds of the largest cross-sectional dimension of any member attached to it. Regarding rivets, the pitch was not to exceed 152 mm or 16 times thickness of the thinnest outside plate, nor be less than three rivet diameters. Rivets were to be either 19.1 mm or 22.2 mm, and the distance from any edge to the center of each rivet hole was not to be less than 31.8 mm. Holes were to be so accurately punched such that a hot rivet of 1.6 mm diameter less than the hole could be inserted without reaming or straining the iron by forcing drift pins to achieve alignment. When driven, the

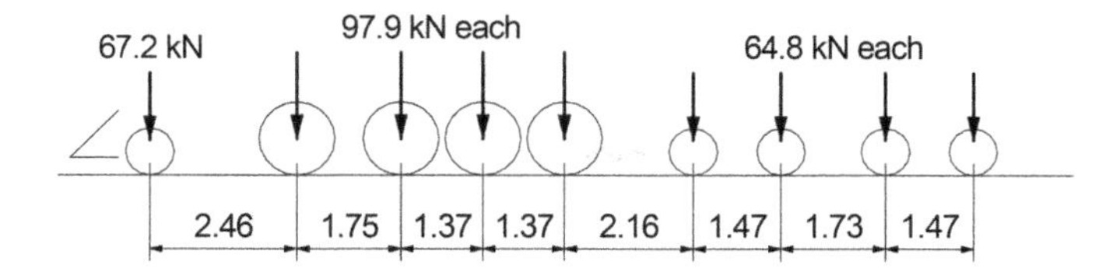

(a) Design locomotive load as stipulated by the 1881 O&W *General Specification for Iron Bridges*.

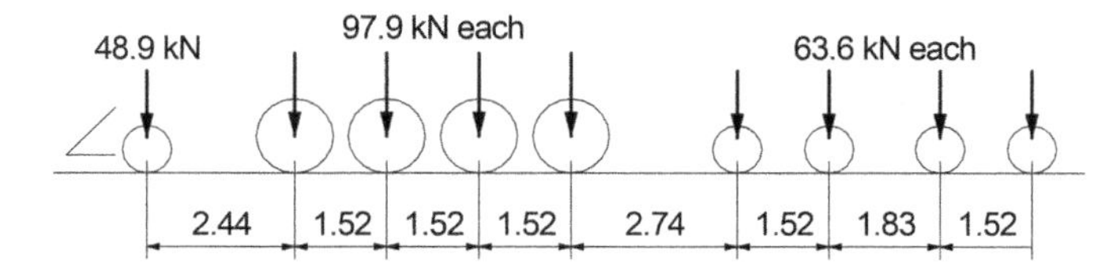

(b) Design locomotive load representative of Cooper E22.

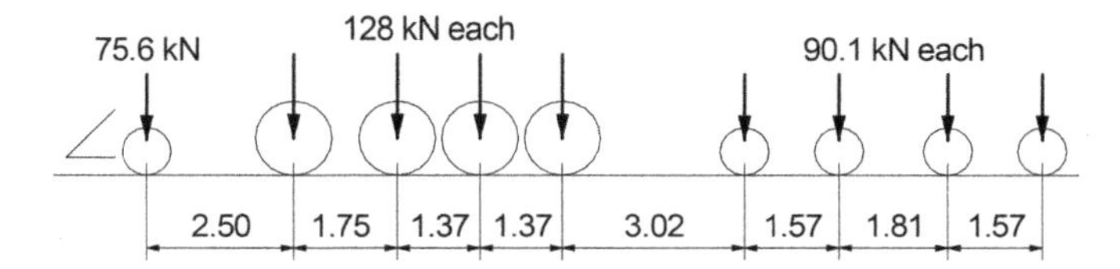

(c) O&W Class S locomotive, introduced in 1890.

Figure 5. Live loads (dimensions in m).

rivets were to completely fill the holes. There was no requirement for sub-punching and reaming rivet holes, so the presumption in general is that all rivet holes were simply punched. However, given the rather large thickness of the eyebar hangers at 31.8 mm, in this case it might seem more likely that their holes were drilled, but again, probably not reamed. The bridge design was also to be such that the floor beam hangers were readily accessible for inspection.

The required design dead and live loads for bridges were also given by the O&W specification. The dead load was to include the actual weight of the iron in the structure, as well as a floor load of 5.84 kN/m of track, which accounted for the weight of the rails, ties, and guard timbers. For the Fish's Eddy Bridge, this resulted in a total design dead load of 19.7 kN/m. The specified live load, consisting of two consolidation-class locomotives (one of which is shown in Figure 5a) and a uniformly-distributed trailing load of 32.7 kN/m representing the rest of the train, was remarkably similar to the Cooper system developed shortly thereafter and still in use today. For comparison, a Cooper E22 loading is shown in Figure 5b, where the driver loads have been set to match those of the O&W specification. In 1890, the O&W introduced the Class S double-cab locomotive, the heaviest locomotive used by the railroad at that time, and eventually becoming the largest single class of locomotive operated on the line (Houck 2011). Figure 5c shows the axle loads for this locomotive as documented by 1904 company records and reported by Houck.

3 O&W'S BRIDGE PROTECTION SYSTEM FOR THROUGH TRUSSES

3.1 *Bridge protection system details*

The tragic collapse of the Fish's Eddy Bridge in 1886, caused by a derailed train striking an end post, was the apparent motivation behind the railroad's development of their unique protection system for

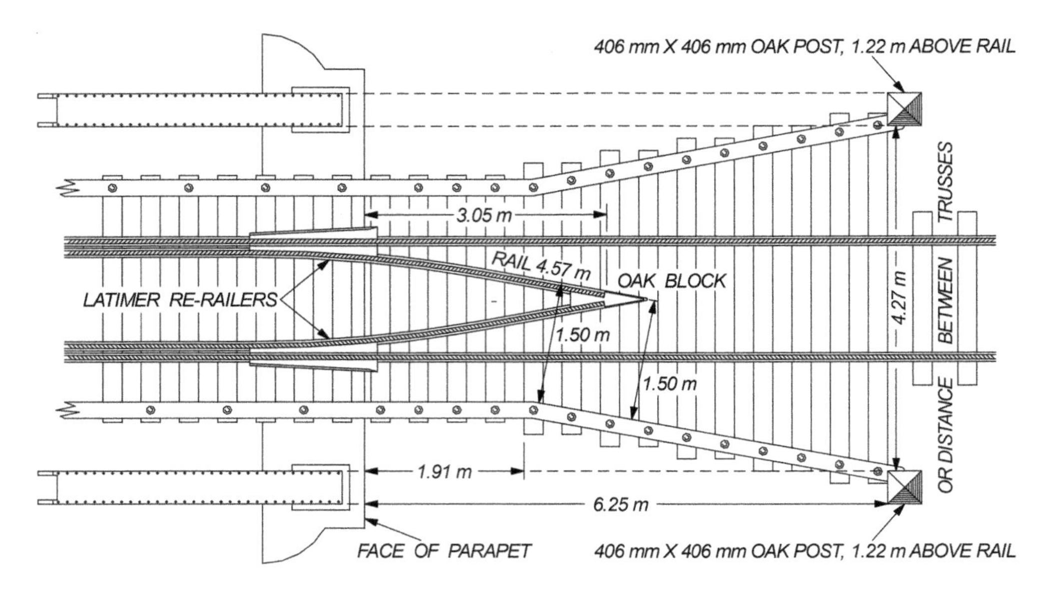

Figure 6. Plan view of O&W standard through-truss bridge-guard system.

through trusses. Details of the invention were first published in 1887 by A.M. Wellington (1906), civil engineer and editor of *Engineering News*, a week after the "Woodstock Disaster" in Vermont. So named for a nearby town, that catastrophe involved a passenger train of the Central Vermont Railroad, whose last four cars derailed due to a broken rail as the train approached a bridge, causing these cars to plunge into the White River and killing 37, with 50 others being injured (Ferguson 2013). Although the O&W invention was specifically intended to protect *through-truss* bridges from being struck by an approaching derailed car, Wellington was convinced that had such a system been in place for the Vermont deck truss, the loss of life would have been far less. Credited to James E. Childs, General Manager and Chief Engineer of the O&W, the invention combined two safety features that, individually, had previously been used by some railroads. Figure 6 illustrates the design of the bridge guard, where one component was the Latimer re-rail device, consisting of cast-iron blocks incorporated into the guard rail system and possessing inclined planes that would lift the derailed wheels back onto the track. The other component was a pair of 406 mm × 406 mm oak bumper posts tied into the bridge approach and aligned with the end posts of the truss. The bumper posts were 3.66 m long, with each post set into the subgrade and leaving 1.22 m extending above the top of the rails. The approach itself consisted of flared guard timbers to help direct a derailed truck toward the center of the track and guide the wheels to the correct side of the guard rail point. In the event that the wheels failed to properly catch the point and instead ended up on the wrong side of the guard rail system, the bumper posts were intended to break the car's coupling from the forward part of the train and strip the truck from the car body, thereby stopping the out-of-control car from striking the bridge.

Wellington (1906) noted that this bridge-guard system for through trusses was standard on the O&W, and that the railroad was in the process of installing these throughout their line when the 1887 Woodstock disaster took place on the Central Vermont Railroad. Given the timing, this further suggests that the very development of this system was indeed motivated by the 1886 collapse of the north span of O&W's Fish's Eddy Bridge. Wellington was convinced that the O&W bridge-guard system was not only the optimal and most cost-effective means available to protect a bridge from such a hazard, but that it also provided the best possible protection to the derailing train as well. With its primary purpose being to place a derailed car back on the track, the system's alternate purpose of stopping or diverting a car that had become too far out-of-control to re-rail was intended to allow the forward part of the train to safely pass over the bridge (and be re-railed if necessary) while providing the remainder of the train (and any passengers aboard) the best chance for survival by being safely stopped by the errant car.

Civil engineer E.E. Russell Tratman (1897) also strongly encouraged railroads to adopt bridge-guard systems of the style used by the O&W, promoting the installation of re-railing devices along with flared guard timber approaches and bumper posts. Specific mention is even made of an improved form of the original Latimer re-rail, manufactured by the Morden Frog & Crossing Company of Chicago and referred to as the "Childs-Latimer re-railing device," an obvious reference to its redesign by Childs of the O&W. Eighteen years later, Tratman (1909) continued to affirm these bridge safeguards, and was quite frustrated by the general reluctance of the broader railroad community to implement them. He observed that flared guard timber approaches were generally not being used, making the "present practice ... decidedly inferior to that of 10 or 20 years ago." He also lamented that re-railing devices were not widely used either, and stated that "their omission is a serious defect in bridge floor design." Interestingly, while both Wellington and Tratman trumpeted the virtues of the O&W's complete system (re-railing devices combined with flared guard timber approaches and bumper posts) and promoted their use on all bridges, it appears that even the O&W itself limited their usage to only through-truss bridges.

3.2 *Subsequent consensus industry standards for bridge protection*

The American Railway Engineering Association's (AREA) Committee on Wooden Bridges and Trestles convened on November 2, 1912 to develop recommended standards for railroad bridge protection (AREA 1913). To support this endeavor, 61 participating railroads were surveyed regarding their individual practices for using guard rails on wooden bridges and trestles, with the O&W being one of the respondents. Interestingly, by this time the O&W had dispensed with the use of re-railing devices on bridges, while several other railroads participating in the survey did indicate that they were using them at least in certain circumstances. A few anecdotes included with the survey even described specific examples of their demonstrated effectiveness, while none of the respondents reported any problems associated with their usage. Wellington (1906) too had commented earlier on the effectiveness of the Latimer re-railer, relating an example where a derailed car on the New York, Pennsylvania & Ohio Railroad was believed to have traveled nearly 0.8 km before it reached and was successfully re-railed by such a device. Although the O&W discontinued their use of Latimer re-railers, by 1912 the railroad did significantly increase the distance that guard rails were extended beyond the end of a bridge, from a little more than 3 m (as shown in Figure 6) to 13.1 m, with these guard rails now terminating at a No. 6 frog. In addition, they still employed bumper posts and flared guard timber approaches to protect their through-truss bridges. However, consistent with the observations of Tratman (1897, 1909) concerning railroads in the U.S., these protection elements were not in general use by most of the other 60 railroads participating in the 1912 survey. One exception was the Illinois Central Railroad, which did use re-railing devices in combination with a guideway that employed flared steel rails, and this guideway was quite similar in form and function to the flared guard timbers used by the O&W. Ultimately, the final conclusions of the AREA committee were as follows:

(1) It is recommended as good practice to use guard timbers on all open-floor bridges, and same should be so constructed as to properly space the ties and hold them securely in their places.
(2) It is recommended as good practice to use guard-rails to extend beyond the ends of the bridges for such a distance as required by local conditions, but that this length in any case be not less than fifty feet (15.2 m); that guard-rails be fully spiked to every tie and spliced at every joint, the guard-rail to be some form of a metal guard-rail.
(3) It is recommended that the guard timber and guard-rail be so spaced in reference to the track rail that a derailed truck will strike the guard-rail without striking the guard timber.
(4) The height of the guard-rail to be not over one inch (25.4 mm) less than the running rail.

These recommendations did not include the distinctly O&W practice of using bumper posts with flared guard timbers on the approaches. In part, this might be attributed to possible skepticism regarding their effectiveness for steel cars that were rapidly coming into common usage at that time. Or perhaps the omission of these posts was simply a reflection of the apparent lack of evidence that such protections actually functioned as intended in the first place. In addition, the committee's recommendations did not include the use of re-railing devices either, even though there was anecdotal evidence that these devices had successful restored cars to the rails on at least several occasions. Obviously, the consensus of the AREA membership was that guard timbers on the bridge deck, combined with a basic guard rail system, provided a sufficient degree of protection against derailments. In the years that followed, the O&W also adopted these recommended practices of AREA. Even the modern railroads of today continue to employ bridge protection practices that are in essence similar to those established in 1912.

4 1897 COLLAPSE OF THE FISH'S EDDY BRIDGE

4.1 *Potential causes*

As shown by Figure 1, all eleven coal gondolas that fell into the river during the 1897 collapse are in the general proximity of the southern pier for the failed middle span, indicating that whatever triggered the collapse, initial failure of the truss most likely took place at the southern end of the span, and within the first few panels. Among the potential causes of the collapse of a through truss bridge such as this are the following:

a) Derailment of train on the bridge
b) Derailment of train approaching the bridge
c) Separated train, with the two portions colliding on the bridge
d) Excessive overload of the structure
e) Pier failure
f) Failure of a component

Taking each of these possibilities in order, collapse due to (a) derailment on the bridge seems unlikely given the slow speed of the train, plus the fact that the guard rails and guard timbers would likely have prevented the empty cars from striking the trusses. Collapse due to (b) derailed cars approaching the bridge is also unlikely, especially given the bridge protection devices employed by the O&W as discussed earlier. Furthermore, from the ensuing investigation, there were no reports of the telltale ruts that car wheels would have gouged into the timber ties had a derailed train traveled over any significant distance. Collapse due to (c) separation of the train followed by a collision of its two parts is also questionable, as it was reported that a brakeman was near the middle of the train when the span collapsed and cars were plunging off the pier, and he was able to safely retreat to the rear of the train while all this was taking place (*The Walton Chronicle* 1897). Thus, had the train initially broken into two parts prior to the collapse, the brakeman would likely have witnessed such an occurrence. It is apparent that there was not (d) excessive overload of the structure either, given that the coal gondolas were empty. Further, had a major structural member failed, such as a chord, post, or diagonal, the entire span would have immediately collapsed, taking with it whatever portion of the train was upon it. That all eleven wrecked cars were piled up towards one end suggests that the collapse progressed slowly, allowing the forward part of the train to clear the span while cars were bunching up to the rear and eventually collapsing the truss. From photographs, it is also evident that (e) the piers remained intact and were not a contributor to the collapse.

This leaves the final potential cause listed: (f) failure of a critical component. To fit the pattern and sequence of events as previously discussed, such a component would need to be located within the first few panels at the south end of the truss, and its failure would need to be such that it did not lead to the immediate collapse of the span, thereby allowing the forward part of the train to safely traverse and exit. A good candidate would be the hanger supporting the floor beam at the first panel point. Unlike the main members of the bridge, this hanger does not provide direct support to

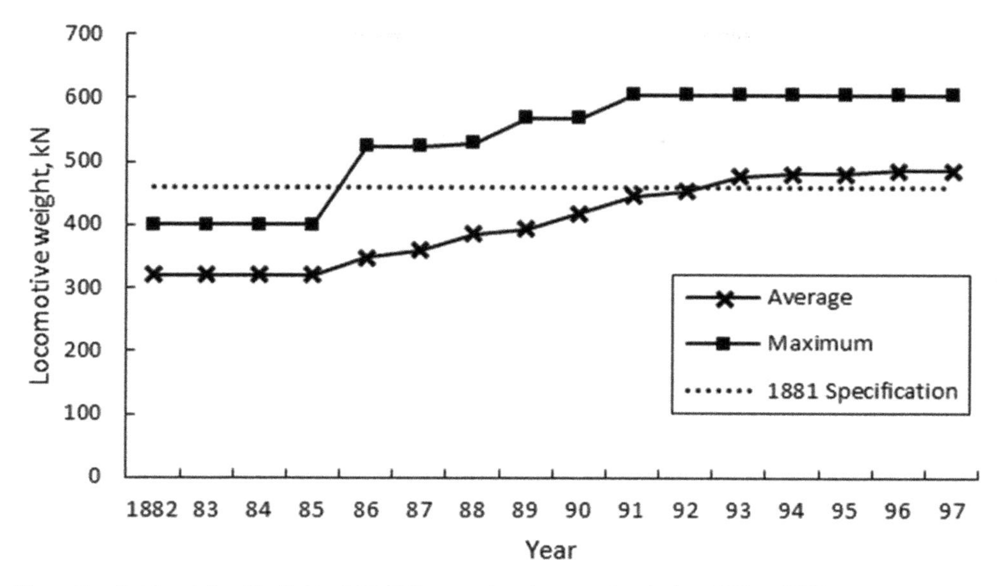

Figure 7. Total weight of individual O&W locomotives in operation during 1882 to 1897.

the truss, that is, the truss itself does not rely on this element in order to remain stable. However, failure of a hanger would cause the end of the floor beam to significantly drop, supported only in a secondary manner by lower chord members tied into that panel point, and by stringers. This would result in a severe vertical misalignment of the track, potentially leading to the derailment of a car and its separation from the rest of the train. This errant car would then cause the following cars to begin to pile up, eventually striking and damaging the truss to a degree that could lead to its collapse. The time delay necessary for all this to happen could conceivably allow the forward part of the train to safely clear the span, as happened in the Fish's Eddy collapse. The possibility of such a hanger failure will be examined in the following sections.

4.2 *Load analysis*

The evolution of the heaviest live loads (i.e., the locomotives) operated by the O&W during the 15-year life of the Fish's Eddy Bridge will be considered first. Using mechanical data compiled by Helmer (1957) for the entire roster of locomotives operated by the O&W, Figure 7 shows the total weight of the heaviest locomotives in operation each year (not including tender loads), as well as the corresponding average weight of all locomotives. Also shown is the total weight of the "standard" O&W locomotive as stipulated in their 1881 specification, and within five years (1886) the heaviest locomotives in use exceeded this standard. However, it was not exceeded by the average weight until 1893, and even then by only a small margin over the remaining five years of the bridge's life. Overall, therefore, while it is evident that the operational loads eventually did surpass those for which the bridge was designed, it might seem that this intensification was slight and its effects relatively modest. However, consideration of individual driver weight reveals a much more severe increase in loading, as shown by Figure 8. In fact, the O&W specification was actually exceeded in this regard from the very beginning, and for most of the bridge's life the maximum weight of individual drivers was nearly double that given in this specification. Thus, relative to certain components, and especially those with a rather small influence area of loading (such as hangers), the ramifications of this increase could be quite significant.

Due to the 53.5° skew associated with the bridge, one of the two hangers in the end panels of each truss supported a larger portion of floor than the other hanger. In performing the live load analyses that follow, the focus will be upon this hanger supporting the greater load. Among the

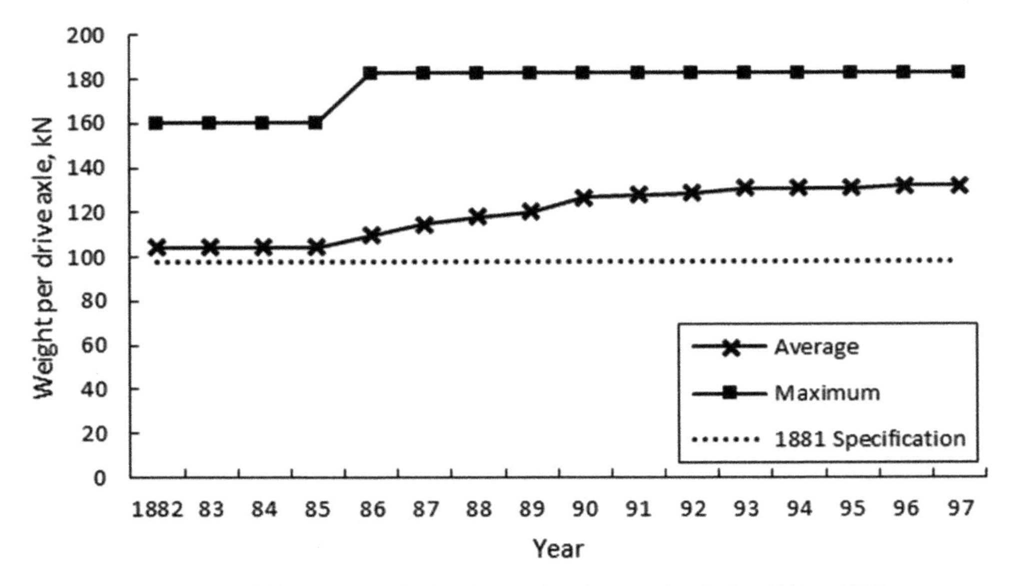

Figure 8. Weight on individual drivers of O&W locomotives in operation during 1882 to 1897.

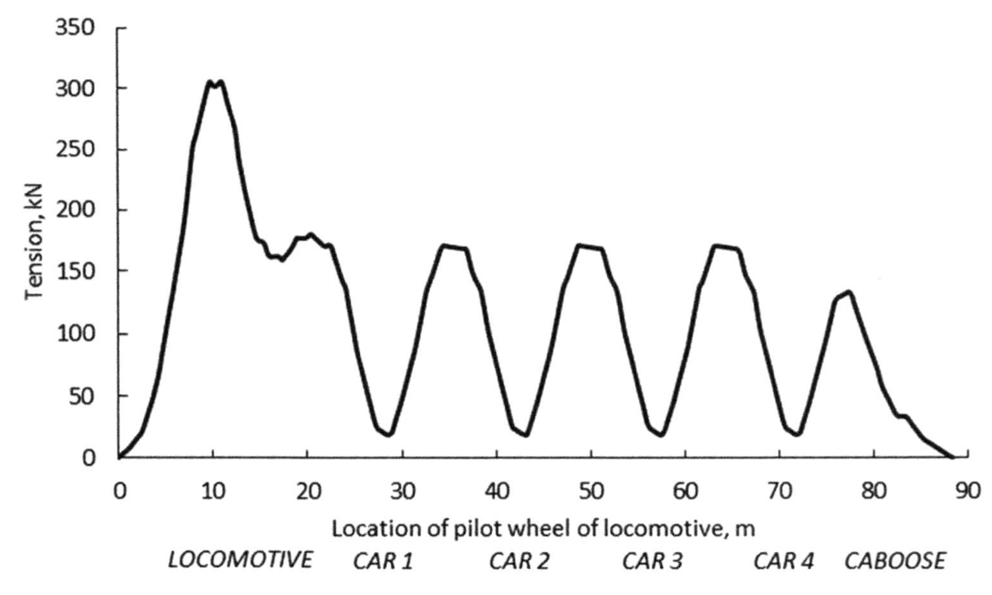

Figure 9. Hanger force developed due to milk train.

worst cases observed was for a Class S locomotive (Figure 5c) pulling a milk train, with the cars consisting of O&W's 267-kN capacity 1001 series (Mohowski 1995). As was done for all freight trains analyzed, a four-wheel caboose was included at the end. The milk train was assumed to travel at 40.2 km/h (which exceeds the synchronous speed for this locomotive, thereby maximizing the applied impact), and based on current recommended practices for steel railroad bridge design and rating (AREMA 2018), impact factors of 59.5% on the drive wheels and 24.0% on the remaining axles were applied, along with a rocking effect of 20%. Figure 9 shows the resulting hanger force

Table 1. Hanger tension stress per O&W 1881 specification, using design live load (no impact added).

Hanger component	Dead load	Stress MPa	41.4 MPa allowable exceeded?
Head, net section	Design	25.3	No
	Actual	24.0	No
Shank, gross section	Design	40.5	No
	Actual	38.4	No
Rivet line, net section	Design	59.0	Yes, by 42.5%
	Actual	55.2	Yes, by 33.3%

developed as a function of the pilot wheel (or lead axle) position, measured from the stringer's pier support for the longer rail.

4.3 *Hanger stress analysis*

As noted earlier, the O&W's specifications for the design of the wrought iron through trusses used at Fish's Eddy (Katte 1881) required that the stresses in floor beam hangers "and other similar members liable to sudden loading" be limited to 41.4 MPa. This was the most restrictive of all the tension member limits given, and while there was no specific provision for impact within the entire specification, the severity of this floor beam limit was motivated to mitigate the effects of impact. There was also a separate provision that stated that "full allowance shall be made for reduction of section by rivet holes" for members subject to tension.

Table 1 examines the critical elements of the hanger, including the eyebar head, shank, and net section through the upper rivet line, and assesses the adequacy of each relative to the provisions of the 1881 specification. In doing this assessment, two dead load cases are considered. The first is the dead load used by the Central Bridge Works in the actual design, which was determined by calculating the total dead load per unit length of truss, and applying this uniformly throughout at 48.0 kN per panel per truss. Thus, the designers also applied this same 48.0 kN on the hanger. However, this appears to be rather high, as the actual dead load acting on the hanger is estimated to have been approximately 35.6 kN acting on the riveted end and 37.8 kN acting on the eyebar head at the top. Thus, in Table 1, both cases are considered. The live load used is that given in the 1881 specification. It is seen in this table that the average stress across the net section of the eyebar head is well below the allowable stress, as well as being far less than the stresses developed in the shank or net section through the rivet line, and thus did not govern the design of the hanger. Examination of the shank stresses shows that especially when using the design dead load, the actual stress just barely satisfies the allowable stress, suggesting that this criterion was treated as the governing basis for the hanger's design. However, consideration of the average stress acting on the net section through the upper rivet line shows the hanger to be highly overstressed, suggesting that this criterion was overlooked. Thus, it could be argued that the hangers were not in total compliance with the very specification that governed their design.

The current rules generally followed throughout North America for the design of steel railway bridges are the recommended practices given in Chapter 15 of the *Manual for Railway Engineering* (MRE), published by the American Railway Engineering Association (AREA) until 1996, since which time it has been maintained by the American Railway Engineering and Maintenance-of-Way Association (AREMA 2018). The basic allowable tension stress, including bending, stipulated by AREA for floor beam hangers with riveted end connections was 96.5 MPa on the net section (AREA 1995). This was considerably less than that the $0.55F_y$ (where F_y = yield strength) permitted for ordinary tension members, as a higher factor of safety was considered necessary for riveted floor beam hangers in light of their many failures experienced over the years as documented by AREA (1950). The 96.5 MPa allowable stress was established based on ASTM A36M steel (and using

Table 2. Hanger tension stress evaluation per MRE, with the live load consisting of a Class S locomotive crossing at 40.2 km/h, and including design impact.

Design criterion	Dead load	Allowable stress MPa	Actual stress MPa	Allowable exceeded?
AREA, net section	Design	96.5	115	Yes, by 19.1%
	Actual	96.5	111	Yes, by 14.9%
AREMA, gross sect.	Design	69.0	76.6	Yes, by 11.2%
	Actual	69.0	73.9	Yes, by 7.2%
AREMA, net section	Design	172	115	No
	Actual	172	111	No

Table 3. Cooper load rating of hanger.

Design criterion	Allowable stress MPa	Allowable live force kN	Cooper rating
O&W spec., gross sect.	41.4	155	E24.4
O&W spec., net sect.	41.4	91.5	E14.4
AREA, net section	96.5	261	E24.8
AREMA, gross sect.	69.0	282	E26.8
AREMA, net section	172	494	E47.0

$0.39F_y$), and it was deemed appropriate to limit all riveted floor beam hangers to this same limit, regardless of actual yield strength. During 2002–2005, AREMA made several significant changes to the provisions for tension members (and hangers), including using two limit states instead of just one: yield stress on the gross section and ultimate stress on the effective net section. While the general philosophy of applying a higher factor of safety to hangers has been retained, there exist certain inconsistencies in the AREMA provisions that are currently being reviewed by committee. Thus, for this paper, the former AREA provision is included as well.

Table 2 examines the hanger's adequacy in light of modern steel railway bridge criteria, considering the case of a Class S locomotive crossing the bridge at 40.2 km/h. Impact was applied as specified earlier, and both the design dead load of 48.0 kN as well as the estimated actual dead load of 35.6 kN were considered as separate cases. Based on a rigid frame analysis of the hangers and the supported floor beam, hanger stresses were increased by 4.99% due to bending. The focus is limited to the gross shank area and net section through the upper rivet line, as these regions are the governing portions of the hanger. The results in Table 2 show that the hanger design is not in general compliance with either version of the MRE Chapter 15. In fact, the only criterion it clearly satisfies is the current requirement for the net section, but the factor of safety for this criterion is much more liberal than that reflected by the gross section limit due to inconsistencies in the AREMA provisions as previously noted. Thus, the modern standards generally indicate that at best, the hanger design was loaded nearly to its full capacity, and at worst, was significantly overloaded. Furthermore, records of scheduled trains suggest that this overloading due to Class S locomotives might have occurred as often as 10 times per day or more.

Based on the provisions of O&W's 1881 specification as well as the MRE, Table 3 shows the results of a Cooper load rating analysis, assuming the hanger to be the controlling structural element. Focusing on the riveted end, the allowable live force shown in the table was determined from the allowable stress by removing the increased tension stress due to bending (4.99%), multiplying by the appropriate area ($A_g = 4840\ \text{mm}^2$, $A_{net} = 3230\ \text{mm}^2$), and deducting the estimated dead load of 35.6 kN. In applying the MRE provisions, driver impact of 59.5% and non-driver impact of 34.4% was included (consistent with a Class S locomotive crossing the bridge at 40.2 km/h), as well as a

Table 4. Maximum hanger tension stresses (from linear-elastic finite element analysis).

Location and condition	Calculated stress MPa
Eyebar head, edge of hole	185
Rivet hole, edge closest to eyebar edge, uniform bearing	249
Rivet hole, edge closest to eyebar center, uniform bearing	232
Rivet hole, edge closest to eyebar edge, 50% bearing	292
Rivet hole, edge closest to eyebar center, 50% bearing	268

rocking effect of 20%. The severity of the net section criterion from the O&W specification results in a much lower rating of E14.4, and as discussed earlier, evidently this criterion wasn't used in the design of the hanger. Further, application of this criterion would seem to be overly conservative. Conversely, the liberality of the current AREMA net section criterion is reflected by its overly high rating of E47.0. The remaining three criteria set the rating of the hanger within a rather tight band, with E24.4 for the O&W's specification (gross section), E24.8 for AREA (net section), and E26.8 for AREMA (gross section). These levels would have been exceeded by equipment that regularly operated over the bridge, such as the Class S locomotive that exerted an E29.0 load. As of 1894 (and prior to the Fish's Eddy Bridge collapse in 1897), Theodore Cooper stated that E40 should be the minimum standard generally used by all railroads (Houck 2011). Based on the standards considered here, the hanger design was not nearly at this capacity either.

Table 4 examines the maximum tension stresses developed in the hanger due to a Class S locomotive crossing at 40.2 km/h, as determined by a linear-elastic finite element analysis. Due to stress concentrations, these maximum stresses develop along the edges of the holes for the eyebar pin and the rivets, occurring in the vicinity of transverse sections taken through the holes. In performing the analysis, the actual estimated dead load of 37.8 kN was also applied to the eyebar head and 35.6 kN applied to the riveted end of the eyebar. Increases due to bending were not included (but would be 4.99% more for the riveted end, and half that for the eyebar head). The table shows that the yield stress of 172 MPa would have been slightly exceeded at the eyebar head under these loads; this excess would have occurred over a very small region, and could have been accommodated by a small amount of localized strain hardening. On the other hand, the stress levels at the rivet holes are much more severe. Assuming uniform bearing on all rivets, the stresses far exceed yield at two locations of each hole, and this excess would have been considerably more widespread than that for the pin hole at the eyebar head. In reality, the bearing upon the rivets would likely be far from uniform, worsening the situation even more. In a study of such connections (AREA 1950), an experiment was conducted for a five-row riveted butt joint. Throughout most of the load range, the results showed that 50% of the total load was taken up by the first row of rivets (i.e., the row closest to the total applied force). For the Fish's Eddy hangers, if 50% of the load is assumed to bear on the first rivet line, Table 4 shows degree that the peak stresses increase (again, based on a linear-elastic analysis). In either case, it is evident that stresses far in excess of initial yield were routinely taking place at four locations (two per hole) across the net section of the upper rivet line, possibly with virtually every locomotive crossing.

5 CONCLUSIONS

Historical background has been presented regarding the O&W's three-span through truss bridge near Fish's Eddy, New York, erected in 1882. Particular attention has been given regarding the circumstances surrounding the collapses experienced in 1886 and 1897.

Due to a derailed car striking the end post of a truss, the 1886 collapse was the apparent motivation leading to the railroad's development of a unique bridge protection system, details of which have been examined. Despite being hailed as an important innovation by two prominent engineers who vigorously promoted its use, the invention was never widely adopted by the railroad industry in general, and eventually was even abandoned by the O&W as well. Besides cost, perhaps a possible reason for this was skepticism regarding the system's effectiveness to protect bridges from the steel cars that were coming into more widespread usage at that time.

With the exact cause of the 1897 collapse having never been established, the possibility of initiation by a hanger failure was examined, as such a scenario is supported by the arrangement of the post-collapse debris. Considering the heaviest loads that regularly operated over the structure, average hanger stresses would have exceeded certain allowable stresses as given by modern standards. It was also observed that the hangers did not completely satisfy the very specifications used to design the structure in the first place. In addition, peak stresses as obtained through linear-elastic finite element analyses were far in excess of initial yield, with these excesses occurring at four places in the critical net section of the hanger. In all, these various examples of overstress show that the hangers did possess a significant degree of vulnerability to failure.

ACKNOWLEDGEMENTS

The authors gratefully acknowledge the invaluable assistance of Jeff Otto and the Ontario & Western Railway Historical Society for access to archived materials.

REFERENCES

AREA. 1913. Report of Committee VII–On wooden bridges and trestles, *Proc. of the 14th Annual Convention* 14: 652–697. Chicago: American Railway Engineering Association.

AREA. 1995. Steel Structures. *Manual for Railway Engineering* Chapter 15. Chicago: American Railway Engineering Association.

AREA Committee on Iron and Steel Structures. 1950. Stress Distribution in Bridge Frames – Floorbeam Hangers. *AREA Proceedings* 51: 470–503. Chicago: American Railway Engineering Association.

AREMA. 2018. Steel Structures. *Manual for Railway Engineering* Chapter 15. Lanham: American Railway Engineering and Maintenance-of-Way Association.

Ferguson, J.A. 2013. The Wrong Rail in the Wrong Place at the Wrong Time: The 1887 West Hartford

Helmer, W.F. 1959. *O. & W. – The Long Life and Slow Death of the New York, Ontario & Western Railway*: 169–190. Berkeley: Howell-North Press.

Houck, M.H. 2011. Center Cabs and Kitchens, *Observer* 45(1): 34–38, 57–58. Middletown: Ontario & Western Railway Historical Society.

Katte, W. 1881. *General Specifications for Iron Bridges*. New York: New York, Ontario & Western and New York, West Shore & Buffalo Railways.

The Middletown Daily Times. 1897a. Railroad Disasters – Erie Cars Piled Up and an O&W Bridge Collapses. April 28. Middletown.

The Middletown Daily Times. 1897b. To Have a New Bridge – O&W Has Made a Contract for One at Fish's Eddy. April 30. Middletown.

Mohowski, R.E. 1995. *New York, Ontario and Western Railway – Milk Cans, Mixed Trains, and Motor Cars*: 74. Laurys Station: Garrigues House.

Orange County Press. 1886. Shocking Casualty – Another Terrible and Fatal Accident on the Ontario & Western Railway. March 3. Middletown.

Tratman, E.E.R. 1897. *Railway Track and Trackwork*: 144–145. New York: The Engineering News Publishing Co.

Tratman, E.E.R. 1909. *Railway Track and Trackwork*, 3e: 169–171. New York: The Engineering News Publishing Co.

Vose, G.L. 1881. *Manual for Railroad Engineers*: 212–214, 481, Boston: Lee and Shepard.

The Walton Chronicle. (1897). "Wreck on the O. & W." Walton, NY, April 29, 1897.

Wellington, A.M. 1906. The Best Safeguard against Woodstock Disasters. *Engineering News* Nov. 8: 489–490 (reprinted from *Engineering News* of Feb. 12, 1887).

Chapter 9

Redesign of collapsed river bridges due to flood and scour

E.E. Laçin, M.C. Dönmez, A. Çiçek, İ.N. Çilingir & Ş. Caculi
Emay International Engineering and Consultancy Inc., Istanbul, Turkey

ABSTRACT: The environmental and climatic conditions are generally reviewed in the early phase of layout plan determination for bridges crossing over streams. In this review, flood flow rate of the stream is obtained and bridge's spanning, vertical clearance and type of piers are determined. If there are piers, they are placed in the stream, piers are exposed to negative effects of water during their service life. The most important problems that occur around the bridge piers are depositions and scouring. Due to these cyclic effects bridge structures may experience stability problems and collapse. In order to decrease these negative effects, stream characteristics around the piers should be investigated and the required data, such as flood flow rate and flow velocity, obtained. These data can be used in determination of stabilities of bridge piers. In this study, twelve bridges located in Ordu, Turkey are investigated which were damaged and collapsed due to the flood that occurred in July 2018. The investigation covers the hydraulic effects which cause the bridges to collapse. Also, the design principles of the new twelve bridges and field precautions to prevent depositions and scouring of piers are studied.

1 INTRODUCTION

Countries need to plan alternative transportation routes due to developing technologies, increasing populations and rising commercial activities. Bridges are in a critical position among engineering structures to be built within the designated route. Damage or destruction of the bridges due to natural disasters such as earthquakes and floods is a condition that affects daily life negatively. In the design phase, structural and hydraulic effects should be considered together in order to ensure continuity of transportation and prevent damage and destruction, especially for bridges with piers in the stream. In this case, piers are exposed to negative effects of water during their service life. In the design of the bridges to be constructed especially in the riverbeds, examination of the climatic conditions, topographic characteristics and stream regime of the region and determination of the pier type is important. For this reason, measurement and observation are performed to obtain precipitation and flow data in order to determine the river characteristics (Yücel & Namlı, 2007). This data is used for determination of precautions to be taken for the bridges and construction methods. Water flows in rivers and streams have two different types of regimes such as river regime and flood regime. Especially in the flood regime, depositions are occurred around the bridge piers and these deposited materials change the water flow characteristics. Sometimes, opposite conditions may happen. That is, flood discharge the materials around the piers, and scouring occurs. Due to these cyclic effects, bridge structures may experience stability problems and collapse. In order to solve this problem, the pier type should be selected according to the stream regime or the stream regime should be adapted to the existing pier type.

In this study, twelve bridges located in Ordu, Turkey are investigated which were damaged and collapsed due to the flood that occurred in July 2018. The investigation covers the hydraulic effects which caused the bridges to collapse. Also, the design principles of the new twelve bridges and field precautions to prevent future pier depositions and scouring are studied.

2 EXAMINED BRIDGES AND COLLAPSING CAUSES

Scouring, which occurs around the bridge piers, causes bridge collapse and financial loss. It happens both around abutments and piers. Bridges in Turkey face negative effects of scouring due to high stream flow, high flood flow rate and thin deposition materials (Yurtseven, 2005). In this section, there is information about twelve bridges located in Ordu, Turkey which were damaged and collapsed because of the flood that occurred in July 2018. The bridges which were observed in situ are listed in Table 1. The location of bridges in satellite view is shown in Figure 1.

Kaleönü Bridge had three spans and was 80 meters long with a deck width of 4.5 meters. The bridge had cast in situ reinforced concrete structure with gerber deck system, which is continuous deck system with intermediate hinges. Settlement occurred in the pier foundation due to scouring and it caused severe damage to the pier and deck as in Figure 2. Displacement occurred in the deck system and it caused transverse cracks in abutments as seen in Figure 3.

Although the geometry of the bridge was suitable for the river bed, it was determined to be unusable due to the fact that during the design (and construction) of the bridge's foundations, the necessary precautions were not taken against scouring.

Kargucak – Geyikçeli Bridge was constructed as a 2.50 m by 2.50 m double cell culvert. Considering the riverbed, the culvert as seen in Figure 4, which had a very small flowing capacity compared to the flood flow rate, was clogged due to deposition materials carried by the flood, causing the culvert to move with approach fillings as in Figure 5 below.

In order to avoid the same situation again, the river passing needs to be re-evaluated in the general layout. The bridge spans to be constructed in a more suitable location must have a suitable

Table 1. Bridges which were observed in situ.

Number	Bridge name
1	Kaleönü Bridge
2	Kargucak Bridge
3	Ünye Cevizdere Söğütlüpınar Bridge
4	Denizbükü Bridge
5	Saraycık Kuşçulu Bridge
6	Yenicuma Bridge
7	Hacıtahir Bridge
8	Kurudere Bridge
9	Kurtluca Bridge
10	Ilıca Bridge
11	Yavaş Bridge
12	İkizce Bridge

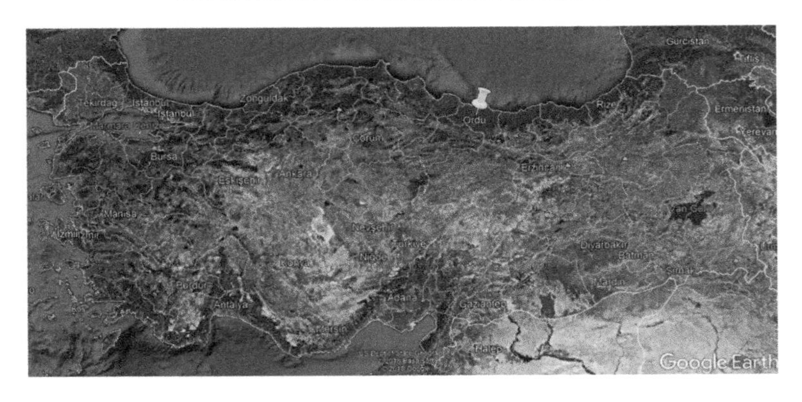

Figure 1. The location of bridges.

Figure 2. Kaleönü Bridge(drone view after damage).

Figure 3. Kaleönü Bridge(drone view after damage).

Figure 4. Clogged Culvert.

Figure 5. Kargucak Geyikçeli Bridge (drone view after damage).

clearance to pass the flood flow rate according to the General Directorate of State Hydraulic Works. Scouring calculations should be taken into consideration in the foundation design phase and deck width should follow highway standards.

Çaybaşı Kurudere Bridge was an 18 meter-long single span bridge with a deck width of 4.5 meters. The bridge was used for transportation to agricultural land. It was observed that the riverbed had a small flood flow capacity compared to the flood flow rate as in Figure 6.

Figure 6. Çaybaşı Kurudere Bridge (drone view after damage).

Figure 7. Çaybaşı Kurudere Bridge – the settlement and rotation occurred in the abutments.

The bridge approach filling moved during the flooding and the bridge could not be used because of the settlement and rotation that occurred in the abutments as in Figure 7.

To avoid a similar situation, it is necessary to obtain the data such as flood flow rate, flow velocity specified for the Kargucak Geyikçeli Bridge according to the General Directorate of State Hydraulic Works and perform the design accordingly.

Hacıtahir Bridge was a 3-span bridge with a cast in situ reinforced concrete system. It was 55 meters long and had a deck 6.2 meters wide. Scouring occurred under the spread foundations which were not suitable for a bridge subject to flooding. Especially, deposition materials located around the piers increased the foundation movements and the bridge superstructure moved in the direction of the piers as shown in Figure 8.

Ünye Cevizdere Söğütlüpınar Bridge was a 70 meter-long, 3-span bridge with a deck width of 12.5 meters. It was a castin-place reinforced concrete bridge. Scouring occurred under the spread foundations which were not suitable for a bridge system subject to flooding. The deck was collapsed due to scouring under the pier foundations. The bridge was completely collapsed as seen in Figure 9 below.

Ünye Kuçculu Saraycık Bridge was a 3-span, 70 meter-long with a 6 meter-wide deck. The bridge was constructed cast in situ reinforced concrete system. The abutment approach fill moved because of the flood. The abutment experienced stability problems and rotation occurred. Horizontal movements happened in the bridge plan. These are due to insufficient bridge length as in Figure 10.

Figure 8. Hacıtahir Bridge (drone view after damage).

Figure 9. Ünye Cevizdere Söğütlüpınar Bridge.

Figure 10. Ünye Kuşçulu Saraycık Bridge(drone view after damage).

3 PRINCIPLES CONSIDERED IN REDESIGN OF BRIDGES

3.1 *General layout*

As it is observed from the examination of the damaged and collapsed bridges, it is concluded that the bridge spans, vertical clearance and pier types were not determined suitable for general layout. Therefore, the pier and abutment rotation and bridge collapse occurred. On the other hand stream characteristics are the most important data for the bridges where piers are placed in the stream. The stream characteristics were taken from the General Directorate of State Hydraulic Works for

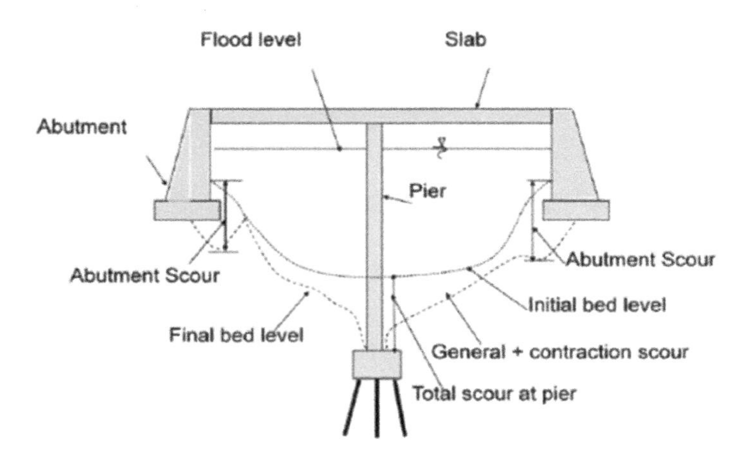

Figure 11. Total scouring mechanism.

reconstructed bridges and the bridge location was determined using these data. The bridge vertical clearance was determined according to the flood flow rate and the longest span was determined in design phase for taken precautions against contraction scour caused by contraction. The bridge skew was determined suitable for stream flow. Wing wall was provided for abutments. In this way, the angular approach of the stream flow will be ensured.

3.2 *Scouring*

Scouring is the concept of removing the backfill materials of the abutments and the sub-base materials of the piers due to the effect of flow rate and causing the elevation changes in streambed because of this transformation (Yanmaz, 2002). For many years, several studies have been carried out to solve this problem. The concept of scouring depends on many factors. It is necessary to examine the velocity and pressure changes around the bridge piers to get an idea of how the scouring is. The total scouring mechanism in a bridge is the interaction of general base scouring, contraction scour in bridge spans and local scouring around the piers and abutments as in Figure 11. In this study, contraction scour and local scouring around the piers and abutments are studied.

Two different types of contraction scour mainly occur in alluvial streams depending on whether the material is carried from upstream. In the case of live-bed scour, the fully developed scour in the bridge cross section reaches equilibrium when sediment transported into the contracted section equals sediment transported out. As scour develops, the shear stress in the contracted section decreases as a result of a larger flow area and decreasing average velocity. For live-bed scour, maximum scour occurs when the shear stress reduces to the point that sediment transported in equals the bed sediment transported out and the conditions for sediment continuity are in balance. For clear-water scour, the sediment transport into the contracted section is essentially zero and maximum scour occurs when the shear stress reduces to the critical shear stress of the bed material in the section. The scouring depth in the bridge is the sum of these scouring conditions. Initially rapidly increasing scour depth, d_s, gradually progresses to a balanced depth, d_{se}. The change in the depth of local scouring occurs around the piers and abutments depending on the time (t) and the average flow rate (u) are shown at Figure 12. u_c is the average critical flow rate.

Equivalent scouring depth after scouring is calculated by the following formulas for live-bed scour (1) and clear-water scour (2);

$$\frac{y_2}{y_1} = \left(\frac{Q_2}{Q_1}\right)^{6/7} x \left(\frac{W_2}{W_1}\right)^{k_1} \tag{1}$$

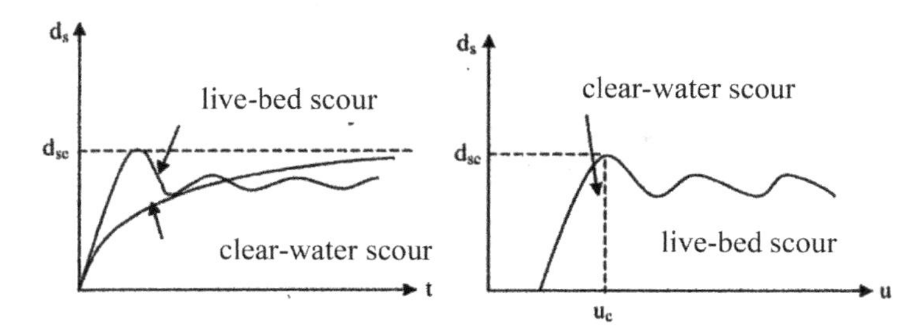

Figure 12. The change of scouring depth occurs around the piers and abutments.

where y_1 = average depth in the upstream main channel (m); y_2 = average depth in the contracted section (m); Q_1 = flow in the upstream channel transporting sediment (m^3/s); Q_2 = flow in the contracted channel (m^3/s); W_1 = bottom width of the upstream main channel depth is transporting bed material (m); W_2 = bottom width of main channel in contracted section less pier width(s) (m); k_1 = coefficient.

$$y_2 = \left(\frac{K_u x Q^2}{D_m^{2/3} x W^2}\right)^{3/7} \tag{2}$$

where y_2 = average equilibrium depth in the contracted section after contraction scour (m); Q = discharge through the bridge or on the set-back overbank area at the bridge associated with the width W (m^3/s); D_m = diameter of the smallest non-transportable particle in the bed material (D_{50}) in the contracted section (m); and W = bottom width of the contracted section less pier widths (m); K_u = 0.025.

The local scouring around the bridge piers is important, as well as contraction scour. The bridge piers placed in the flow area increase the average velocity by reducing the flow area in bridge span because the sediment transportation capacity increases in the contracted section and the scouring begins because of the vorticity occurs in around the piers. The pier scouring is calculated by the following formula (3) according to FHWA Hydraulic Engineering Circular 18 (FHWA HEC 18 2012);

$$\frac{y_s}{y_1} = 2.0 K_1 K_2 K_3 \left(\frac{a}{y_1}\right)^{0.65} Fr_1^{0.43} \tag{3}$$

where y_s = scour depth (m); y_1 = flow depth directly upstream of the pier (m); K_1, K_2, K_3 = correction factors for pier nose shape, angle of attack of flow, bed condition; a, L = the pier width and length ; and Fr_1 = Froude number directly upstream of the pier.

The water level increases in piers upstream due to sudden drop in flow rate in piers. In this case, high pressure occurs in surface and decreasing pressure occurs towards the base. Thus a vertical flow occurs from the high pressure zone to the low pressure zone, in other words it occurs from the surface to the base. It causes vertical vorticity in upstream side as seen in Figure 13 below.

The abutments extend towards the flow area to provide the required structural balance. In this case, the bridge span reduces. The average flow rate increases from the abutments upstream side and vortices occur in base with vertical flow as in Figure 14. The maximum vortices occur in upstream side. When the lateral flow and the stream flow act together, the total scouring in abutment increases.

In the abutment design phase, the scouring depth is calculated by the following formula (4) according to Froehlich approach;

$$\frac{y_s}{y_a} = 2.27 K_1 K_2 \left(\frac{L}{y_a}\right)^{0.43} Fr^{0.61} + 1 \tag{4}$$

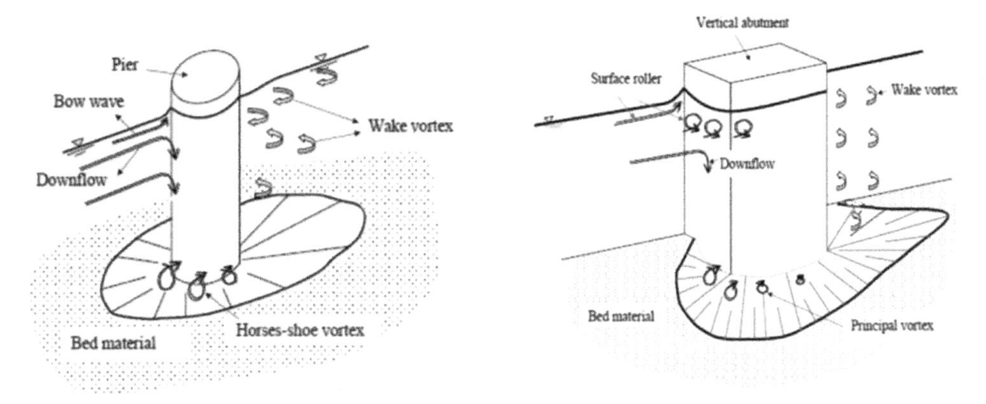

Figure 13. Definition sketch for pier scour.

Figure 14. Definition sketch for abutment scour.

where K_1, K_2 = coefficient for abutment shape, angle of embankment to flow; L = length of active flow obstructed by the embankment (m); and y_a = average depth of flow on the floodplain (m); Fr_1 = Froude number of approach flow upstream of the abutment.

3.2.1 *Calculations in this study:*

In this section, the scouring analysis for Kaleönü Bridge, which is examined and redesigned within this study, is shown. In the design phase, while determining the location of the bridge, the climatic conditions, topographic characteristics and stream regime of the region were examined and the pier types were determined according to this data. In order to avoid a similar situation, the river passing was re-evaluated in the plan. The bridge span was determined according to the flood flow rate taken from the General Directorate of State Hydraulic Works. In the foundation design phase, the scouring depth, which was obtained from the scoring analysis, was taken into account in the SAP2000 model.

The scouring analysis is performed according to FHWA HEC 18 2012 and the scouring depth analysis can be performed using a software which is called HEC-RAS (Hydrologic Engineering Center – River Analysis System). This software is developed by the United States Army Corps of Engineers. Its algorithm has been prepared in accordance with the FHWA HEC 18 2012 criteria. Hydraulic analysis can be performed for stream bed using HEC-RAS software. This software was used in the analysis of the scouring depth of Kaleönü Bridge. For Kaleönü Bridge, topographic specialties and design flow rates were determined, bed material information and bridge geometry data were defined in the program, and the scouring depth was calculated. Kaleönü Bridge modeled in the program and the Elekçi Stream, in which the bridge piers are located, is shown in Figure 15.

While determining the scouring depth, the bridge was designed for Q_{100}, and the design was checked for the Q_{500} (Evaluating Scour at Bridges, FHWA). Program output for the scouring depth of Kaleönü Bridge is shown in Figure 16. The scouring depth for Kaleönü Bridge is 8 meters according to HEC-RAS analysis.

3.2.2 *Precautions against scouring:*

Although the structural design of the bridge is compatible with the hydraulic and geotechnical details, there should be many countermeasures in the bridge span, piers and abutments. There are various structures for decreasing the scouring around the piers, protecting the slope and controlling the flow. There are no generalized criteria in the countermeasures since the local conditions of each region are different. The choice of countermeasures depends on the engineer's experience. Also, when necessary, the adequacy of the countermeasures should be checked by physical models. The precautions against local scouring around the piers are specified in FHWA HEC 23 2009 criteria.

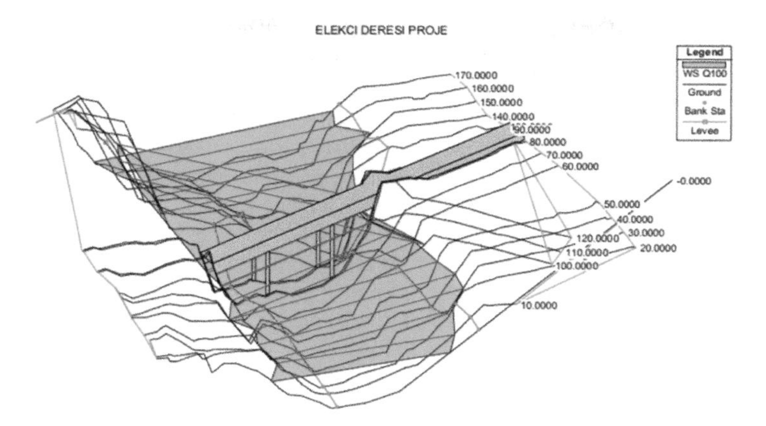

Figure 15. Kaleönü Bridge and Elekçi Stream model in HEC-RAS.

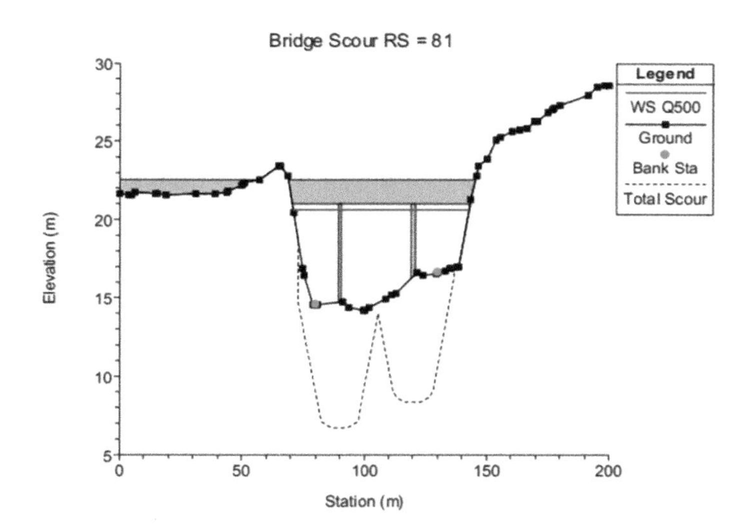

Figure 16. Kaleönü Bridge scouring depth output.

These are hydraulic and structural precautions. The hydraulic precautions include flow-control and slope protection. The structural precautions are measures to reduce the scouring around the piers. These are arrangements for bridge foundation. The main protective coating layers used under the base level around the piers are riprap, concrete block matrix, grout filler matrix, and gabion. Design criteria for these materials are studied in FHWA HEC 23 2009.

Spurs are used to regulate stream flow. Also, the longest span is determined in the design phase for given precautions against contraction scour. In this way the interference of scouring are prevented. If the spread foundation is used, the foundation bottom should be carried below the scouring depth according to the scouring depth determined in analysis. If the scouring depth is greater, then the piled foundation should be adopted.

The first precaution against scouring was the pile foundation design for Kaleönü Bridge. The scouring depth was taken into account in the foundation design. Grouted riprap in abutments and coursed rockfill in piers were applied to prevent movement of the base and backfill materials during flooding. Wing walls were also provided in the abutments.

4 CONCLUSIONS

In this study, bridges damaged and destroyed due to the flood in Ordu, Turkey are examined, and the causes of their failure are determined. For the reconstructed bridges, some information is given about structural and hydraulic calculations which is taken into consideration during the design phase and recommended construction methods during the construction phase. In the design of the bridges to be reconstructed in the riverbeds, data such as climatic conditions, topographic specialties, and the region's stream regime were taken from the General Directorate of State Hydraulic Works. The bridges' layout plans and types of piers were determined according to these data. Scour analysis was performed and scour height was determined and this scour height was taken into consideration in the static design phase. In the construction phase, physical recommendations were applied in the field to prevent scouring in bridges service life.

REFERENCES

FHWA HEC18 Hydraulic Engineering Circular No.18 (2012). U.S. Department of Transportation Federal Highway Administration.

FHWA HEC23 Hydraulic Engineering Circular No.23 (2009). U.S. Department of Transportation Federal Highway Administration.

Yanmaz, A. Melih. 2002. *Köprü Hidroliği*. Ankara: ODTU Geliştirme Vakfı Yayıncılık ve İletişim A.Ş.

Yurtseven, M. Levent. *Köprü Ayakları Arkasında Oluşan Oyulmaların İncelenmesi.* Master Thesis, Istanbul Technical University, 2005.

Yücel, A. & Namlı, R. 2007. Su ve diğer faktörlerin köprü ayakları etrafındaki bozulmalara etkisinin araştırılması. *Doğu Anadolu Bölgesi Araştırmaları.*

Chapter 10

Explicit collapse analysis of the Morandi Bridge using the Applied Element Method

D. Malomo & R. Pinho
University of Pavia & Mosayk Ltd, Pavia, Italy

N. Scattarreggia
Istituto Universitario di Studi Superiori (IUSS), Pavia, Italy

M. Moratti
Studio Calvi Ltd., Pavia, Italy

G.M. Calvi
Istituto Universitario di Studi Superiori (IUSS) & Eucentre, Pavia, Italy

ABSTRACT: Explicit collapse analysis of bridges still represents an open challenge in numerical modeling. In this work, an innovative micro-modeling approach, the Applied Element Method, is used to investigate potential triggering factors that might have contributed to the collapse of the Morandi Bridge, in Genoa, Italy, which took place on August 14, 2018. This paper explores the influence of several parameters, including deterioration of cables, and loading effects on the collapse mechanism. The observed and predicted debris were also compared to assess the possible collapse mechanism of the bridge.

1 INTRODUCTION

In the immediate aftermath of a bridge collapse event, it is not always straightforward to identify the reasons behind the structural failure, also because the extent and quality of in-situ collected data can be limited, as noted by Deng et al. (2016). On the other hand, performing full-scale and field tests is very challenging due to both costs and safety concerns issues (Piran Aghl et al. 2014). Within such context, therefore, forensic investigations may benefit significantly from the use of reliable numerical methods able to represent explicitly damage propagation up to collapse, accounting for various input conditions and for the influence of the adopted construction technique. In this work, a recently developed rigid body and spring discrete model, known as the Applied Element Method (AEM), is used to investigate numerically potential causes and triggering factors which may have contributed to the collapse of the viaduct over the Polcevera river (Genoa, Italy), also known as the "Morandi Bridge". The structure was designed in the early 1960s by Riccardo Morandi, and opened to traffic in 1967. On August 14, 2018, the bridge collapsed, claiming the lives of 43 people. Recent applications have shown that the AEM, originally conceived by Meguro and Tagel-Din (2000) to simulate controlled structural demolition, does appear to be able to capture adequately the progressive failure of large-scale bridge systems (e.g. Salem and Helmy 2014). Thus, making use of only material publicly available online, an AEM-based model was built for assessing explicitly, through a detailed sensitivity study, the influence that several parameters, including reduction of cables cross-section (potentially induced by corrosion), as well as both local and global effects of impulsive loading, might have had on the observed structural failure.

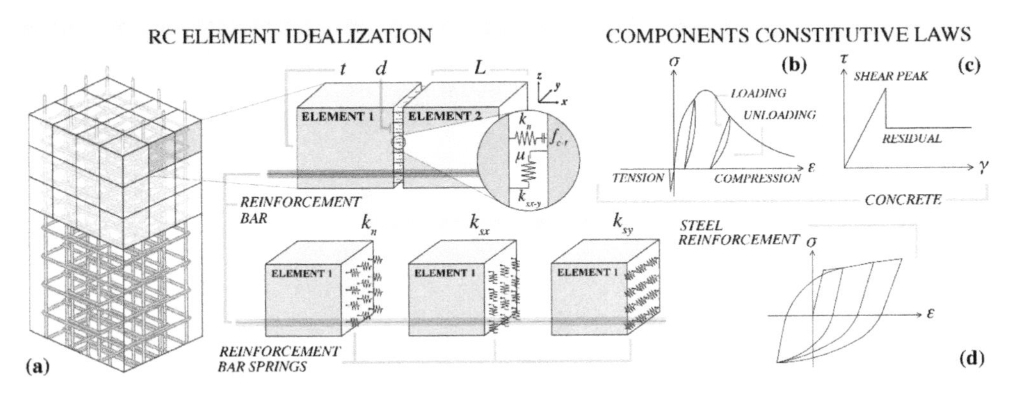

Figure 1. AEM modeling of RC structures (a), material models for concrete (b,c) and steel (d).

2 THE APPLIED ELEMENT METHOD FOR RC STRUCTURES

In the framework of the AEM, a given solid body is modeled as a virtual assembly of small 6-DOF rigid units connected to each other by means of zero-thickness spring layers, characterized by normal (k_n) and shear (k_s) stiffnesses, in which the material properties are lumped. This scheme can be easily adapted to the modeling of reinforced concrete (RC) structures by explicitly coupling the mechanical contribution of concrete elements, whose elastic and inelastic properties are assigned to interface springs, with those of the embedded steel bars, represented by equivalent "reinforcement springs" which can be placed exactly in the same position of their real counterparts, as depicted in Figure 1a below:

In the current version of the AEM-based software employed in this endeavor – Extreme Loading for Structures (ELS, Applied Science International LLC. 2018), the Maekawa and Okamura (1983) model is adopted for modeling the response of concrete under uniaxial cyclic tension-compression loading (see Figure 1b). To this end, the Young's modulus of concrete (E_c), compressive strain and fracture parameters are used to define an initial elastic bound, beyond which k_n is assumed as a minimum value (1% of initial k_n in this work) to avoid stiffness matrix singularities. Tension cracking is accounted by neglecting the contribution of tensile strength right after reaching the peak stress σ_t. When considering shear-compression stress-states, a simplified Mohr-Coulomb failure envelope is typically employed (Figure 1c). The hysteretic model proposed by Menegotto and Pinto (1973) is assigned to normally-loaded reinforcement springs (Figure 1d).

3 BRIEF DESCRIPTION OF THE MORANDI BRIDGE

The Morandi bridge, schematically depicted in Figure 2a, represented a key infrastructure for the Italian road network. While the entire bridge and its foundation scheme are comprehensively described in Morandi (1967a; 1967b), in the following special attention will be given to the symmetrical "balanced system" that actually collapsed (i.e. pier 9), as shown in Figure 2b.

With reference to the nomenclature reported in Figure 2b, the considered balanced system essentially comprised the following main elements:

1. A pier with eight inclined struts (with cross-section varying between 4.5 × 1.2 to 2.0 × 1.2 m) that props the deck over a distance of about 42 m.
2. An antenna with two A-shaped structures (element cross-section varying between 4.5 × 0.9 to 2.0 × 3.0 m) that converge about 45 m above the deck level.

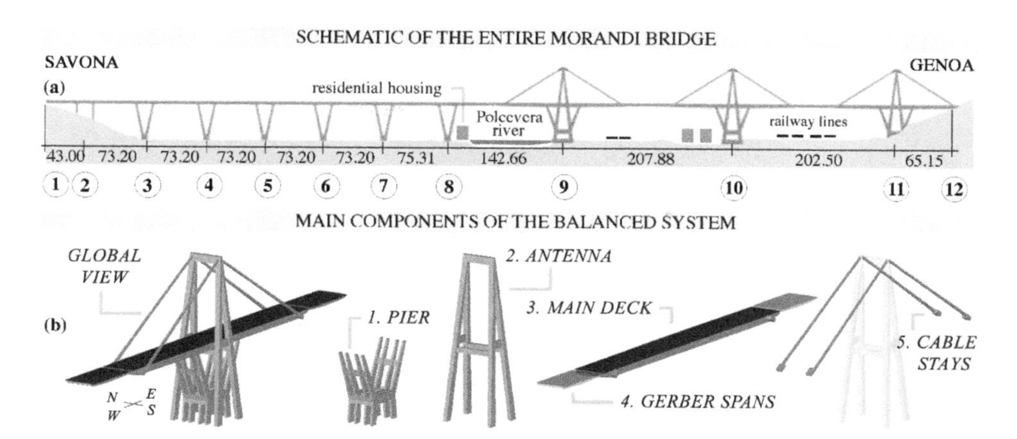

Figure 2. Schematic of the entire bridge (a), balanced-system sub-components (b).

3. A main deck with a five-sector box section of depth variable between 4.5 and 1.8 m, an upper and lower slab 16 cm thick, and six deep webs with thickness varying between 18 and 30 cm. In its final configuration, the deck of the balanced system 9 was 172 m long and supported at four points: from below by the piers at the aforementioned spacing of 42 m and from above by the cable stays at a distance of 152 m. Two 10 m cantilevers thus completed the deck length.
4. Four transverse link girders connected stays and pier trusses to the deck.
5. Two simply-supported Gerber beam spans connecting the balanced system to the adjacent parts of the bridge. Each span was 36 m long and comprised six precast pre-stressed beams, with a variable depth equal to 2.20 m at mid-span, sitting on Gerber saddles protruding from the main deck.
6. Four cable stays, hanging from the antenna's top and intersecting the deck at an angle of about 30°.

4 POST-COLLAPSE DEBRIS DISTRIBUTION AND OBSERVED DAMAGE

In this section, the actual configuration of the in-situ collapse debris, as well as the post-collapse observed damage (see Figure 3), is discussed on the basis of the documentation made publicly available through various national and international media after of the events of 14 August 2018.

As can be gathered from Figure 3, the bottom portion of the pier (i.e. below the eight inclined struts) did not collapse, somehow withstanding the impacts of both antenna and main deck debris. Eyewitnesses reported that the antenna structure, as shown in many videos, was the last component to collapse. Looking at the image above, it seems that most of the antenna debris collided with the top of the pier, covering a relatively reduced area. While a clear identification of debris is not possible when considering the West supported-span (which impacted the underneath residential and commercial buildings), it appears that the East Gerber span impacted the ground on its back side. On the other hand, the post-collapse configuration of the West portion of the deck (oriented towards northwest), would indicate that a torsional failure mechanism could have occurred. The North-side stay-to-deck connection, as well as the extremity of the transverse link girder, although severely damaged, appears still connected to the deck. With respect to the East portion of the deck, given that no asphalt is visible (contrary to the West-side deck), one could assume that it collapsed on its back side.

Figure 3. Actual configuration of debris after the collapse of the Morandi bridge (adapted from Malomo et al. 2019, original images from tg24.sky.com, video.repubblica.it, huffingtonpost.it and umbriajournaltv.it).

5 NUMERICAL IDEALIZATION OF CONSTRUCTION DETAILS

While a traditional bottom-up construction approach was employed for erecting both pier and antenna, the main deck was constructed by progressively casting-in-place 5 m-long RC segments, which were temporarily supported. Then four cable stays, each of which containing a total of 464 strands with nominal diameter of 1/2 inch, were added. More precisely, 352 strands were located first and connected to the deck to bear its dead weight, after which concrete shells of variable section (0.61 to 1.22 m) were cast around, post-compressed using the remaining 112 strands and injected with cement mortar. At this stage the main deck experienced upward bending, which was afterward made negligible through the introduction of supported spans and dead loads (assumed equal to 2.4 kN/m^2 plus 18 kN/m for three lines of New Jersey barriers). With a view to simplify this collapse modelling effort, live loads, which amounted to less than 20% of the deck self-weight, were not considered in the analyses, since they were deemed unlikely to have played a role in the triggering of the bridge failure (Calvi et al., 2018).

Construction phases were faithfully reproduced numerically, as shown in Figure 4a. For each stage predicted deck displacements and stays responses were also monitored, and results compared to those obtained in other modelling endeavors (Calvi et al. 2018; Bellotti 2019; Orgnoni 2019). For instance, the initial tension strain to obtain zero vertical displacements after both the application of dead load and the construction of the supported spans predicted by ELS was 148 mm (whilst in the works mentioned above values varying between 140 and 145 mm were obtained), the vertical reaction at the pier base was 165 MN (values ranging from 168 to 170 MN were reported in the publications above), the axial force in the 352 tendons and the post-compressed concrete element was 20,800 kN (values between 22,360 and 22,600 kN were predicted by the aforementioned researchers).

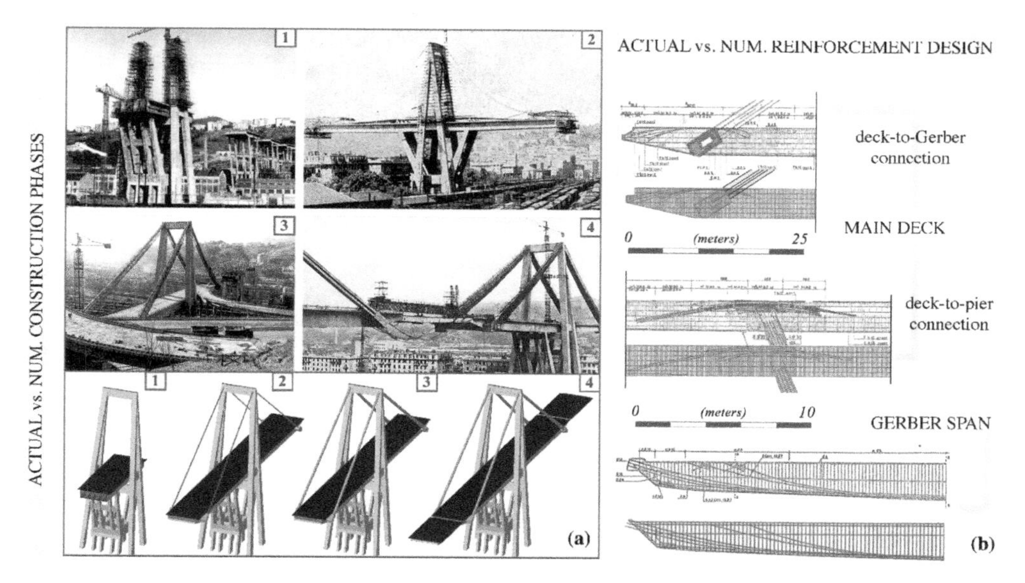

Figure 4. Actual vs. numerical construction phases (a) and reinforcement design (b) (adapted from Malomo et al. 2019).

As illustrated in Figure 4b, where original drawings are compared with numerical counterparts, both passive and active reinforcement schemes were explicitly implemented in the AEM model; it is noted that, since no online-available details regarding the morphology of transverse link girders were found, a minimum reinforcement level was again considered herein, as well as visually-deduced section properties (i.e. hollow, thickness from 0.5 to 1 m, approx. 4.5×2.0 m). The stays were modeled as an assembly of two different elements working in parallel; beam elements to represent the post-compressed concrete components, and bilinear links to represent the pre-tensioned tendons (making sure that zero vertical displacements were obtained after the addition of the supported spans and the dead load, as per original design and construction). According to Morandi (1967a), stay cables are continuous elements passing over a saddle at the top of the antenna, without any local restraint. Thus, and for instance, the removal of the southwest (SW) stay would imply also the subsequent loss of its southeast (SE) counterpart. However, from a numerical viewpoint, representing explicitly the strands curving over the saddle is cumbersome and was beyond the scope of the paper. Thus, link/beam elements were interrupted at the interface with the top of the antenna, while the failure of two consecutive stays was externally-induced or precluded depending on the investigated failure mechanism.

6 SENSITIVITY STUDY

In this section, a number of modeling scenarios are presented and discussed with the aim of investigating potential triggering factors and possible causes that might have induced the collapse of the Morandi bridge. Their selection was guided by both analytical and numerical outcomes reported in Calvi et al. (2018), where it is shown that the Morandi bridge, as conceived and designed, seemed to have had significant capacity reserves with respect to deck flexure, shear and torsion mechanisms, whilst the complete loss of a stay could have resulted in the type of complete collapse that was observed.

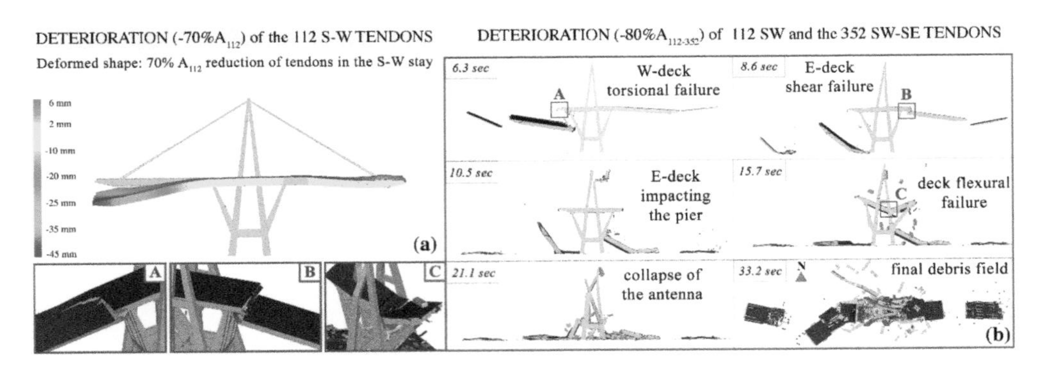

Figure 5. Deformation-induced by a 70% are reduction of the 112 SW tendons (a), collapse in the case of 80% of area reduction of 112 (SW) and 352 (SW-SE) tendons (b) (adapted from Malomo et al. 2019).

6.1 *Scenario 1 – Progressive deterioration of stays reinforcement*

A progressive reduction of the cross-sectional area of the 112 tendons implies a simultaneous decrease of the stays' stiffness and their consequent elongation, while inducing a migration of shear and torsional actions in the deck that could potentially lead to the failure of the SW stay and the consequent collapse of the bridge. Analyses results, however, seem to indicate that, even in the case of unrealistically low values of residual cross-sectional area of the post-compression tendons of the SW stay, the bridge was unlikely to collapse.

For instance, considering a 50% tendons cross-sectional area ($A_{112-352}$) reduction leads to only a -19 mm additional vertical displacement at the connection between the SW stay and the deck (and naturally even smaller vertical displacements on the NW, SE and NE stays-deck connections). Considering instead a reduction of 70% (see Figure 5a) leads to a maximum displacement of -45 mm on the SW side (and -17 mm NW, -10 mm SE, $+6$ mm NE), which is a condition still far from inducing collapse. Given that the reduction of cross-sectional area of the 112 post-compression tendons of the SW stay alone did not lead to collapse, a number of additional cases have been modelled assuming cross-sectional area reduction also for the 352 pre-tensioned cables, both in the SW stay alone, as before, as well as in the other three stays. In order to be able to obtain an explicit collapse of the structure, an area reduction in the range of about 80% of both the 112 post-compression tendons (SW stay) and the 352 pre-tensioned cables (SW and SE stays) would need to be introduced. In such a case, as depicted in Figure 5b, where the progressive collapse of the bridge is shown, a torsional failure of the West-side of the deck was predicted.

However, when comparing predicted and actual (see Figure 3) debris configuration, several dissimilarities can be observed. First, while the actual West-side deck is oriented towards NW, the model predicted the opposite (i.e. the deck is oriented towards SW). Further, the predicted debris generated by the collapse of the antenna extends ca. 30m from the pier, whilst the actual debris configuration seems to indicate that the antenna collapsed 5–10 m far from the pier. Finally, because of the large extent of tendons deterioration for which the model predicted collapse, and since conspicuous signs of significant structural distress would have had to appear well in advance, it can be thus concluded that whilst a progressive reduction of tendons cross-sectional area and related post-tensioning force could perhaps have contributed to the observed collapse, it could not by itself alone be the cause of the collapse of the bridge.

6.2 *Scenario 2 – Impulsive load acting on critical sections*

This modeling scenario explores the possibility of a global collapse induced by a hypothetical case of an impulsive load acting on critical sections, possibly weakened by some loss of post-tensioning. Although the simply-supported Gerber span appears to possess sufficient strength to withstand the

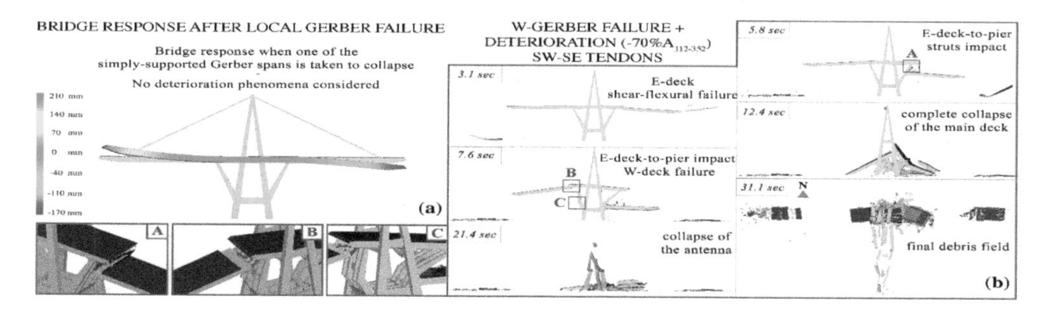

Figure 6. Bridge response when one of the simply-supported Gerber spans is taken to collapse with (a) and without (b) considering deterioration phenomena (adapted from Malomo et al. 2019).

considered hypothetical accidental impulsive load, the effect of its potential failure on the global dynamic response of the bridge was nonetheless investigated, through the sudden removal (after the application of the static loads) of one, and then two of its six constitutive Gerber beams.

In the first case, no explicit collapse of the supported span was obtained, while the simultaneous removal of two of the Gerber beams did lead to a collapse of the supported span, which induced on the main bridge system a flexural deformation producing vertical displacements at the connection between the SW stay and the deck of +160 and +170 mm towards NW and SW respectively, whilst on the NE and SE sides −135 and −145 mm. As shown in Figure 6a below, however, such a scenario does not lead to the collapse of the bridge.

Thus, with a view to induce global failure in case of the collapse of a Gerber span, several combinations of deterioration percentage of the 112 + 352 tendons were considered. As depicted in Figure 6b, explicit collapse was first obtained assuming a widespread deterioration of stay elements, i.e. corresponding to a reduction of cross-sectional area of both 112 and 352 tendons of 70%. During collapse, the East-side deck first collapsed, impacting the pier inclined struts and inducing global failure of the system. As in the previous case, the antenna was the last element to collapse, even though it experienced an out-of-plane collapse towards the southside (maximum distance from debris generated by the antenna and the pier equal to approximately 120 m).

With respect to the actual debris distribution, significant differences are clearly visible, including the antenna debris and the position of the main deck, located at a maximum distance of 5–8 m from the pier (actual was 10–25 m). It is therefore concluded that no reasonable level of impulsive loading could cause the collapse of the bridge, unless in combination with other problems, such as a concurrent loss in the stay capacity.

6.3 *Scenario 3 – Failure of the deck/antenna-stay connection*

As depicted in Figure 7a below, two scenarios are herein considered. The first is failure at the interface between the SW stay and the antenna (possibly related to fatigue in the tendons).

The second scenario is the sudden loss of connection between the same SW stay and the main deck (as previously discussed, the limited knowledge about the transverse link details cannot exclude this possibility).

The collapse sequence (as induced by the antenna-to-stay interface failure), depicted in Figure 7b, comprised a torsional collapse of the deck in a section next to the West-side pier strut and the subsequent falling to the ground of the west supported span.

Then, the consequent release of the SW stay and flat collapse to the ground of the west deck and supported span was predicted and followed by the collapse of the South and North A-shaped structures of the antenna, Finally, the central span collapsed when hit by the falling antenna debris. A similar collapse mode was obtained for the case of a deck-to-stay interface failure, as shown below. The progressive collapse sequence described in Figure 7b,c seems to be consistent with the actual evidence, as may be gathered also from Figure 8 below, where observed and predicted debris

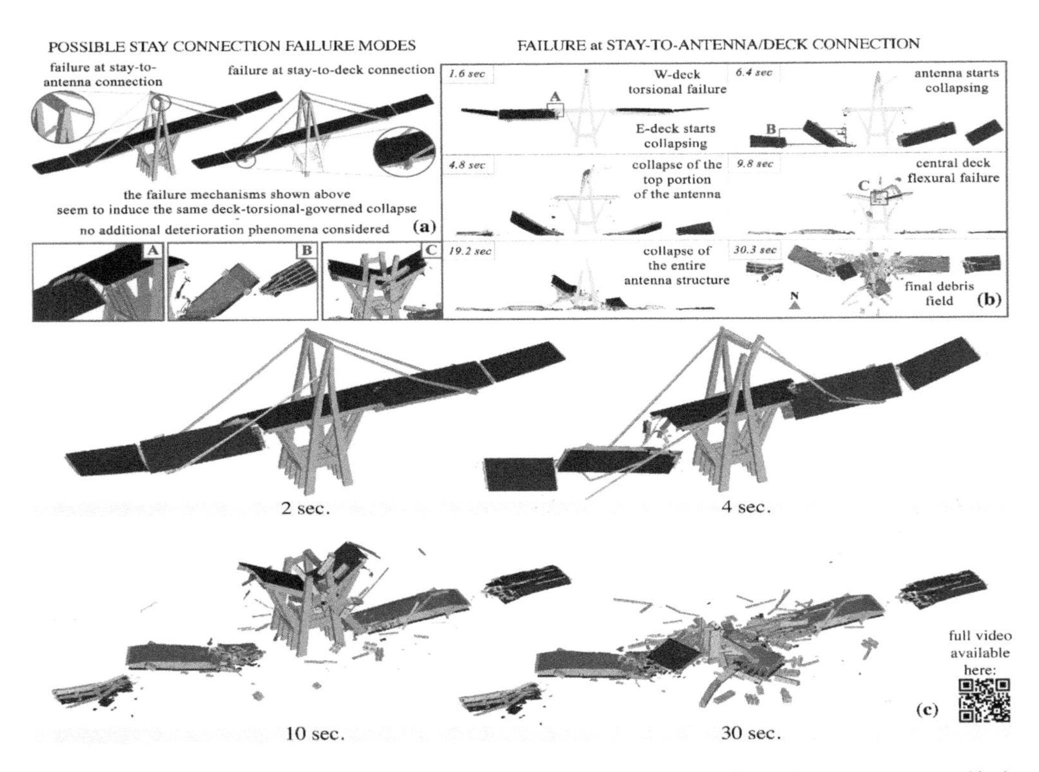

Figure 7. Possible SW stay connection failures (a), collapse induced by the failure at SW stay-to-antenna/deck connection (b) and failure sequence up to complete collapse (c) (adapted from Calvi et al. 2018)

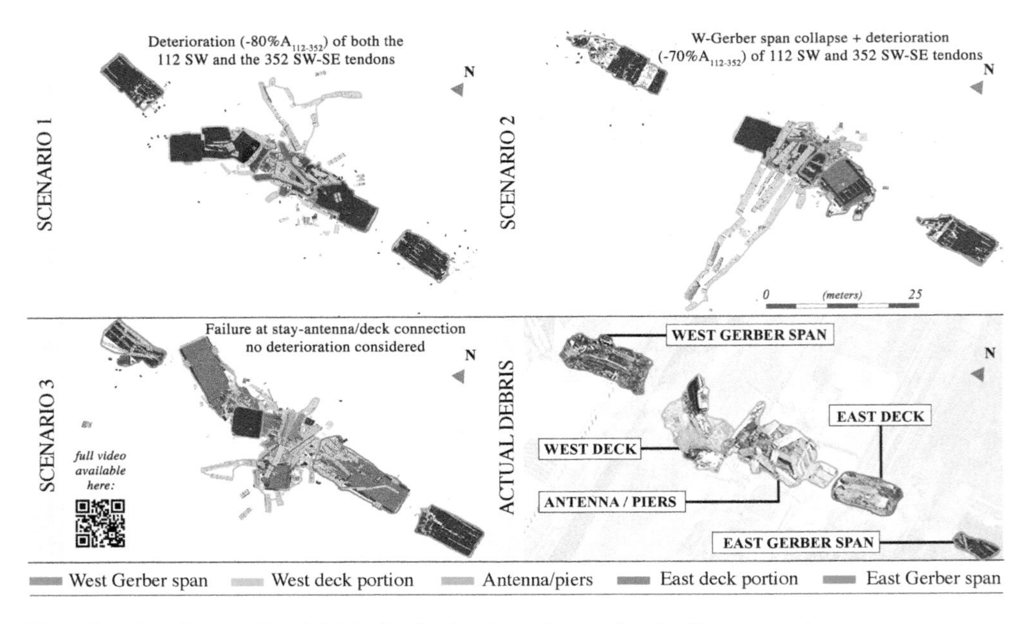

Figure 8. Actual vs. predicted debris distribution for each considered collapse scenario.

are compared. Indeed, the observed East deck orientation, as well the actual position of the debris generated by the collapse of the Gerber spans, were realistically predicted by the model. As for the previous scenarios 1 and 2, the bottom portion of the pier did not collapse, and the antenna collapsed last.

This notwithstanding, the model marginally overestimated the extent of the antenna debris, which extends towards the southside. However, it recalled that the buildings surrounding the bridge were not modeled in this work. The satisfactory agreement found seems to lend further weight to the possibility that the collapse of the bridge was indeed triggered by a failure of the deck/antenna interfaces of the SW stay.

7 CONCLUSIONS

On August 14, 2018, due to presently unknown reasons, a portion of the viaduct over the Polcevera river (Genoa, Italy), designed by Riccardo Morandi in the early 1960s, collapsed killing 43 people. In this work, the Applied Element Method (AEM), which explicitly represents the structural behavior passing through all the stages of loading, from the elastic domain up to full collapse, was employed for investigating numerically potential triggering factors and possible causes that may have induced the observed failure, which involved one of the balanced systems of the Morandi bridge. Several different scenarios, whose selection was guided by the work of Calvi et al. (2018), including the progressive deterioration of stays reinforcement, as well as the effects of both impact loads acting on critical sections and a sudden failure of the deck/antenna-stay connections, were considered.

The numerical results obtained seem to indicate that a progressive reduction of the steel section of the stays was unlikely to have played a triggering role in the collapse, given that unrealistically low values of residual cross-sections were found to be necessary to induce collapse. This conclusion is also supported by the fact that in such case a progressive elongation of one or more stays, which would result in increased local displacements and significant deck level irregularities, should have been evident long before the reaching of near collapse conditions.

Similarly, it was observed that, even though an impulsive load acting on a critical section of the supported span may induce local element collapse in presence of relevant corrosion phenomena, no global collapse modes were predicted. Indeed, it was found that the simultaneous shear failure of at least two Gerber beams of the supported span may induce the collapse of the span, but not of the bridge itself.

On the other hand, when considering the failure at the stay-to-deck/antenna connection, which could have been induced e.g. by either fatigue problems or by the progressive deterioration of the stay-to-transverse link connection, the numerically-inferred collapse sequence and corresponding debris distribution seemed to be consistent with their actually-observed counterparts (see Figure 8), especially if compared with the results obtained in the case of Scenario 1 and 2, in which the model predicted rather different debris distributions. Thus, it can be concluded that, regardless of the initiation of the collapse sequence, the loss of a stay seems to be the most likely cause of collapse.

It is however recalled that variations, with the respect to the available online drawings that were herein employed to develop the numerical models, e.g. of tendons location and layouts, could introduce variations in the locations of damage concentration that could in turn change the post-collapse debris distribution. We will thus continue to endeavor to acquire additional certainties regarding the actual construction details of the bridge, by trying to secure access to the hard-copies of the final as-built drawings of the bridge (even if potential changes in reinforcement will not necessarily be critical for what concerns the conclusions above).

ACKNOWLEDGEMENTS

The authors gratefully acknowledge the assistance and collaboration from Applied Science International LLC (ASI) in the use of Extreme Loading for Structures (ELS).

REFERENCES

Applied Science International LLC. (2018). "*Extreme Loading for Structures*." Durham, USA.

Bellotti, F. (2019). "*Seismic analysis of the Morandi Bridge*." MEng Dissertation, Civil Engineering and Architecture Department, University of Pavia, Italy.

Calvi, G. M., Moratti, M., O'Reilly, G., Scattarreggia, N., Malomo, D., Calvi, P., Monteiro, R., and Pinho, R. (2018). "*Once upon a time in Italy: the tale of the Morandi bridge*." Structural Engineering International, DOI: 10.1080/10168664.2018.1558033.

Deng, L., Wang, W., and Yu, Y. (2016). "*State-of-the-Art Review on the Causes and Mechanisms of Bridge Collapse*." Journal of Performance of Constructed Facilities, 30(2).

Malomo, D., Scattarreggia, N., Pinho, R., Moratti, M., and Calvi, G. M. (2019). "Numerical study on the collapse of the Morandi bridge." *Journal of Performance of Constructed Facilities* – submitted for publication

Maekawa, K., and Okamura, H. (1983). "*The Deformational Behavior and Constitutive Equation of Concrete using Elasto-Plastic and Fracture Model*." Journal of the Faculty of Engineering, 37(May 1985), 253–328.

Meguro, K., and Tagel-Din, H. (2000). "*Applied element method for structural analysis: Theory and application for linear materials*." Structural Engineering/Earthquake Engineering, Japan Society of Civil Engineers, 17(1).

Menegotto, M., and Pinto, P. (1973). "*Method of analysis for cyclically loaded RC plane frames including changes in geometry and nonelastic behavior of elements under combined normal force and bending*." Proc. of IABSE symposium on resistance and ultimate deformability of structures acted on by well defined repeated loads, Zurich, Switzerland, 15–22.

Morandi, R. (1967a). "*Il viadotto sul Polcevera per l'autostrada Genova-Savona*." L'Industria Italiana del Cemento, 12, 849–872.

Morandi, R. (1967b). "*Il viadotto del Polcevera dell'Autostrada Genova – Savona*." Unpublished report.

Okamura, H., and Maekawa, K. (1991). "*Nonlinear analysis and constitutive models of reinforced concrete*." Giho-do press, Tokyo (Japan) – ISBN: 978-4-7655-1506-1, 10(2), 42.

Orgnoni, A. (2019). "*Detailed finite element modelling of the construction sequence of the Morandi Bridge*." MEng Dissertation, Civil Engineering and Architecture Department, University of Pavia, Italy.

Piran Aghl, P., Naito, C. J., and Riggs, H. R. (2014). "*Full-Scale Experimental Study of Impact Demands Resulting from High Mass, Low Velocity Debris*." Journal of Structural Engineering, 140(5), 04014006.

Salem, H. M., and Helmy, H. M. (2014). "*Numerical investigation of collapse of the Minnesota I-35W bridge*." Engineering Structures, Elsevier, 59, 635–645.

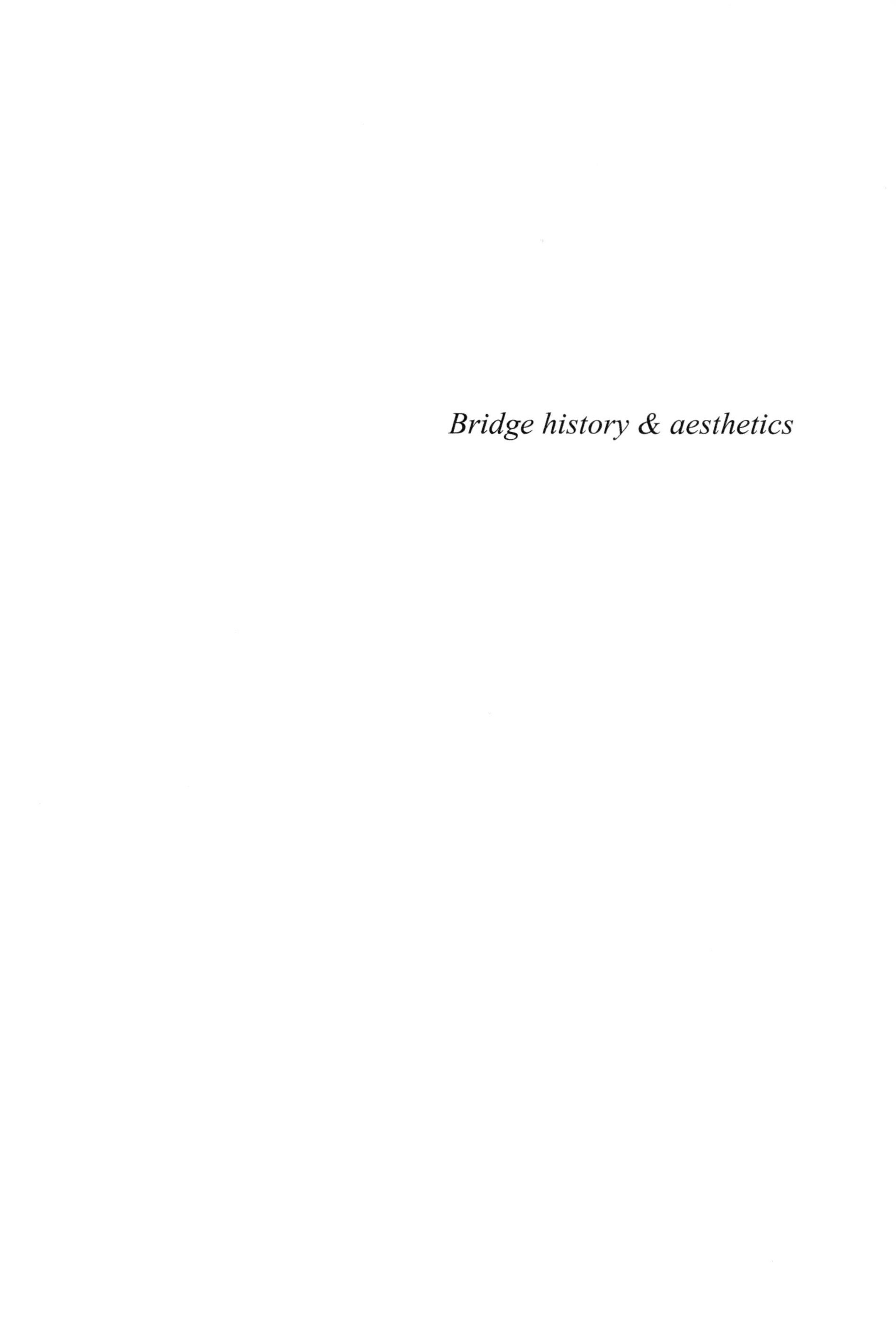

Bridge history & aesthetics

Chapter 11

The Walter Taylor Bridge – Florianopolis Australis

S. Rothwell
Stuart Rothwell & Associates, Brisbane, Australia

ABSTRACT: The Walter Taylor Bridge over the Brisbane River at Indooroopilly in Queensland Australia, completed in 1936, is a rare example of the Florianopolis type, originated by David B Steinman. His bridge, however, and those by others, used eye bar suspension chains, while the cable at Indooroopilly consists of wire ropes salvaged from the tie-back cables used in construction of the Sydney Harbour Bridge. The bolted clamps that connect the stiffening trusses to the wire rope cables were developed specifically for the bridge at Indooroopilly, and these two features make the bridge unique. When it was opened the bridge had the second longest span in Australia, only exceeded by the then new bridge over Sydney Harbour. Design provenance is clouded by uncertainties, although it is now generally accepted that R. J. McWilliam designed the pylons and W. J. Doak was responsible for the suspended steelwork. At the time, Doak was Bridge Engineer for Queensland Railways, which may explain the reluctance to publically acknowledge his involvement at the time. The bridge is well maintained and remains in service.

1 PREAMBLE

In 1936 a private company incorporated to perform a public duty completed a Florianopolis type suspension toll bridge (see Figure 1) over the Brisbane River at Indooroopilly in Queensland Australia. Initially known simply as the Indooroopilly Toll Bridge, the name was eventually changed to the Walter Taylor Bridge to honor its principal promoter and builder.

The bridge, which is still operating as originally intended, has a main span of 600 feet (182.9 m). When it was completed this was the second longest clear span of any bridge in Australia, and the longest for a suspension bridge. The back-span lengths, from piers to anchorage are 185 feet (56.4 m) on the Indooroopilly (left) side and 198 feet (60.4 m) on the Chelmer side.

The Florianopolis type is named for the eponymous Brazilian city, where the first bridge of this style, the Hercilio Luz Bridge, was completed in 1926. David Barnard Steinman (1886–1960) of the USA consulting practice *Robinson and Steinman* designed that bridge in Brazil, so the relatively few bridges completed in this style have also been referred to as Steinman types.

The principal characteristic of a Steinman / Florianopolis type suspension bridge is that the suspension cable forms a part of the top chord for the stiffening truss. Steinman believed this was

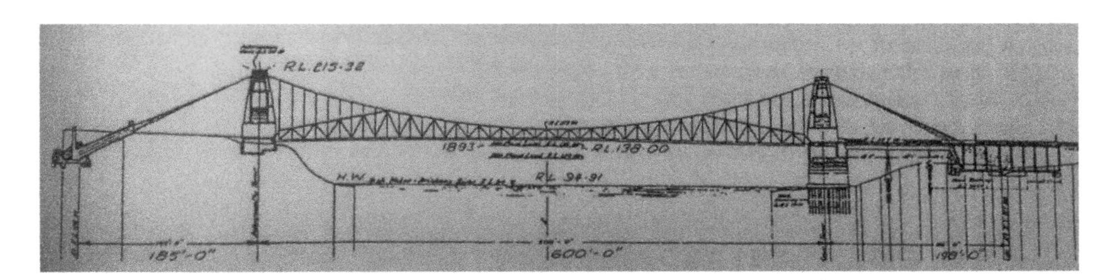

Figure 1. Walter Taylor Bridge elevation – looking downstream with Indooroopilly on the left (O'Connor 2003).

Figure 2. View of Walter Taylor Bridge, looking north to the Indooroopilly tower (O'Connor 2003).

Figure 3. Joint on the cable of the Walter Taylor Bridge between end chord and cable (O'Connor 2003).

a more efficient arrangement that could yield cost savings (refer Section 2). Steinman's bridge at Florianopolis, constructed by American Bridge Company, used steel eye bar chains for the suspension cable. At Indooroopilly, however, as shown in Figure 2, wire rope cables were adopted, which were salvaged from falsework used to construct the Sydney Harbour Bridge (refer Section 5).

To connect these cables to the trusses a bespoke clamp detail (see Figure 3) was developed and tested locally. These two distinct particulars, the wire cables and the bolted clamp joints, make the bridge an internationally unique Steinman type.

Figure 4. Schematic of a design (not adopted) for a suspension bridge at Florianopolis (Steinman 1929).

Figure 5. Schematic of the adopted design for the suspension bridge at Florianopolis (Steinman 1929).

2 THE STEINMAN TYPE DESCRIBED

The principles characteristics and attractions of a Steinman type suspension bridge, like the one adopted at Indooroopilly, are best explained by reference to the two bridges David Steinman developed for the Florianopolis site and his published comments about them.

Steinman's first (eventually superseded) design for that bridge, circa 1920, was a conventional form of suspension bridge with a stiffening truss of uniform depth suspended from separate cables (see Figure 4). Over the central half of the span each cable was in close proximity to the top chord of the truss. Steinman figured that as the cable was in tension, and the top chord in compression, the "juxtaposition of two principal members carrying opposing stresses represents a waste of materials, or rather, a neglected opportunity for economizing". He determined that by "combining the two opposing structural elements, one member is made to take the place of two". As he remarked, "the result is a subtraction of stresses instead of an addition of sections" (Steinman 1929).

Steinman reckoned that it wasn't efficient to continue this combination of elements outside the central half of the span. As his revised and eventually adopted design shows (see Figure 5), this led to a fortuitous coincidence. He asserted that the moments in his suspension bridge were greatest at the quarter points, while the shears were greatest at the ends and middle of the span. So, he postulated, "the economic profile of a stiffening truss is one having a maximum depth at the quarter-points and minimum depth at mid-span and at the end". As he observed, this result is "automatic" if the combination of cable and top chord is "limited to the middle half of the span" (Steinman 1929).

Steinman also calculated that the revised arrangement was more efficient in reducing deflection, again because the stiffening truss was deepest where it was most necessary (Steinman 1929).

The redesign at Florianopolis was undertaken because of an innovation at the time; high-tension, heat-treated carbon steel eyebar cables, which suited the new configuration. (Steinman 1929).

Nevertheless, Steinman was at pains to record that "the new form of stiffening construction can also be used in conjunction with wire cables" and he intimated that he and Robinson had developed "the necessary connections between truss members and wire cables" (Steinman 1929). These are important observations, discussed further in Section 5, because they relate directly to the decisions taken at Indooroopilly.

Other contemporary Florianopolis style bridges to use eye bar chains were the ill-fated Silver Bridge, built over the Ohio River at Point Pleasant (WV) in 1928, and a sister bridge built over the same river at St Marys (WV), also in 1928.

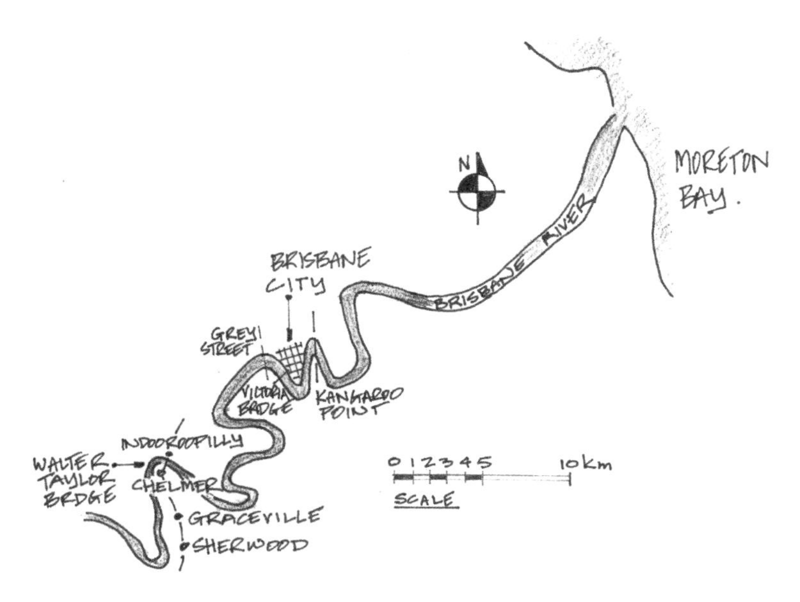

Figure 6. Map of Brisbane and surrounds.

3 HISTORICAL BACKGOUND

The first colonial settlement at the site of Queensland's current capital, Brisbane, was in 1825. A hundred years later the Victoria Bridge was the only connection over the Brisbane River able to link road transportation between the burgeoning city and the expanding region to the south (see Figure 6).

As the centenary of settlement approached citizen groups and progress associations south of the river had been calling for a road bridge crossing to Indooroopilly for 40 years. A railway bridge connecting Chelmer and Indooroopilly did serve those southern citizens wishing to travel to Brisbane by train, but road traffic was forced to cross via a ferry.

These citizen initiatives had generally been impeded by the fragmented arrangement of the surrounding local governments and the expense of bridging over such a relatively wide river. In 1925, however, those numerous and small local authorities around the city's commercial centre were amalgamated into one entity, the greater Brisbane City Council (BCC), and an almost immediate decision of the new council was to appoint a Cross River Commission, to "investigate what further facilities for crossing the Brisbane River are necessary and also as to the best sites for same" (Hawken et al 1926).

The Commission received numerous submissions about a bridge at Indooroopilly and Walter Taylor (1872–1955), a local building contractor who lived in nearby Graceville, appeared before it on behalf of the Indooroopilly-Chelmer Memorial Bridge League. In its final report the Commission gave conditional acceptance of the general site at Indooroopilly, as its fifth priority location, although it gave no clear recommendation as to the form to be adopted, nor the precise position of the bridge.

The first priority recommended by the Commission, a bridge at Kangaroo Point, was deemed too expensive, so BCC decided to construct the second priority bridge at Grey Street instead, using public funding. In 1930, however, several private companies approached the Queensland State Government for approval to build a toll bridge at the Kangaroo Point site. To accommodate these approaches, and those at other locations, the Queensland state government of Premier A. E. Moore, enacted the *Tolls on Privately Constructed Road Traffic Facilities Bill*, enabling it to award Franchises. The enterprising Walter Taylor saw his chance and in June 1931 he made an application

Figure 7. Walter Taylor.

to the Premier of Queensland for a Franchise. In November 1931 an Order-in-Council was gazetted, giving approval for the construction "of a road suspension bridge" across the Brisbane River at Indooroopilly. The Franchise was limited to 35 years (O'Connor 2003).

The Prospectus of *Indooroopilly Toll Bridge Ltd* was drawn up soon afterwards, in which Taylor assigns to the company all his rights and interests as Franchisee. The prospectus also notes an agreement between Taylor and the company "for the construction of the said bridge". As well as becoming its builder, he was also a Director of the company, and eventually its Secretary (O'Connor 2003).

4 WALTER TAYLOR

Walter Taylor (see Figure 7) was born in Sheffield, England in 1872 and emigrated to Australia with his parents in 1882. His father was a builder and Walter joined him in that industry. Taylor senior died tragically in a construction accident, after which Walter returned to England, for a decade from 1902, where he was involved in the design and construction of fairground attractions. Whilst in Europe he also took an interest in reinforced concrete, during its fledgling years. (Davis 2016)

He returned to Australia in 1912 and spent the rest of his life as a successful construction contractor, in and around Brisbane, specializing in reinforced concrete structures, and building warehouses, apartment blocks, factories, schools, churches and hospitals (Davis 2016).

Taylor also constructed several bridges. None were as substantial as his achievement at Indooroopilly, but they were not, at the time, inconsequential, and he apparently learned some valuable, albeit painful lessons from them. Principal amongst his early bridge contracts was the Abbotsford Road Bridge. When opened, in 1928, the *Architecture and Building Journal of Queensland* reported that "The reinforced concrete bridge now being constructed over the Breakfast Creek, Albion is the longest bridge of its kind attempted in Brisbane" (Davis 2016).

All these earlier bridges were over water, and the prospectus for *Indooroopilly Toll Bridge Ltd* indicates this may have presented difficulties for Taylor. The prospectus extols the virtues of the suspension bridge proposed in it, particularly the avoidance of works within the river, and refers to its builders "experience with the enormous cost of Under-Water Work" (Davis 2016).

5 THE SYDNEY HARBOUR BRIDGE CONNECTION

In January 1924, tenders closed for the detailed design and construction of the Sydney Harbour Bridge. Dr John Job Crew Bradfield (1867–1943), the project's chief engineer, had called tenders "for cantilever and arch bridges only". However, as he noted at the time, "tenders ...have been

Figure 8. Perspective drawing of English Electric Company of Australia tender design (Bradfield 1924).

submitted for suspension bridges", which Bradfield also considered on their merits (Bradfield 1924).

One of the unsuccessful suspension bridge bids was submitted by the English Electric Company of Australia to a design by "Messrs Robinson and Steinman of New York City, U.S.A." (Bradfield 1924). The perspective elevation in Figure 8, which is reproduced from Bradfield's tender report, indicates a similar type to the one eventually adopted by Steinman at Florianopolis, but which predates it.

Bradfield understood this when he reported that "Messrs Robinson and Steinman have produced a suspension bridge of novel design" (Bradfield 1924). Bradfield's team undertook check analyses of the computations provided by Steinman and proclaimed that "this suspension bridge is remarkably rigid". In this task he was grateful to "Mr G A Stuckey ... who has assisted me in checking any calculations and technical matter in connection with the tenders" (Bradfield 1924). This was Gordon Stuckey (refer Section 8), who worked on the Sydney Harbour Bridge project through to completion (Raxworthy 1989).

Later, after the Florianopolis Bridge was completed using eyebar cables, Steinman wrote that "in their design for the Sydney Harbour Bridge, Robinson and Steinman developed plans and details for combining wire cables with a continuous stiffening truss of the Florianopolis principle" (Steinman 1929). The use of steel wire cables in the tender design appears to be confirmed by Bradfield, who reported that amongst the material proposed to be sourced locally by the English Electric Company of Australia was "4,722 tons of galvanised steel wire and cable wrapping" (Bradfield 1924).

There is an important observation to make from all this. In 1924, Australian bridge engineers were aware that Steinman had developed a new type of suspension bridge, they understood its principal design philosophy, they admired its rigidity, they had an example of the design methodology adopted by Steinman, and they were quite capable themselves of carrying out independent analyses. And, depending on the interpretation given to Steinman's later comments (refer Section 2), they may well have had some preliminary joint details.

The successful tenderer for construction of Sydney Harbour Bridge was *Dorman Long & Co Ltd* who had proposed a two-pin arch solution that involved falsework at each end to hold back the two halves of the arch as they reached out to each other. The eventual, rather neat solution was to use looped wire rope cables deeply entwined into rock tunnels, as shown in Figure 9.

In August 1930 the two halves of the arch were closed, allowing the tie back cables to be released and removed.

In early March 1931, Walter Taylor enquired whether the cables were available for purchase, and they were.

Interviewed about the Indooroopilly Toll Bridge, in 1952, for an article in *The Telegraph*, a Brisbane newspaper, Taylor maintained that the "basis of the design had come to him in a flash", when he "had been thinking about ... the Sydney Harbour Bridge ... and the great steel cables that had been taken down when the arms were joined" (Davis 2016).

6 SELECTING THE BRIDGE TYPE AT INDOOROOPILLY

On 26 March 1931 an article appeared in the *Brisbane Courier*, a local newspaper, which announced "Mr Walter Taylor, of Graceville, advises that he has designed a suspension bridge for erection at

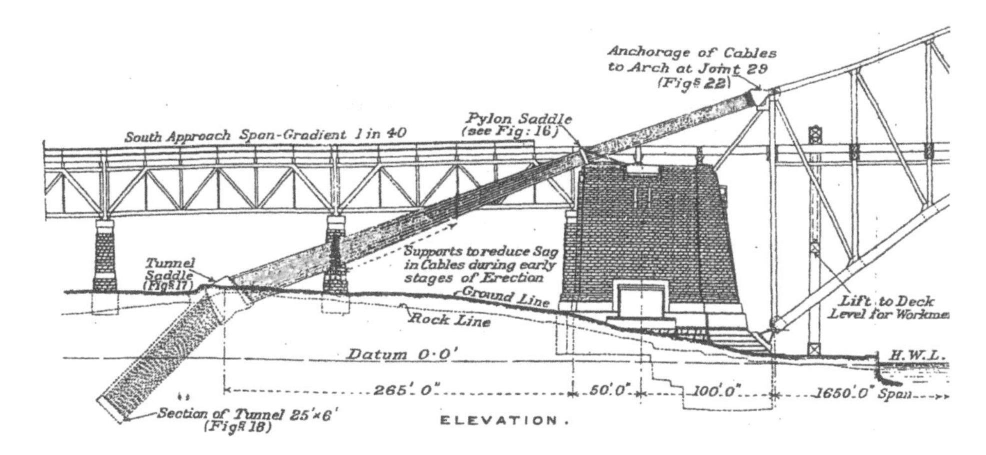

Figure 9. Detail of arrangement for the wire rope cable tie-backs used on the Sydney Harbour Bridge.

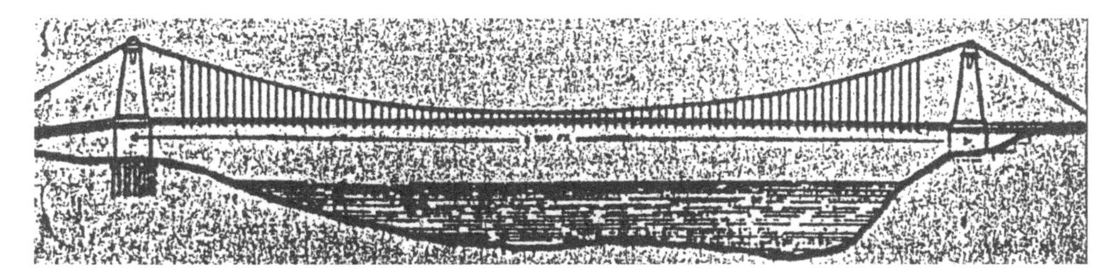

Figure 10. Taylor's initial concept for a suspension bridge across the Brisbane River (O'Connor 2003).

Indooroopilly" (O'Connor 2003). The article included a sketch of the proposed bridge, which is shown in Figure 10.

This timing can be compared with information provided later that "on 7 March 1931, Mr W. Taylor communicated with Messrs. Dorman Long & Co Ltd, Sydney, as to whether it was possible to purchase some of the Cables used in the erection of the Sydney Harbour Bridge". Some form of understanding, including price, had been reached by 17 March 1931, and "Mr Taylor at once set to work to design the bridge with the intention of using two groups of 12 Cables" (O'Connor 2003).

This initial design consideration appears to have produced the concept drawing that appeared in the newspaper ten days later. One thing is obvious though; at that stage it wasn't a design for a Steinman type.

The Order-in-Council from the Queensland Government awarding a franchise to Walter Taylor, to construct a toll bridge at Indooroopilly, was issued on 19 November 1931, and he wasted no time in issuing the prospectus for the formation of a company to finance the project, which is dated less than a week later, on 24 November 1931. Interestingly, the prospectus's title page contained an artist's impression of the proposed toll bridge, shown in Figure 11, which is unmistakably of a Steinman type suspension bridge, located adjacent to the pre-existing Albert Rail Bridge. Sometime between March and November 1931, Taylor had amended his concept.

Taylor had applied for his franchise in a letter to the Queensland Premier, dated 15 June 1931, and the change may have been made by that time, as the 1932 annual report of the Main Roads Commission (MRC), recording the Indooroopilly franchise, advised that the "design submitted by the franchisee is for a suspension bridge under the Steinman system" (O'Connor 2003).

Taylor did travel to Sydney during this time, to meet with Dorman Long & Co representatives "when the whole scheme was discussed, and the cables and other suitable materials inspected". It is

Figure 11. Extract from a watercolour sketch by D. Webster used on Prospectus frontispiece

probable that he was introduced to the Florianopolis style at this stage by exposure to the following sources.

i) Bradfield's Report on Tenders (see Section 5)
ii) The 1929 second edition of Steinman's book *A Practical Treatise on Suspension Bridges*, which includes an Appendix devoted to the bridge at Florianopolis. There is evidence that Taylor possessed a copy of this book (Davis 2016).

Taylor was probably enticed to change from his original traditional suspension bridge concept by cost considerations. In his 1929 book, Steinman did extol the cost saving potential of the new type constructed at Florianopolis, and Taylor needed to effect economies wherever he could find them. In the souvenir booklet, issued for the official opening day ceremony, mention was made of the imperative "to keep down costs in order to make the proposition a financial success", and the necessity of preparing a design "to get the maximum amount of efficiency with the minimum amount of expenditure". Steinman could hardly have put it better himself, as an expression of his own motivation (refer Section 2).

It's also possible that during this time he was introduced to Gordon Stuckey, the assistant engineer to Bradfield (see Section 5), who had been responsible for checks on the temporary tie back cables and would have been a valuable source of information on their properties and capabilities. Stuckey was a highly regarded analytical structural engineer, who was involved in the checks of Dorman Long & Co.'s detailed arch design and would have been in a position to explain the finer points of analysis required for the Steinman type (Raxworthy 1989).

7 DESIGN DETAILS

The Walter Taylor Bridge is considered unique for two principal reasons.

i) It is the only known Steinman type suspension bridge in the world where wire ropes were adopted for the suspension cables. Elsewhere, at Florianopolis and over the Ohio River at Point Pleasant and St Mary's, eye bar chains were used.
ii) Because it uses wire ropes as the suspension cables, it required a bespoke node connection system at the stiffening truss joints.

Each cable at Indooroopilly comprises 12 of the wire ropes salvaged from Sydney (refer Section 5). Each rope has a diameter of 2.76 inches (70.1 mm), made up of a central wire of 0.2 inch (5.1 mm) diameter, and 216 wires of 0.16 inch (4.1 mm) in 8 layers, spirally wound in alternate directions in successive layers. The number of wires in the layers increases regularly from six in the first to 48 in the outer layer, an increase by six in each layer (O'Connor 2003).

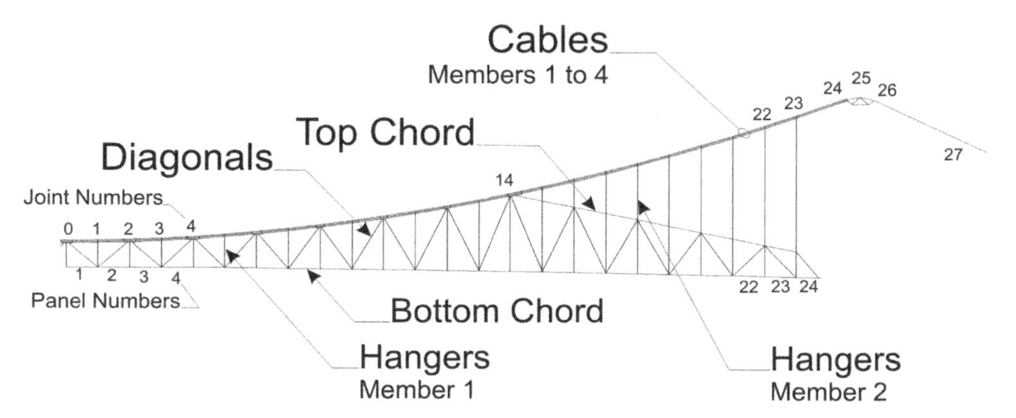

Figure 12. Half elevation of main span indicating node and panel designations (Stevens 2012).

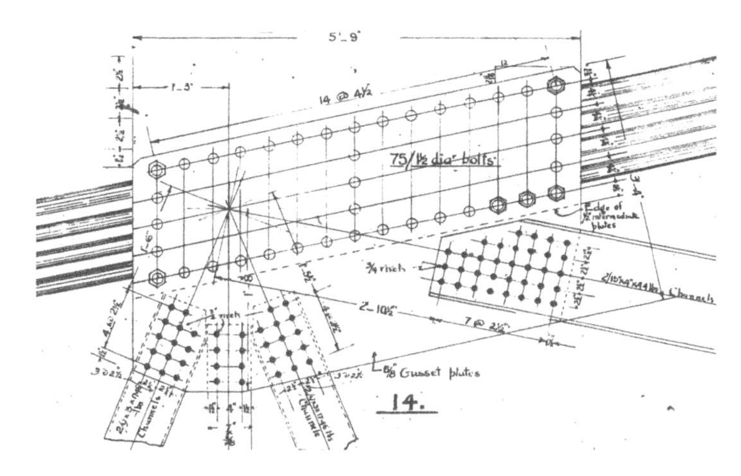

Figure 13. As constructed detail for Joint 14 (from project Drawing No 20A).

In the stiffening truss for the main span there are 24 panels on either side of the centerline. In the original drawings the upper chord joint nodes were numbered sequentially from mid-span; 0, 1, 2, 3 etc., with the pier ends of the bridge being node 24 (see Figure 12). The bottom chord nodes were signified alphabetically, from A at mid-span to Y at the pier ends. The suspension cable joins the upper chord of the stiffening truss at nodes 14, the 0.208 points of the span, 10 panels in from the pier end.

The joint detail at node 14, probably the most significant joint in the bridge, is shown in Figure 13, which is the fabrication detail for the in-situ joint shown Figure 3.

Examining the various drawings archived by Brisbane City Council and the Department of Transport and Main Roads (previously the MRC), it can be observed that detailing of the joints was a work in progress throughout the project. This can be appreciated most readily by comparing the joint detail for node 14 in Figure 13 above, which is from a drawing (No. 20A), dated February 1935 (amended in May and June), with the superseded detail shown in Figure 14, from a drawing (No 8A) dated 15 May 1934.

It is obvious from Figure 3 that the detail in Figure 18 is the one adopted. But the difference is substantial. The as constructed detail has adopted seventy-five, $1^1/_2$ inch (38 mm) bolts while the superseded joint has forty-five, 1 inch (25 mm) bolts.

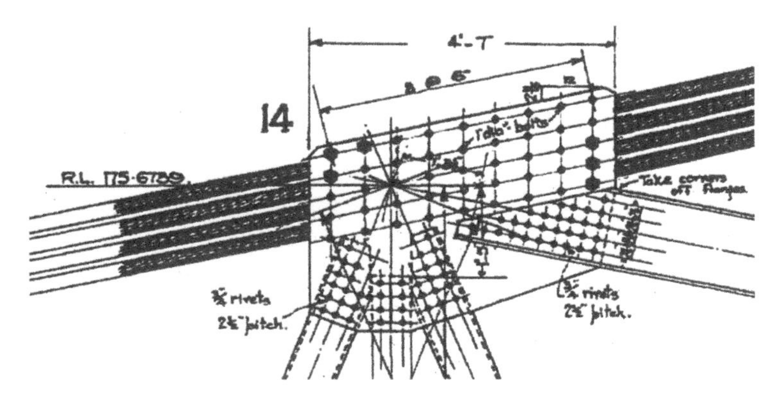

Figure 14. Superseded detail for Joint 14 (from project Drawing No 8A).

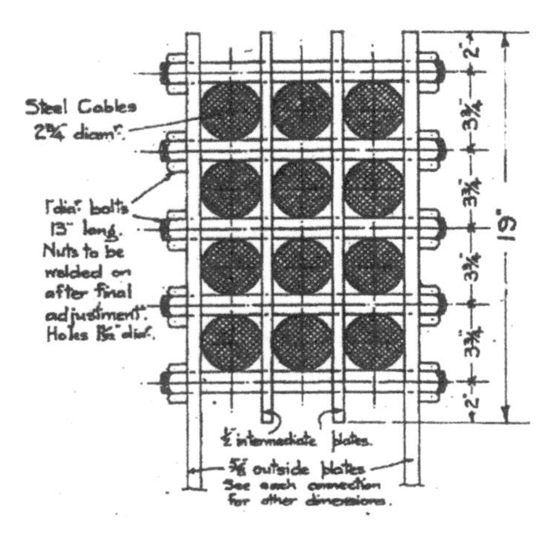

Figure 15. Superseded section for Joint 14 (from project Drawing No 8A).

A typical section through the superseded joint is shown in Figure 15. Although the as constructed bolt sizes are different, the clamp arrangement is otherwise the same. There are two outer gusset plates and two inner packer plates clamping onto the wire rope cable. Transfer of forces is by friction, and it is important that this be properly achieved.

There is an even earlier drawing dated 30 November 1933, which shows another variation of Joint 14 with only twenty-five, 1 inch (25 mm) bolts.

Circa May 1934, before the final revision, a seemingly exasperated government bridge engineer wrote that "Mr Taylor, I think, realizes the importance of joint 14, and said that he could find a satisfactory solution for the problem" (Davis 2016). While the form of the joint remained essentially the same throughout, the number and size of bolts increased markedly, and serious reservations about the capacity of the joints, were – it would appear – held by MRC and BCC engineers responsible for approving the design on behalf of these two government bodies, both with a vested interest in the outcome.

Eventually a decision was made to carry out tests. The *Indooroopilly Toll Bridge Ltd* alludes to this issue in the souvenir booklet, issued for the official opening day ceremony, where it is admitted that "In order to ascertain the correct number of bolts required for each set of plates, a clamping device was constructed to grip six cable ends and tested at the Queensland University with highly satisfactory results". The result sheet for these tests is available (refer Figure 16).

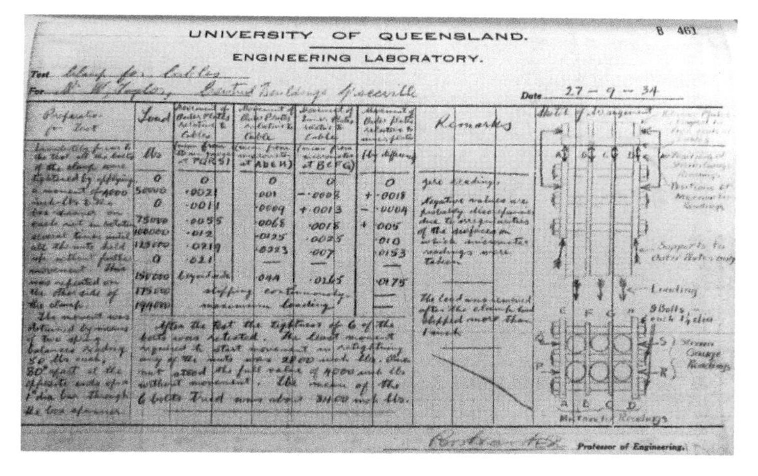

Figure 16. University of Queensland 1934 cable clamp test results (O'Connor 2003)

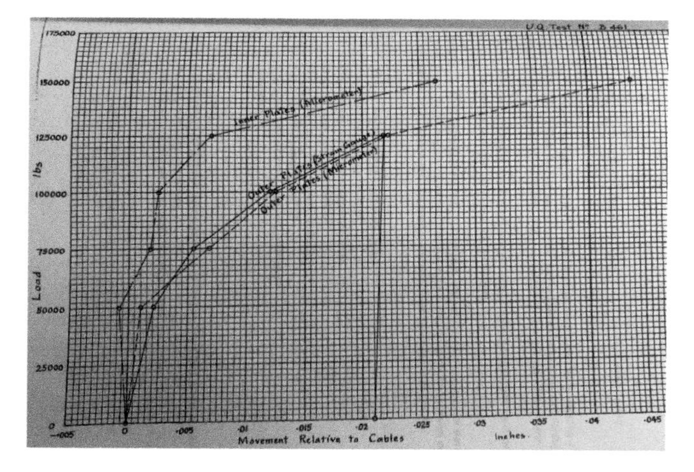

Figure 17. University of Queensland graph of 1934 cable clamp test results (O'Connor 2003).

The graph of plate movements versus applied load derived therefrom is shown in Figure 17.

While recent investigations undertaken for Brisbane City Council (refer Section 10) suggest that these tests were not a completely accurate indication of the in-situ joint's performance, they provided important and sufficient information to Taylor and his advisor(s) about the number of bolts required and the required way to tighten them.

8 DESIGN PROVENANCE

There is no doubt that Walter Taylor was a remarkable man, but he was not a trained, or experienced bridge design engineer. The bridge arrangement he adopted is indeterminate, and he needed help with the details. Here the history of the bridge becomes a little blurred.

The *Indooroopilly Toll Bridge Ltd* acknowledged that Taylor was assisted by the involvement of expert engineers. In the souvenir booklet, issued for the official opening day ceremony, it is commented that, "Several technical engineers were consulted and reports obtained on different aspects of the work, and the information furnished with the plans to the Main Roads Commission" (O'Connor 2003).

Nevertheless, the only signature on the extant drawing is Walter Taylor's. Although it is now well established that other engineers were involved, their role was never formally acknowledged.

Colin O'Connor, Emeritus Professor of Civil Engineering at Queensland University, has studied the records. Based on his extensive research (O'Connor 1998 & 2003), and other sources, the following observations are made.

i) Walter Taylor himself can probably be given the credit for deciding to adopt a suspension type bridge.
ii) Taylor probably adopted a stiffening truss of the Florianopolis type after discussions with engineers in Sydney including those from Dorman Long & Co Ltd, and possibly others, like Gordon Stuckey, familiar with an unsuccessful tender submission for the Sydney Harbor Bridge based on a Steinman design (see Section 5).
iii) Walter James Doak (1874-1962), then Chief Engineer for the Queensland Railways department, is principally responsible for detailing the steelwork.
iv) Russell John (Jack) McWilliam (1894–1991) carried out the reinforced concrete design for the piers and anchorages.
v) Various engineers employed by the Main Roads Commission and Brisbane City Council were involved in carrying out independent structural checks and construction inspections.

Jack McWilliam graduated with a Bachelor of Engineering from the University of Queensland in 1920, and then gained experience in companies specializing in reinforced concrete. By 1924 he was in private practice as an architectural and structural engineer and thrived until the Great Depression intervened. He later remarked that "In February 1930, I had more work than I could handle, but by April 1930, I had nothing to do". Walter Taylor made him an offer to design and detail the towers and anchorages for his bridge at Indooroopilly, and as McWilliam recalled, he "was pleased to have the job" (Cossins 1999).

Walter Taylor would have known McWilliam from previous projects. In 1928, when Taylor was constructing the Abbotsford Road Bridge (refer Section 4), he was also engaged on an eight-storey apartment building (Davis 2016), constructed with a reinforced concrete frame, for which Jack McWilliam provided structural engineering services to the architect (Cossins 1999).

Jack McWilliam, went on to establish McWilliam and Partners, a highly successful and respected structural engineering consultancy, which continued after his death. During his research, Professor O'Connor interviewed several members of that practice, who told him that "in casual conversation, McWilliam would say that he designed the towers of the Indooroopilly Bridge" (O'Connor 1998).

Further confirmation is contained in the archives from the Main Roads Commission (M.R.C.), where a memo to the Chief Engineer, dated 27 April 1932, reported that, "Mr Taylor has asked Mr Doak to confer with the M.R.C. regarding the details of the steel design, and is willing to ask Mr McWilliam who did the concrete design" (O'Connor 1998).

This neatly introduces, W. J. Doak, who was Bridge Engineer for Queensland Railways, a position he had attained in 1919. Doak graduated with a Bachelor of Engineering from Sydney University in 1895, where he was a brilliant student, being awarded the University Medal. He worked for Queensland Railways until 1939. He retired from that position at the standard retirement age of 65 but continued working in a variety of other private and public engineering organizations until 1952.

There is no doubt that his bridge engineering abilities were widely recognized, and other agencies frequently sought his advice. In 1923 he was seconded to assist Brisbane authorities effect bearing repairs on the Victoria Bridge, at the time the only cross-river road bridge in the city. In 1926, assisting the Cross River Commission (see Section 4) he prepared designs and estimates for a possible cantilever bridge over the Brisbane River at Kangaroo Point, the Commission's recommended first priority. In 1930 he reported to Rockhampton City Council on the condition of the Fitzroy River Bridge.

All this extra-curricular activity, however was for public agencies, or relevant to his government position. His apparent involvement in the Indooroopilly project, however, was for an ostensibly private venture, and this may explain the reluctance to publically acknowledge his involvement at the time. In November 1932, a new Labor Party Government, led by W. Forgan Smith, replaced the government of Premier A. E. Moore. The new government opposed construction of privately owned toll facilities, but honoured any agreements reached before it took office. As the terms of the Franchise allowed for the eventual transfer of the bridge to public ownership, it was in the government's interest that expert engineers, like Doak, were involved in the design. O'Connor has surmised that "the Labor Government of the time preferred that the public did not know of his apparent aid to a private company" (O'Connor 2003).

This is reinforced by archival material that shows his involvement was known to senior engineers within the Main Roads Commission (MRC), including the Commissioner and Chief Engineer. This view is supported by a 1934 letter from the *Indooroopilly Toll Bridge Ltd* to the MRC in which it is stated, "When the company was formed it was deemed advisable to have the original design checked before going to allotment. The Commissioner for Railways and Chief Engineer made it possible to secure the services of Mr W. J. Doak" (Davis 2016).

Then there is Doak's apparently modest demeanor. He occasionally collaborated in research with R.W.H. Hawken, then Professor of Civil Engineering at the University of Queensland, who provides a telling insight into Doak's self-effacing character. In 1925 they developed a method of assessing whole-of-life costs, called *Economy of Purchase*, a form of discounted cash flow. In the publication describing his methodology, Hawken acknowledged that the idea for his paper originated from the solution of a problem "given him by Mr W. J. Doak" who, Hawken continued, "revised the manuscript at various stages". Such was Doak's contribution, Hawken had wished to include him as joint author, an offer that, Hawken recorded, Doak "generously" declined (Hawken 1941).

Even more elusive than Doak and McWilliam's involvement is what might be called the Sydney connection. It is known that Taylor visited Sydney on several occasions during design (see Section 7). A company minute of 9 December 1933, during construction, records that, "Mr Taylor stated that a previous minute gave him authority to consult with an independent engineer relative to any difference between himself and the Main Roads Board and Mr Taylor stated that when in Sydney he was placing his plans before an independent authority" (Davis 2016). It is intriguing to speculate who this might have been, and Gordon Stuckey (see Section 7) is the most probable candidate. Prof O'Connor spoke with former colleagues of Stuckey from the Sydney Harbour Bridge project, and "is inclined to believe that he may have given Taylor advice" (O'Connor 1998).

9 CONSTRUCTION

Early in 1932 the necessary resumptions were taken, site workshops were established and foundations investigations took place. Putting down the first borehole, gold was stuck, and Taylor was required to take out a Mining Lease to protect his site.

By November 1932 the high-level footings for the Indooroopilly pier were approved by the Main Roads Commission and construction of the reinforced concrete tower commenced. Foundation work for the Chelmer pier then began but was more involved; rock was not found at a high level, and 168 ironbark timber piles were driven, some to a depth of 61 feet (18.6m). The anchorage on the south side was also more extensive. On the Indooroopilly side a compact anchorage was achieved in rock. No such opportunity was available on the Chelmer side and the anchorage was achieved by constructing a large, multi-cell reinforced box, infilled with sand. The cables were then taken down to the anchorage points via inclined shafts, formed in the side-walls of the box. Work on the pier towers and preparation of the anchorages was complete by March 1935.

From about April 1935 until February 1936 John Stephens Robertson was the inspecting engineer at the Indooroopilly bridge site for the Main Roads Commission. Robertson was a very experienced bridge construction engineer and kept a chronological photographic record. According to

Figure 18. John S Robertson photograph, dated 11 November 1935, of deck steelwork in place.

Robertson, stringing of the ropes began on 10 April 1935. All 12 ropes were in place by 26 June 1935 (O'Connor 2003).

By 30 July 1935, work was well advanced attaching temporary hangers to the cable joint gusset plates from which a temporary access structure was attached. This temporary bridge was used to erect the steelwork. Apart from riveting the cross girders to the bottom chord gussets, all steel fabrication took place on-site.

By mid-October 1935 erection of the vertical and bottom chords, and cross girders was well underway and a month later, Robertson took a photograph with the deck steelwork erected (see Figure 18). Note only the vertical chords of the eventual stiffening truss are in place at this stage, which suggests the stiffening truss was not primarily intended to support the dead loads (O'Connor 2003).

In January 1936, Robertson took a photograph of two boilermakers drilling holes in a stiffening truss diagonal, who appear to be using the gusset as a template for the holes, providing further evidence for that supposition.

10 CONSERVATION

The franchise to operate the bridge as a toll facility expired in October 1965. Photographs from around that time suggest that maintenance on the bridge had been minimal. The government removed the toll and transferred the bridge to Brisbane City Council (BCC). Since that time BCC has been assiduous in both understanding the operational characteristics of the bridge, and in developing necessary maintenance activities.

Initial attention was given to the cables. Circa 1968 suspicions were raised that some of the cable wires may have fractured, particularly at the joints, and in subsequent years considerable effort was made to quantify the issue. Small holes were introduced in the gusset plates to allow inspection of the wire ropes using a fibre optic light source and cystoscopes. Widespread corrosion was observed, and broken wires were visible in 9 of the 17 joints inspected.

Subsequently, circa 1971 a falsework arrangement was developed that allowed removal of the bolts and plates at selected joints (see Figure 19).

The No 6 joints (see Figure 12), were dismantled using this falsework. Pitting of the plates (see Figure 20) and the cable wires was observed. Pronounced decarburation of surfaces was also noticed. Both these defects raised considerable concerns about the fatigue strength of the cables, and

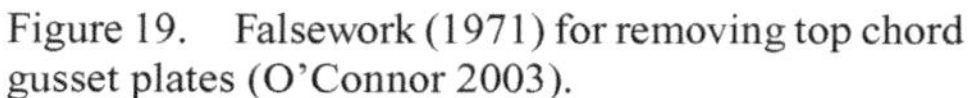

Figure 19. Falsework (1971) for removing top chord gusset plates (O'Connor 2003).

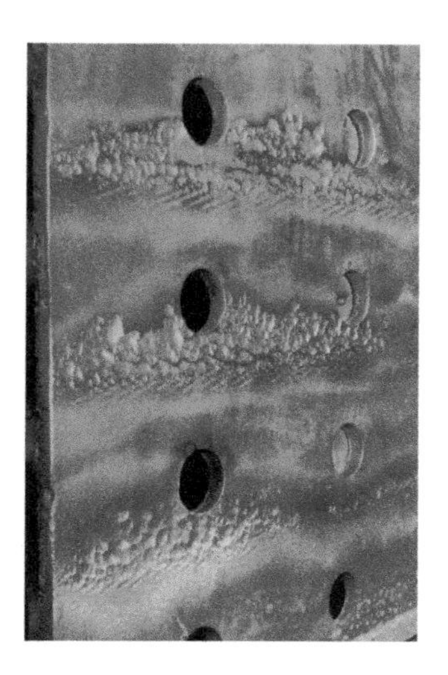

Figure 20. Pitting of original top chord gusset plates (O'Connor 2003).

fatigue testing of corroded wire samples, again by the University of Queensland, was undertaken at both minimum and maximum stresses. These tests indicated the risk of fatigue failure was low.

Nevertheless, this selective investigation alerted BCC engineers to the need for development of a corrosion protection strategy, and subsequently all joints were treated with a corrosion inhibitor. It was also appreciated that the joints would require further, regular inspections and, ultimately, refurbishment.

In January 1993 the original timber deck was replaced with lightweight, precast concrete panels. The original deck comprised 9 inch (225 mm) × 4 inch (100 mm) hardwood planks, with an asphalt surface varying in thickness from 2.5 inch (64 mm) to 1.5 inch (38 mm). The timber planks were supported on steel RSJ stringers, which connected to the cross girders. While the stringers were retained as far as possible, some corroded members were replaced. This work considerable improved ride quality over the bridge and reduced road noise. Most importantly, however, it involved waterproofing of the deck, with long-term benefits for durability of the supporting steelwork.

During 2004–2008, when the bridge was about 70 years old, BCC carried out a major investigation to provide information necessary for ongoing bridge management (Stevens 2008), including determination of:

i) its load carrying capacity
ii) forces in strands and truss members under dead load
iii) live load capacity after consideration of corrosion damage
iv) required strengthening of truss members
v) a joint refurbishment strategy, involving opening, inspection, cleaning and corrosion protection.

Analysis of the bridge's dead load condition was modelled with 3D frame idealization using the LUSAS finite element package. A key finding of the analyses was that the 1993 deck modifications had increased the dead load of the bridge by 16%, increasing the load carried by the diagonals and top chords of the stiffening truss. The original construction intended no dead load contribution to the truss (see Section 9).

The live load response of the bridge was analysed using a Spacegass model. A review of historical data during the BCC investigation (Stevens 2008) suggested the original design live load was 70 pounds per square foot (3.35kPa). This was assessed as being equivalent to 75% of T44 and L44

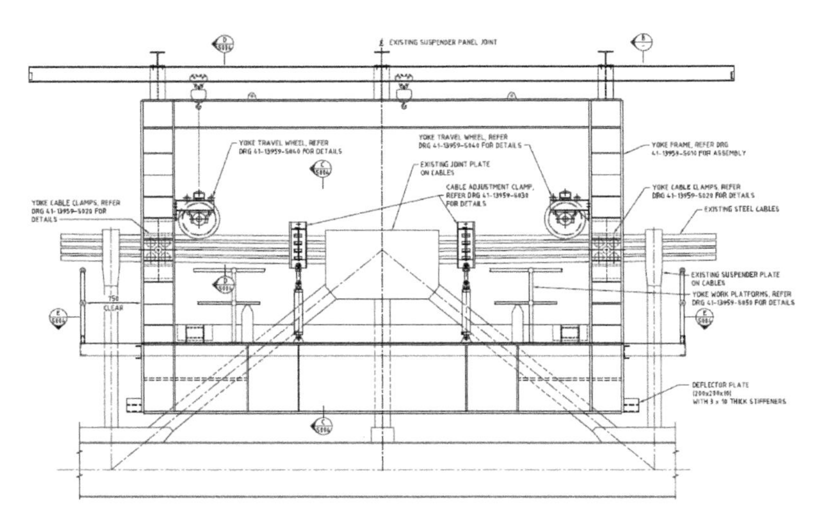

Figure 21. General arrangement of the joint maintenance yoke (Stevens 2008).

loadings (NAASRA 1976). BCC decided that live load capacity should be assessed using three lane T44 and L33 configurations, where L33 denotes 75% of the L44 standard. As a result, it was found (refer also Figure 12) that:

a) diagonals in panels 8 to 14 had insufficient capacity
b) top chords from joints 14 to 22 had insufficient capacity
c) allowing for corrosion previously observed in the wires, minimum factors of safety for cables were 3.54 for the serviceability limit state and 2.66 for the ultimate limit state, which was considered satisfactory
d) the critical locations for the cable capacity are in the vicinity of Joint 14
e) additional bolts were required at joints 8, 10 and 12
f) adopting the proposed corrosion protection system (Amerlock 400), and allowing for a lower coefficient of friction, higher bolt clamping forces were required, necessitating high strength bolts, thicker outside plates and reduced clearance at bolt holes.

Initial refurbishment took place between 2008 – 2012, during which the plates were replaced at all the even numbered joints in Figure 12, between 0–14, in both directions, and the original bolts were also replaced, with M36, Type 8.8 high strength bolts. The original chord connecting rivets, evident in Figure 3, which had to be removed to dismantle the joints were also replaced with bolts (Steven 2012).

A yoke (see Figure 21) was designed to enable joints to be opened. The yoke had to:

i) accommodate the different geometries of the various joints
ii) transfer all necessary loads to allow the bridge to remain in service once the yoke is installed
iii) be sufficiently stiff to avoid joint displacement or load redistribution
iv) be able to move from joint to joint without bridge closure.

Strengthening of the top chords of the stiffening truss also took place, between the piers and the number 14 joints, and those diagonals considered overstressed by the new deck, as presaged in a) and b) above.

Current work, commenced in 2019, involves refurbishment of all the remaining minor 'hanger' joints, which can also be seen in Figure 3 on either side of the more heavily bolted stiffening truss joint. While this will mean that, eventually, no suspension cable joints will be original, the essential 'Florianopolis' characteristics of the bridge will be maintained.

11 AN EPILOGUE

The Sydney Harbour Bridge opened on 19 March 1932.

After the Indooroopilly Toll Bridge opened on 14 February 1936. Walter Taylor continued as a builder. The bridge at Indooroopilly was his greatest achievement, although he is also remembered for the various other structures he contributed to the city.

David Steinman became one of the great American bridge engineers. His achievements include the Mackinac Bridge, in Michigan, which opened in 1957.

On 15 December 1967, the Silver Bridge, over the Ohio River collapsed, killing 46 people.

12 CONCLUSIONS

The Walter Taylor Bridge at Indooroopilly in Queensland, Australia is a rare extant example of a Florianopolis style bridge, a type originally conceived by D B Steinman, a celebrated American bridge engineer.

While the bridge is a testament to the vision, industry, capability and public mindedness of its principal promoter, Walter Taylor, the role of other eminent engineers was unacknowledged at the time, and the reasons for this have never been satisfactorily established.

The bridge's principal claims to fame, as an internationally unique example of the Steinmann type, are the adoption of wire rope suspension cables for the top chord of the stiffening truss, and a specially devised clamp plate arrangement to effect load transfer at node points between the cable and the stiffening truss chords.

The bridge is still in normal operation as a road transport bridge, with pedestrian access, as originally intended, and is well maintained by its custodian, Brisbane City Council.

ACKNOWLEDGEMENTS

The Author is indebted to Dr Nick Stevens for generously providing opportunities to discuss his technical investigations into the bridge (Stevens 2008, 2012). He is also grateful to Mr Phil Cutler, of Brisbane City Council, who arranged permission to view the reports of those investigations.

REFERENCES

Bradfield, J. C. C. 1924. *Sydney Harbour Bridge: Report on Tenders.* Sydney: NSW Government Printer.

Cossins, Geoffrey (ed.) 1999. *Eminent Queensland Engineers, Volume II.* Brisbane: I.E.Aust Qld. Div.

Davis, Noel 2016. *The Remarkable Walter Taylor.* Brisbane: Oxley-Chelmer History Group Inc.

Hawkin, R. W. H., Nelson, W. I. M. & Wilson R. M. 1926. *Report of the Cross River Commission ap- pointed by Brisbane City Council.* Brisbane: R. G. Gillies & Co. Ltd.

Hawkin, R. W. H. 1941. Economy of Purchase with Tables. *University of Queensland, Papers, Faculty of Engineering* v.1, n.6.

NAASRA, 1976. *Bridge Design Specification.* Sydney: National Association of Australian State Road Authorities.

O'Connor, Colin 1998. The Walter Taylor Bridge. *I.E.Aust Qld. Div. Engineering Update* v.6, n.1, p. 3–8.

O'Connor, Colin 2003. *Walter Taylor Bridge Conservation Plan.* Brisbane: Brisbane City Council.

Raxworthy, Richard 1989. *The Unreasonable Man, J. J. C. Bradfield.* Sydney: Hale & Ironmonger.

Steinman, D. B. 1929. *A Practical Treatise on Suspension Bridges, 2nd Edition.* New York: John Wiley & Sons.

Stevens, Nick 2008. *Report on Walter Taylor Bridge Investigation.* Brisbane: Brisbane City Council.

Stevens, Nick 2012. *Report on Verification of Structural Adequacy of Walter Taylor Bridge after completion of Joint Maintenance.* Brisbane: Brisbane City Council.

Chapter 12

Design and construction of the new Frederick Douglass Memorial Bridge

K. Butler & N. Porter
AECOM, Richmond, Virginia, USA

ABSTRACT: This project showcases how District Department of Transportation (DDOT) is undertaking the largest transportation project in its history in an extremely challenging environment. The presentation will describe the procurement, design and construction of this one-of-a-kind signature arch bridge in the Nation's Capital. The challenges associated with developing a signature bridge in a price competitive design build procurement will be presented. The presentation will share with the audience the extreme level of effort and bid phase cost that goes into pursuing and delivering signature bridges under a design build format. The presentation will also discuss the risk of delivering these types of projects and how the owners have successfully transferred almost all risk to the design build teams.

1 PROJECT LOCATION AND BACKGROUND

South Capitol Street was a primary corridor in Major Pierre L'Enfant's 1791 Plan (*L'Enfant Plan*) of Washington D.C., which developed South, East and North Capitol streets to extend directly from the U.S. Capitol, and become prominent gateways to the city's Monumental Core. As shown in Figure 1, underscoring this historic plan is the replacement of the existing Frederick Douglass Memorial Bridge (FDMB) with a new multiple arch bridge that will transform the South Capitol Street corridor with an iconic entry route to the District's federal area. The ultimate design goal is to improve connectivity, traffic mobility, safety and operational characteristics in the project corridor while ensuring compliance with all environmental requirements related to the protection of all natural and historic resources.

As part of the overall *Anacostia Waterfront Initiative (AWI)*, the South Capitol Street Corridor Phase I Project is part of the District's strategy to support neighborhood revitalization and economic development on both sides of the Anacostia River. As shown in Figure 2 the bridge project will help frame public spaces, such as the South Capitol Street Ovals and extend the Riverfront Park to the Yards Park and the Navy Yard.

The 70-year-old existing Frederick Douglass Memorial Bridge is being replaced with a signature three-arch bridge designed to last 100 years. DDOT established an Aesthetic Review Committee (ARC), along with fifteen (15) specific Project Design Appearance Goals (PDAGs), to ensure the visual quality of the final structure. This high-profile, prestigious project underscores the legacy of Frederick Douglass, who was a believer in dialogue and making alliances across divides. The new, elegant structure will reflect his ideology of unity and integration.

The $442 million project is the largest construction project in the District's history. In addition to replacing the existing Frederick Douglass Memorial Bridge, the project also includes improvements to South Capitol Street, I-295 and Suitland Parkway. The multifaceted project consists of the refinement/redesign of the horizontal and vertical alignments of 11 local roads; Interstate I-295 widening including reconstruction of three bridges over Howard Road, Suitland Parkway and Firth Sterling Avenue; five Interstate ramps; the I-295 and Suitland Parkway Interchange; and two traffic ovals to maximize the corridor's efficiency as shown in Figure 2.

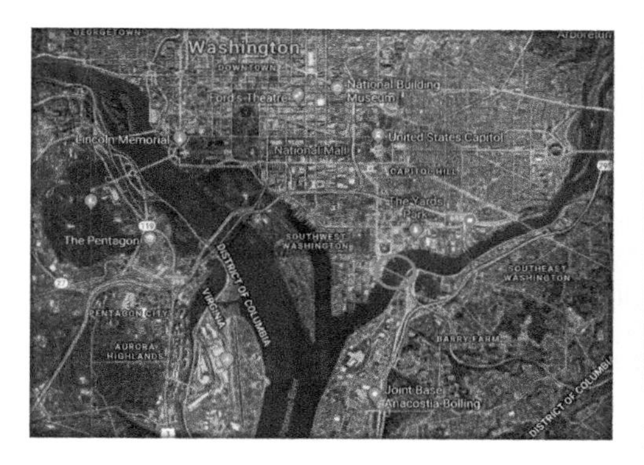

Figure 1. Project location.

Figure 2. Creation of public spaces.

AECOM is the lead designer for South Capitol Bridgebuilders (SCB) a joint venture team of Archer Western Construction and Granite Construction Company. This design build team, in partnership with DDOT, has a once-in-a-century opportunity to redefine the existing corridor, help shape the character of development along the Anacostia Waterfront and introduce impressive civic architecture that forms a worthy river crossing in the nation's capital.

2 DDOT'S PROCUREMENT STRATEGY FOR ENSURING A SIGNATURE BRIDGE

For the project to be successful, DDOT's procurement strategy for the largest transportation project in its history had to be meticulously thought out. Delivering any project in Washington D.C. can be challenging but creating an iconic bridge with a grand urban boulevard to serve as the gateway to the Nation's Capital offered unique challenges. From the early stages of the Federal Environmental Impact Statement/Record of Decision (SFEIS/ROD), DDOT focused on how to reduce risk for the design build teams by taking on activities such as United States Coast Guard (USCG) permits, right of way (ROW) acquisitions and preliminary engineering. DDOT had to strike the right balance of determining what efforts needed to be done in advance of the design build procurement. They also needed to strike the right balance between budget, schedule, and the desire for a signature bridge given the oversight of the U.S. Commission of Fine Arts (CFA) and the National Capital Planning Commission (NCPC). The SFEIS included a Section 106 MOA to ensure signatories and consulting agencies compliance. Ultimately DDOT developed unique Visual Quality goals and the ranking criteria that drove creativity and ingenuity.

All this development had to be done with the understanding that the existing Frederick Douglass Memorial Bridge, one of the District's busiest commuter gateways, is 70 years old and past its service life.

DDOT spent from 2013 to 2016 developing the preliminary engineering for the project. This intense effort culminated in a comprehensive set of Reference Information Documents (RIDS) that were provided to the design build teams to aid in their project understanding and ultimately their bidding process. DDOT developed and provided the following information to the design build teams:

- Surveys & Mapping
- Geotechnical Design including Preliminary Geotechnical Reports and Subsurface Data
- Environmental including USCG permits, Unexploded Ordnance (UXO) identification, Lead Based Paint Analysis, Section 106 MOA requirements
- Utilities including location plans and conflict matrix

- As-built bridge plans including inspection reports
- Streetscape and Bridge Aesthetics including a Visual Quality Manual to be used as a guide
- Traffic signals and ITS concept designs
- Roadway Interchange Modification Report (IMR)
- ROW Plans

Perhaps the most important challenge that DDOT overcame was securing the USCG's acceptance to change the existing bridge from a moveable swing span to a fixed span bridge with a defined navigational clearance envelope. In addition to the base engineering information provided, DDOT also developed a comprehensive Request for Proposal (RFP) including project specific Technical Provisions and Specifications.

In tandem with the technical development of the project, DDOT set their project goals and Visual Quality Concept (VQC) requirements to further define the project:

1. Design and construct improvements within the corridor which transforms South Capitol Street into a grand urban boulevard consistent with the Anacostia Waterfront Initiative:
   - Replace the existing Frederick Douglass Memorial Bridge with a new iconic structure that reflects the traditions of great civic design in the District.
   - Develop public spaces within the ovals which provide opportunities for future commemorative works integrated with a network of urban open spaces throughout the project.
2. Improve connectivity, traffic mobility, safety and operational characteristics in the Project corridor:
   - Maintain worker and public safety while minimizing impacts to local residents, businesses and the traveling public during construction.
3. Incorporate aesthetically pleasing and sustainable materials and elements throughout the corridor reflecting the natural and historic resources of the area:
   - Ensure compliance with all environmental requirements related to the protection of all natural and historic resources in the area.
4. Achieve Project Final Completion on, or prior to, December 31, 2021.

To ensure the PDAGs were met by the design build teams, both pre-award and post-award, DDOT developed a VQC process. The intent of the VQC process was to:

- Allow proposers to have flexibility in developing their proposed designs, while not compromising DDOT's Project Design Appearance Goals, Technical Provisions, or other goals related to the appearance of the project.
- Allow DDOT and the proposer to review, on a confidential basis, the proposers' VQCs through one-on-one meetings, to provide clarification feedback on such VQCs and clarify, by addendum, any misinterpretations or ambiguities in the RFP relating to PDAGs before the proposers' submission of their Technical Proposals.

DDOT required two mandatory VQC submissions in conjunction with two one-on-one meetings. Optional third and fourth submissions, as well as one-on-one meetings, were offered if needed. Each VQC submission had to address 15 PDAGs.

DDOT established an Aesthetic Review Committee (ARC) comprised of representatives from DDOT, CFA, NCPC and State Historic Preservation Office (SHPO). The ARC reviewed each VQC and provided comments indicating whether each of the 15 PDAGs were acceptable or unacceptable. All 15 PDAGs had to be "accepted" by the ARC for the proposal to be compliant. This pass/fail criterion ensured DDOT, and all the key stakeholders, that the design build teams had to create and deliver an elegant and iconic bridge along with the urban design and landscape architecture.

For the final Visual Quality Plan (VQP) that was submitted with the Technical Proposal, DDOT required the proposers not to include any information that disclosed their identity other than a cover sheet. This was done to ensure an unbiased review of the designs by the selection committee. The scoring was based on a total score of 1000 points, which the Technical Proposal counted towards 400 points and the price proposal 600 points. The visual quality accounted for 160 of the 400 points.

Figure 3. Creation of the elegant and iconic FDMB.

Figure 4. Modern application of ancient arch form.

Ultimately the VQP was used to create the Record of Recommendations (ROR), post-award, to ensure that the project is built according to the design build team's proposal.

DDOT's goal was to provide the Design Build teams enough information to ensure they had a thorough understanding of the project challenges, goals and objectives, as well as to help reduce their risk registers. The culmination of their efforts is reflected in Figure 3.

3 ARCHITECTURAL VISION AND APPROACH TO MEETING DDOT'S 15 PDAGS

SCB's approach throughout the bidding phase of the project was to develop a design that truly reflected DDOT's requirements for an iconic bridge and grand boulevard for the South Capitol Street Corridor, all while staying within DDOT's overall budget and schedule. Moreover, SCB's goal during the proposal phase was to exceed all requirements for each of the 15 PDAGs. SCB proactively reacted to the ARC's comments on the VQC submissions. SCB adjusted their design significantly from VQC1 to VQC2 and made further refinements to the VQC3 submission. At the end of DDOT's process, the SCB team produced an exceptional design which exceeded all the PDAGs.

The Frederick Douglass Memorial Bridge is a unique above-deck arch design that captures the essence of Washington D.C., both historically and in consideration of future context. When built, it will be the only one of its kind in the United States and around the world. The new bridge crossing the Anacostia River, depicted in Figure 4, uses the ancient structural form of an arch and marries it to modern technology.

The design is an integrated engineering and architectural approach, with aesthetic and symbolic intent as key drivers. The three-span arches result in only two v-piers in the river. The v-piers are designed to appear as if they "spring" off the water to create a visual effect of continuity and flow from the deck to water surface. The arches are raised high above deck level to the maximum elevation of 168 feet to achieve a better visual impact for the drivers as they enter and exit the Nation's Capital. The central arch is slightly taller (20 feet) than the flanking arches to help signify the center of the river while at the same time creating unity and equality to both sides of the river.

As SCB considered how the design and vision should respond to the goals and objectives identified in the *South Capitol Gateway Corridor and Anacostia Access Studies*, some of the criteria we challenged ourselves with included:

- FDMB must have iconic value (a signature bridge that is that is visually unique)
- FDMB, waterfront esplanades, and the ovals should encourage a renaissance of the surrounding area; evoke public interest and community pride; and create a sense of destination
- Project should be of post-card quality and must fit into the natural and built setting; and
- Design solutions must be respectful of the cultural legacy of Frederick Douglass and the historic context of Washington, D.C.

Since the river channel is the threshold between the home district of Frederick Douglass and the seat of authority with which he both struggled and served, SCB believed it was important that the structure that carries his name is a memorialization of his legacy. Mr. Douglass was a believer in dialogue and making alliances across divides. On a physical level the bridge design strives to achieve a strong, confident and aesthetically eloquent structure that will be redolent of Douglass's clarion call for unity and integration.

Additionally, the architecture of Washington D.C. conveys a sense of gravitas and timelessness that informed the bridge design. Key aspects of FDMB design that will help transform South Capitol Street into a grand urban boulevard and gateway to the District's Monumental Core include:

- Multiple arch structure that references familiar DC bridge typology
- Dramatic drive-through structure that enhances the commuter experience
- Vertical structure that creates skyline aesthetic comparable with evolving context
- Rhythmic sequence of arches that create strong visual signature and silhouette
- Distinctive contemporary arch shape (driven by structural behavior)
- Dramatic scenic overlooks at the arch connections that enhance a rich pedestrian/cyclist environment and
- Structurally reliable, security conscious, resilient and robust design.

4 KEY ELEMENTS OF THE BRIDGE DESIGN

In addition to aesthetic challenges, the project includes technical complexities as well. Unlike conventional arch bridges, the three-arch system is designed to allow the superstructure to freely move through the arches with expansion joints only at the beginning and end of the structure. The bridge expands and contracts through the arches, like a glider chair. The unusual variable-depth "kite" shape of the arch, unbraced parallel arches and internal splice connection details (needed for aesthetics) all added challenges to both design and erection. The design and construction had to consider all the environmental site constraints as well as creating a signature bridge that the National Capital Planning Commission, US Historic Preservation Office and the US Commission of Fine arts would embrace.

As illustrated in Figure 5, the bridge is a total of 1,445 feet in length, with a main span length of 540 feet centered with the existing federal navigation channel (492′-6″ clear span opening) and two side spans measuring 452′-6″ in length. The heights of the arches are set in proportion to the spans at 148′–168′–148′ above mean sea level, with the height of the center arch set at the restricted limit in the Project Technical Provisions.

The arches comprise two primary components: concrete v-piers bearing on pile supported footings and steel arch ribs connected to the v-piers above deck level. The arches and hangers are in

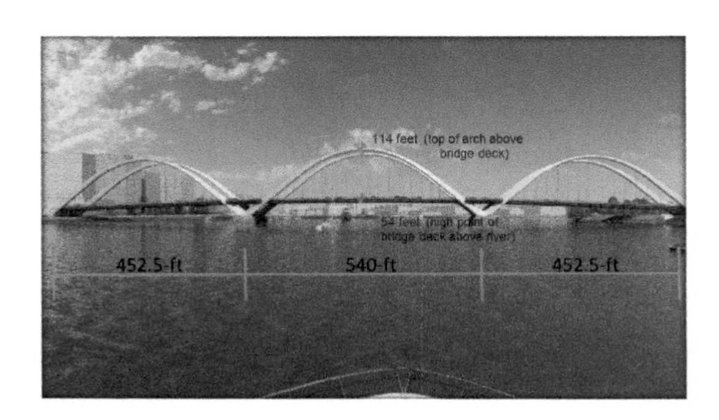

Figure 5. Bridge elevation *Arches – Steel Ribs*.

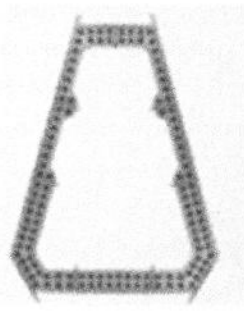

Figure 6. Steel arch rib.

Figure 7. Concrete V-Piers.

a vertical plane centered 5-feet outside of the deck edge, with no lateral bracing or part of the structure extending out over the traffic lanes or split-use path. With no bridge elements directly above the deck, this not only reduces the potential of falling ice from being a safety hazard for motorists and pedestrians, but also eliminates the need for any maintenance operations directly over traffic. This is a significant added value to the project not only in terms of public safety, but also in reducing long term maintenance costs. The use of vertical arch ribs also significantly reduces the amount of erection falsework required compared to a design with inclined members.

As shown schematically in Figure 6, the arch ribs are a hexagonal box section that tapers in depth from the base to the crown. The steel plates composing the arch rib vary in thickness and are stiffened along their length. All structural steel for the arches will be painted weathering steel, which provides a multi-level corrosion protection system. These multiple levels of protection are an added value to ensure that the structure will meet the 100-year service life requirement. The outside of the ribs will be fully protected with a two-part paint system an organic zinc prime, and polysilaxane top coat and the inside of the ribs will be protected with a two-part system consisting of a organic zinc prime and white epoxy top coat.

Each steel arch rib is composed of multiple sections that are spliced together internally with bolted connections. The use of internal bolted splice connections between members eliminates the need for field welds that are costly and have greater quality control challenges. Structural details were incorporated in the final design of the splice connections so as to maintain a consistent appearance along the outside surface of the arch ribs (i.e. external bolted field splices were not used).

4.1 *Arches – Concrete V-Piers*

The v-pier portions of the arches below the steel arch ribs are constructed of post-tensioned, cast-in-place concrete. As illustrated in Figure 7, the v-piers seamlessly continue the geometry of the steel arch ribs to the base and are constructed integrally with the pile supported footing. Concrete "strongbacks" built integrally with the v-piers extend out transversely beneath the bridge deck to support the longitudinal edge girders for the superstructure on bearings. The v-piers will be constructed of high-strength, low-permeability concrete and will be post-tensioned to inhibit cracking, which will ensure the 100-year service life requirement can be achieved. The v-piers are supported from below on architecturally sculpted waterline concrete footings in the river, with the footings founded on groups of 60-inch diameter, open-ended steel pipe piles. Additionally, the upper portion of the pipe piles are filled with reinforced concrete to increase the stiffness of the pile and provide moment connection with the footing. For added corrosion protection, the exterior surface of the piles within the river immersion zone are coated with coal tar-epoxy system.

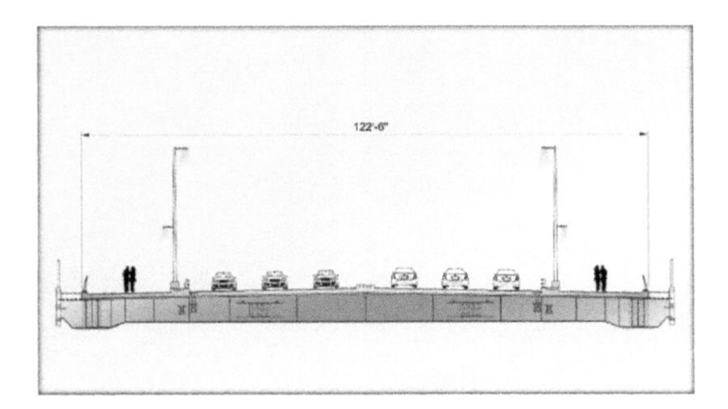

Figure 8. Transverse deck section.

4.2 *Hanger system*

The hanger system supporting the superstructure below the arches consists of 14 hangers for each of the side arches and 16 hangers for the center arch (for a total of 88 hangers). The hangers are spaced along the deck at 24-foot intervals. The hangers use stay-cable technology and are comprised of individually polyethylene sheathed, parallel 7-wire strands with an ultimate tensile strength of 270 ksi. Each individual 7-wire strand is filled with a corrosion inhibiting blocking compound and the bundle of strands is contained within ultra-violet resistant co-extruded HDPE pipes. This system comprising nested layers of corrosion protection delivers a service life greater than 75 years and has been proven durable by both fatigue and leak testing in accordance with the latest Post-Tensioning Institute (PTI) standards. The hangers along with the associated anchors at each end are being supplied by Schwager Davis, Inc. (SDI). The hanger sizes vary from 25 strands near the ends of the arches to 18 strands for the majority of the hangers. In line with requirements in the Project Technical Provisions, the hanger anchor and outer pipe will allow for the future addition of at least 5% additional strand capacity and each hanger will have one reference strand that may be removed by DDOT for corrosion inspection.

4.3 *Superstructure*

As illustrated in Figure 8, the typical out-to-out width of the deck is 122′-6″ and accommodates three 11-foot traffic lanes in each direction, two 18-foot wide split-use paths, and required medians and shoulder widths. Further, the bridge cross section can accommodate a reduced split-use path width for the future inclusion of an additional 11-foot traffic lane in each direction. The superstructure for FDMB comprises a steel composite deck section. This reliable structural form has proven to be extremely durable over time and has been used on the majority of arch and cable-stayed bridges in North America over the last 30 years. The use of I-shaped plate girders for the edge girders and floorbeams, together with precast concrete panels for the deck, is a design that takes these simple, durable elements and creates a bridge that simplifies fabrication and removes a lifetime of complex inspection associated with more elaborate structural systems.

The primary load path for the superstructure loads is through the transverse floorbeams, to the edge girders, and then to the hangers. In the unlikely event of a transverse floorbeam or edge girder fracture, the system will remain stable due to alternate load paths provided by the relatively closely spaced floorbeams and continuity of the deck in combination with the multiple edge girder supports provided by the hangers and bearings. The edge girders are steel I-girders with 6-ft deep webs and 2′–6″ top and bottom flanges. The thicknesses of the top and bottom flanges are both a constant 2″. There are two bays of transverse floorbeams between each hanger location, which results in a typical floorbeam spacing of 12′–0″. Similar to the edge girders, the floorbeams are also steel I-girders with a varying web depth to align with the crowned slope of the deck. Both the top and bottom flanges are 2′–0″ wide. Openings in the webs of the floorbeams accommodate the required utilities. Similar to the steel arch ribs, all structural steel for the edge girders and floorbeams will be

Figure 9. Aeroelastic wind tunnel model.

Figure 10. Temporary erection trestles.

painted weathering steel, which provides a multi-level corrosion protection system. These multiple levels of protection ensure that the structure will meet the 100-year service life requirement. The steel for the edge girders and floorbeams will be fully protected with a two-part paint system similar to the arch exterior.

The deck is primarily composed of 10-inch thick precast panels. The use of precast panels ensures that a large portion of the bridge deck is cast within a high quality, climate controlled environment that will enhance its durability. After the panels are placed on the steel framework, the spaces between the panels are closed by placing concrete between all gaps along with reinforcement. These cast-in-place closures will both connect the panels to each other and to the steel framework. The deck will be reinforced with a combination of stainless and epoxy-coated reinforcing with sufficient concrete cover to ensure the required 100-year service life is satisfied. Additionally, a 1-inch thick polyester polymer concrete (PPC) overlay with a high molecular weight methacrylate (HMWM) sealer prime coat will be provided along the entire length of the bridge to enhance rideability and further armor the deck against abrasion and chloride attack.

Rowan Williams Davies & Irwin Inc. (RWDI) performed the wind tunnel testing and analysis for the bridge to confirm that the superstructure meets aerodynamic performance requirements of vortex shedding response, galloping, and flutter instability. Sectional model tests were completed to ensure aerodynamic stability of the transverse cross section. A pedestrian comfort study was also performed. Figure 9 shows the aeroelastic model being tested during a critical arch erection stage. The bridge was tested in its completed condition, with and without the existing bridge, as well as for the critical erection stages.

5 ERECTION AND CONSTRUCTION

SCB's erection of the new FDMB began with erection of two temporary trestles. As shown in Figure 10, one trestle extends from the west abutment to V-Pier 1 while the other trestle spans from the east abutment to V-Pier 2.

The trestles are designed to support a 300-ton crane, which is used to drive foundation piles, place the concrete footings, and ultimately erect the arches and bridge superstructure.

After the trestles were installed, 60-inch diameter spiral welded steel pipe test piles were driven. Two piles were Statnamic load tested to confirm the 1800-ton design capacity. As of the end of March all foundation piles have been driven with the exception of the east abutment.

Following pile driving, SCB began constructing the waterline footings. On March 2, 2019 SCB poured 1,700 cy of concrete in the first v-pier footing (Figure 11).

After the lower concrete v-piers have been completed, erection of the arch steel will commence on both outer spans. The 300-ton crane will be used to erect the steel arch on the west span while another 300-ton crane is concurrently erecting the east span. The general sequence of arch erection for each of the spans is depicted in Figure 12. Arch erection consists of the following steps:

- Install temporary shoring towers to provide stability of the arch sections during the erection process. Temporary towers are designed to handle appropriate construction and wind loading.

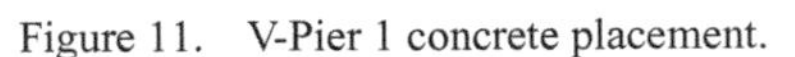

Figure 11. V-Pier 1 concrete placement.

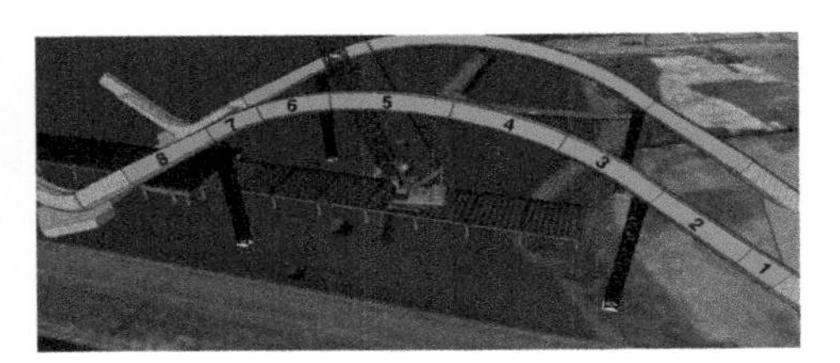

Figure 12. Erection of side arches.

Figure 13. Arch base cross section.

Figure 14. Channel span arch erection.

- Transport arch sections to the crane via the trestle. Each arch section will be approximately 50-feet in length and weigh 50,000 to 80,000 lbs.
- Erect arch section # 1, 2, 3, 7, and 8.
- Erect arch section #4 and 6.
- Erect the arch keystone piece #5.

The steel arches were being fabricated as of the end of March 2019. Figure 13 shows the base connection where the steel arch will connect to the concrete v-pier.

The most challenging aspect of the erection sequence is maintaining the 150-foot navigation channel during construction of the center span. As shown in Figure 14, the sequence of erecting the center span is like the outer two spans with the following exception:

Figure 15. Steel edge girder & floor beam erection.

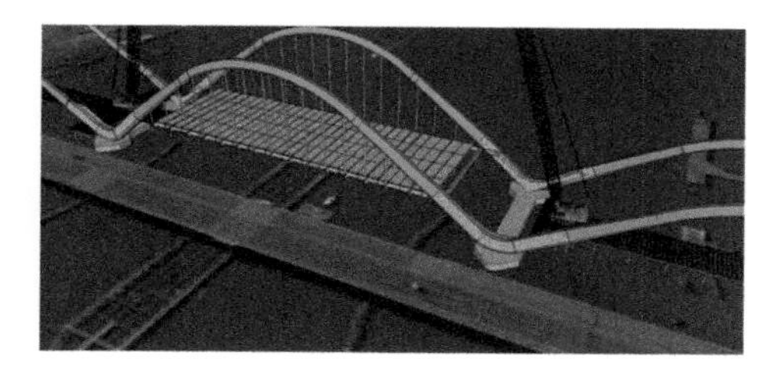

Figure 16. Precast deck panel erection.

Figure 17. Aesthetic lighting.

Due to the weight of the arch sections and the distance across the navigation channel, both 300-ton cranes will be needed to pick arch sections #8 and #9. All other arch sections can be lifted into position with a single crane. To ensure the 150-foot navigation channel is maintained open, SCB will use temporary horizontal guy wires in lieu of shoring towers to support arch sections #8 and #9.

Once all arch sections have been erected in the center span, SCB will start erecting the composite deck system which is comprised of edge girders, floorbeams, and precast deck panels. As depicted in Figure 15, the erection process starts with erection of edge girders using the permanent cable stays to support the edge girder. Erection of the floor-beams follow, also being supported by the permanent cable stays. Following the steel erection, precast deck slabs are placed as shown in Figure 16. The erection of the composite deck will start with the section over the 150-feet navigation channel. Cast-in-place concrete closure pours between the precast deck panels complete the process.

After the deck section above the 150-foot navigation channel is complete, the process will continue with one of the 300-ton cranes installing the composite deck from the navigation channel moving towards the east abutment. The second 300-ton crane will start at the navigation channel and move towards the western abutment.

Bridge PPC overlay and furnishings (i.e. barriers, roadway lighting, aesthetic lighting, etc.) will then be placed. Figure 17 illustrates the high-tech aesthetic lighting features that the bridge will be equipped with.

SCHEDULE

Design: August 2017 to November 2018
Bridge open to traffic: May 2021
Project Completion: December 31, 2021

CREDITS

Owner: District Department of Transportation
(HNTB Owner's PM/CM for Contract Compliance)

DESIGN BUILD TEAM:
South Capitol Bridgebuilders (JV: Archer Western Construction, LLC/Granite Construction Co.)
AECOM (Lead Bridge Design, Blast Security, Corrosion Protection Plan)
ECS (Geotechnical Engineering)
RWDI (Wind)
BeAM (Bridge Architect)
RBLD (Bridge Aesthetic Lighting)
Systra/IBT (Independent Design Check)
McNary Bergeron (Erection Engineering & Temporary Trestle)
SDI (Cables Stays & Post-tensioning)

6 CONCLUSIONS

Cities across the nation more than ever are demanding signature bridges. Common themes we always hear are:

- We want a Gateway to our city that represents our citizens
- We want you to create an elegant and iconic new bridge and
- We want a structure that is a destination that will facilitate urban development.

The challenge is also always the same: how do we create a vision that will be embraced by the community; and at the same time do it within the budgetary constraints of the owner. The key is DON'T overpromise during the bid phase and develop sufficient design and construction details to adequately bid the project while at the same time minimizing risk. The design build team must strike the right balance between appropriate engineering solutions and over-the-top extravagant solutions. They must interactively work together as contractor, designer, and bridge architect to ensure this occurs. The design build team must be passionate regarding their understanding of local cultures, and then weaving this culture into the bridge design. However, they must always be pragmatic keeping one eye on cost and budget as well as risk.

REFERENCES

L'Enfant P.: *L'Enfant Plan*, August 1791.
DDOT: *Anacostia Waterfront Initiative - Transportation Master Plan*, February 2014.
DDOT: *South Capitol Gateway Corridor and Anacostia Studies*, September 2003.

Chapter 13

The failure of the Tacoma Narrows Bridge

K. Gandhi
Gandhi Engineering, Inc., New York, NY, USA

ABSTRACT: The original design of the Tacoma Narrows Bridge developed by Washington State engineer Clark Eldridge included a suspension bridge with a center span of 2,600 feet (792 m), two side spans of 1,300 feet (396 m) each, trusses and cables 39 feet (11.9 m) center to center, stiffening trusses 22 feet (6.7 m) deep, and two travel lanes and sidewalks. However, when the Washington State Toll Bridge Authority (WSTBA) requested federal assistance from the Public Works Administration (PWA), the PWA agreed to a grant of 45% of the construction cost on the condition that the WSTBA hire Leon Moisseiff of New York for the design of the superstructure. Moisseiff told the PWA that his design would reduce estimated construction cost from $11 million to $7 million. The foundation design was to be performed by another New York firm, Moran, Proctor, and Freeman. Moisseiff increased the center span length to 2,800 feet (853.4 m) and reduced the side span lengths to 1,100 feet (335.3 m). The 25-foot deep (7.6 m) stiffening trusses were replaced by 8-foot-deep plate girders. On July 1, 1940, the bridge was opened to traffic. It was the third longest bridge in the world after the Golden Gate and George Washington Bridges. On November 7, 1940, torsional oscillations caused the failure of the bridge. The wind speed was about 42 miles per hour. Due to World War II, construction of the new bridge was delayed, and it was finally opened in 1950 with one twist: the new bridge had stiffening trusses 33 feet (10 m) deep or more than four times the depth of the plate girders designed by Moisseiff. This paper covers the construction and destruction of the original bridge and the people connected to it.

1 INTRODUCTION

Puget Sound, in the vicinity of Tacoma, is restricted at its narrowest point to a width of about 4,600 feet (1,402 m) in what is termed the Tacoma Narrows. The bridging of the sound at this location had been discussed since 1923. However, due to the great depth of water and the swiftness of the tidal currents, the cost of the bridge was prohibitively high for financing by private individuals.

In 1937, the Washington State Toll Bridge Authority (WSTBA) was created with full power to finance, construct, and operate toll bridges. Immediately thereafter, Pierce County, in which the bridge project was situated, contributed $25,000 to the State Highway Department for such a study. The feasibility study was conducted by its Chief Engineer, Clark Eldridge. His design included a suspension bridge with a center span of 2,600 feet (792 m), two side spans of 1,300 feet (396 m) each, trusses and cables 39 feet (11.9 m) center-to-center, stiffening trusses 22 feet (6.7 m) deep, and two travel lanes and sidewalks.

However, when the WSTBA requested federal assistance from the Public Works Administration (PWA), the PWA agreed to a grant of 45% of the construction cost on the condition that the WSTBA hire Leon S. Moisseiff of New York for the design of the superstructure. The foundation design was to be performed by another New York firm, Moran, Proctor, and Freeman.

Both Moisseiff and Moran, Proctor, and Freeman made revisions to the design prepared by Eldridge. The contractor working on the foundations of the Tacoma Narrows Bridge stated that the revisions proposed by Moran, Proctor, and Freeman were not buildable and the original design by Eldridge was adopted. The comparison between the two superstructure designs is listed below:

Item	Bridge by WSDOT	Bridge by Moisseiff
1. Main Span	2,600 feet (792 m)	2,800 feet (853.4 m)
2. Side Spans	1,300 feet (396 m)	1,100 feet (335.3 m)
3. Height of Towers	East: 476.5 feet (145 m) West: 463.5 feet (141 m)	425 feet (129.5 m) 425 feet (129.5 m)
4. Depth of Stiffening Trusses	22 ft. (6.7 m) truss	8 ft. (2.4 m) plate girder
5. Distance between Center Lines of Cables	39 ft. (11.9 m)	39 ft. (11.9 m)
6. No. of Lanes	2	2
7. Max. Vertical Deflections	Not Available	6 ft. side spans 10.1 ft. center span
8. Max. Lateral Deflections	Not Available	3.4 ft. side spans (1 m) 20 ft. center span (6.1 m)

The failure of the Tacoma Narrows Bridge resulted in multiple investigations, their findings, lawsuits, examples of past successes and failures, and the resolve to build a new bridge based on wind tunnel testing validating the design. There subjects are covered in this paper.

2 THEORIES FOR THE DESIGN OF SUSPENSION BRIDGES

2.1 *Intuition and rule of thumb*

Charles Ellet, Jr. and John Roebling, two early pioneers of the design and construction of suspension bridges in the U.S., designed their bridges based on their intuition and basic laws of statics (such as *stress* = *load/area*) and sized their structural members accordingly.

2.2 *The elastic theory*

According to Gimsing (1997), Maurice Levy developed the Elastic Theory in 1886 to analyze the interaction between the cable and the stiffening girder in a suspension bridge. It was derived from simple consideration of elastic equilibrium of the system. The Elastic Theory was based on the following five assumptions (Steinman 1929):

1. The cable is supposed perfectly flexible, freely assuming the form of the equilibrium polygon of the suspender forces.
2. The truss is considered a beam, initially straight and horizontal, of constant moment of inertia and tied to the cable throughout its length.
3. The dead load of truss and cable is assumed uniform per lineal unit, so that the initial curve of the cable is a parabola.
4. The form and ordinates of the cable curve are assumed to remain unaltered upon application of loading.
5. The dead load is carried wholly by the cable and causes no stress in the stiffening truss. The truss is stressed only by live load and by changes of temperature.

The last assumption is based on erection adjustments, involving regulation of the hangers and riveting-up of the trusses when assumed conditions of dead load and temperature are realized.

2.3 *The deflection theory*

Clark Eldridge, who first designed the Tacoma Narrows Bridge for the Washington State DOT (WSDOT), had a 22-foot (6.7 m) deep stiffening girder. Then how was it that Moisseiff was able to reduce the stiffening girder to an 8-foot (2.4 m) deep plate girder and still considered the Tacoma Narrows Bridge to be safe? The answer is that Moisseiff based his design on Deflection Theory, a term he coined himself.

The "More Exact Theory," or Deflection Theory, was first published by Josef Melan of Prague, then a part of Austria, in 1888. Steinman translated Melan's theory into English and published it in 1913 in a 300-page book titled *Theory of Arches and Suspension Bridges* (Steinman 1913). The fundamental assumptions of this theory are:

1. The initial curve of the cable is a parabola, and
2. The initial dead load (w) is carried by the cable (producing the initial horizontal tension (Hw)) without causing stress in the stiffening truss.

Gimsing has provided a simple and intuitive explanation of the Deflection Theory without any mathematical formulas in his book (Gimsing, pp. 12–13).

3 REQUEST FOR INVESTIGATION OF BLACKWELL'S ISLAND AND MANHATTAN BRIDGES

Moisseiff was in charge of the design of the Manhattan and Blackwell's Island Bridges. Many engineers who worked on the design of these bridges, but who were either not familiar or comfortable with the Deflection Theory, were concerned about the safety and sufficiency of these bridges. Some of them were members of an organization called Technical League of New York City and the identities of its members were kept secret.

The League arranged three lectures by three separate speakers on the design of the Blackwell's Island Bridge. These presentations were published in *Engineering-Contracting* in 1908. Based on these presentations, the League and the *Engineering-Contracting* journal requested the appointment of an independent Board of Engineers to "make a thorough investigation of the contract and specifications, the stress sheets and drawings, the methods used in calculating the stresses and, in fact, all details regarding the bridge." These doubts were raised by three papers read before the Technical League of New York. These papers showed that there were points of controversy on the loading of such mammoth bridges, and in one paper the starting statement was made that on the original stress sheets on which the contract was let, there was "not a single stress correctly calculated."

3.1 *Safety and sufficiency of the Blackwell's Island Cantilever Bridge*

The Department of Bridges retained the services of Boller & Hodge of New York City and Professor William H. Burr of Columbia University to investigate the Blackwell's Island Bridge.

Two independent reports prepared by Boller & Hodge and Professor Burr recommended reducing the dead load and limiting future live loads (Engineering-Contracting 1908a).

3.2 *Safety and sufficiency of the Manhattan Suspension Bridge*

Based on the demands for investigation by the journal *Engineering-Contracting*, and the City Club of New York, Commissioner James W. Stevenson of the Department of Bridges retained the services of Ralph Modjeski of Chicago on January 26, 1909. Modjeski was also involved in the investigation of the failure of the Quebec Cantilever Bridge (Gandhi 2017).

The design calculations for the Manhattan Bridge were performed by Leon Moisseiff using Josef Melan's theory which Moisseiff had named the "Deflection Theory." Since no one in Modjeski's office knew about the deflection theory, Modjeski retained Professor F.E. Turneaure of the University of Wisconsin as his consultant to work with his staff in Chicago. Modjeski also sent his assistant Mr. W.R. Weidman to New York to work in the Bridge Department's office to supervise calculations of stresses in the stiffening trusses and towers performed by the employees of the Bridge Department. The teacher (or *guru* in Sanskrit) Moisseiff converted his students (*shishyas*) F.E. Turneaure and W.R. Weidman to placing their full faith in the Deflection Theory. No wonder that when Modjeski submitted his report on September 16, 1909, it gave the Manhattan Bridge a clean bill of health.

4 THE FAILURE OF THE TACOMA NARROWS BRIDGE

The newly completed bridge before it opened to traffic is shown in Figure 1. On the morning of November 7, 1940, the wind was blowing through the Narrows at 42 mph (68 kph) and one of the bridge's new diagonal stays broke. The bridge began not only to ripple up and down, but also to twist from side to side in a torsional motion so violent that the bridge was closed to traffic (Figure 2) (Scott 2001).

Leonard Coatsworth, a copy editor for the Tacoma News Tribune, was the last motorist permitted on the bridge. In his car was his daughter's black spaniel. When Coatsworth was about halfway across, the roadway tilted sideways so violently that he lost control of the vehicle. He decided to abandon the car, but the dog did not want to go with him (Figure 3). When he reached the east shore, Coatsworth saw Professor Farquharson of the University of Washington filming the bridge's gyrations. Then the bridge began to whip up and down so violently that the roadway was rising and falling as much as 28 feet (8.5 m) and was tilting up to 45 degrees sideways.

Suddenly, the vertical cables that held up the roadway began to snap, their ends whipping through the air, and the roadway along the 2,800-foot (853.3m) center span began to break up. Huge chunks of concrete broke off and crashed into the water 200 feet (61 m) below, taking along with it Coatsworth's car and dog (Figure 4) (Scott 2001).

The bridge continued to writhe and shatter and by the time wind had subsided, about 90% of the center span was lying in pieces on the bottom of the Puget Sound, with the rest hanging down in the air and the side spans sagging 60 to 70 feet (18 to 23 m). The steel towers were bent backwards towards the shores and were damaged beyond repair (Figure 5) (Scott 2001).

Figure 1. Tacoma Narrows Bridge, 1940 (Scott 2001).

Figure 2. Torsional oscillation, Tacoma Narrows Bridge (Scott 2001).

Figure 3. Leonard Coatsworth (*shown here*) achieved notoriety as the last person to drive onto the bridge–and for abandoning his car to escape the collapse. The last person off the bridge was Professor Farquharson, of the University of Washington, who had been observing the structure's misbehavior (Petroski 2009).

Figure 4. Collapse of the main span, Tacoma Narrows Bridge (Scott 2001).

Figure 5. Side span after collapse, Tacoma Narrows Bridge (Scott 2001).

Moisseiff told reporters in New York that a "peculiar wind condition" had hit at an unfortunate angle. "I am completely at a loss to explain the collapse," Moisseiff said according to a November 8, 1940 Associated Press report. He said that testing a model of the bridge at the University of Washington had convinced him that the span would be permanent despite its early shakes. Moisseiff came to Tacoma days later to examine the collapse, which no doubt haunted him the rest of his life.

5 REPORT OF THE BOARD OF ENGINEERS AND ITS KEY FINDINGS

The PWA, which had made a grant of $2.88 million to finance the Tacoma Narrows Bridge project, appointed a Board of Engineers within weeks after the collapse of the bridge in November 1940. The board members were:

1. Othmar H. Ammann, Director of Engineering, New York Port Authority
2. Theodore von Karman, Director of the Daniel Guggenheim Aeronautical Laboratory at the California Institute of Technology
3. Glen B. Woodruff, Engineer of Berkley, CA who had been employed on the design of the San Francisco-Oakland Bay Bridge

The Board submitted its report with 15 key findings in about four months, dated March 28, 1941 (Ammann 1941). The summary of their findings is also given in the *Engineering News Record* (May 8, 1941, pp. 74–76). The key findings of the Board of Engineers were:

1. The bridge failed due to excessive oscillations caused by wind action
2. The excessive vertical and torsional oscillations were made possible by the extraordinary degree of flexibility of the structure and of its relatively small capacity to absorb dynamic forces

6 MOISSEIFF'S ROLE IN THE CONSTRUCTION OF THE MANHATTAN BRIDGE

The Manhattan Bridge was constructed under seven major contracts as follows (Johnson 2010):

1. Brooklyn Tower Foundation, let in 1901, completed in 1902
2. Manhattan Tower Foundation, let in 1902, completed in 1910
3. Brooklyn Anchorage, let in 1905, completed in 1910
4. Manhattan Anchorage, let in 1905, completed in 1910
5. Steel Tower, Cable, Suspenders and Suspended Superstructure, let in 1906, completed in 1909
6. Manhattan and Brooklyn Anchorages, let in 1907, completed in 1910
7. Roadway and Footwalk Pavements, Electrical Equipment, and Lower Deck Railway Tracks, let in 1909, completed in 1910.

The author did not see any credit being given to Moisseiff in the technical journals published during these years. The details about the engineers who were responsible are given below.

Year	Commissioner	Chief Bridge Engineer	Chief Engineer-Manhattan Bridge
1902–3	Gustav Lindenthal	Gustav Lindenthal	Richard S. Buck (resigned) L.A. LaChicotte
1904–5	George E. Best	Othniel F. Nichols	Richard S. Buck (returned) Holton D. Robinson (Engineer-in-Charge)
1906–7	James W. Stevenson	C.M. Ingersall Jr. Othniel F. Nichols (consulting engineer)	Richard S. Buck Holton D. Robinson (Engineer-in-Charge) D.E. Baxter (Resident Engineer)
1908–9	James W. Stevenson	C.M. Ingersall Jr. Alexander Johnson	Alexander Johnson (consulting engineer) A.I. Perry (Engineer-in-Charge, Design) D.E. Baxter (Engineer-in-Charge, Construction)

The above table indicates that Moisseiff may have been an occasional or a periodic visitor to the Manhattan Bridge during its construction but was never an inspector assigned full-time to the Manhattan Bridge. Besides, he was also the designer of the Blackwell's Island Bridge during the same time. Rather than learning from the failures of past bridges, he trusted his deflection theory which ultimately led him to believe that a stiffening truss was unnecessary for the stability of the Tacoma Narrows Bridge and to his eventual downfall along with the bridge.

7 WIND FAILURES OF SUSPENSION BRIDGES

Between March 15, 1934 and October 31, 1935, W. Watters Pagon wrote eight informative articles on the design of structures to resist wind starting from "What Aerodynamics Can Teach the Civil Engineer" and ending with "Using Aerodynamic Research Results in Civil Engineering Practice." In the introduction of his first paper, he pointedly observed that civil engineers had shown very little interest in the subject of wind forces and had been content to abide by the results of tests made 75 to 100 years prior by such men as Duchemin.

After the failure of the Tacoma Narrows Bridge, Professor Finch of Columbia University presented the results of his research (Finch 1941) on the failures of previous suspension bridges and concluded that "in all cases, it was lack of stiffening or rigidity that caused the difficulty. When heavy stiffening trusses were used, the oscillations ceased."

The principle of strengthening the suspension system with stays was originally introduced by John A. Roebling during the construction of the Railways Suspension Bridge across the Niagara Gorge. This bridge was constructed between 1851 and 1855 (Gandhi 2006).

Although Roebling did not know Josef Melan's Theory, he had a superb grasp and understanding of torsional rigidity and the need for a stiffening truss in a suspension bridge, more than any other engineer of his time. The below comparison between the cross-sections designed 90 years apart (1847 vs. 1937) will prove this point (Figures 6 and 7).

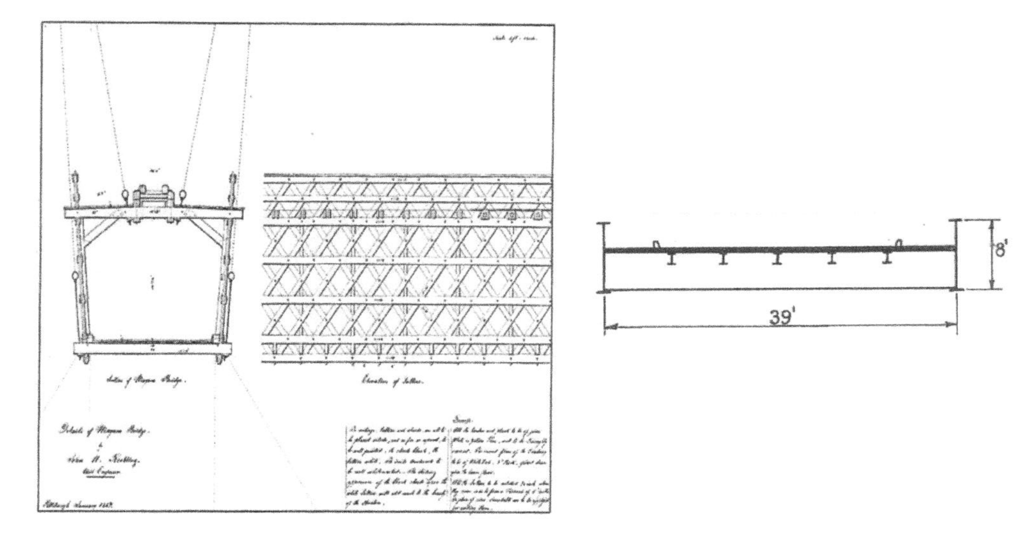

Figure 6. Comparison of Cross Sections of Niagara and Tacoma Narrows Suspension Bridges. Please note that the Niagara Bridge, besides being a box, also had above floor and under floor stays to stabilize the bridge against wind (Finch 1941).

Figure 7. The suspension bridge across the Niagara showing underdeck stays (Scientific American 1867).

8 INSURANCE CLAIMS AND SETTLEMENT

The Tacoma Narrows Bridge was insured for over $5 million and 23 different insurance companies were involved to spread the risk. One of the insurance agents had kept the premium for his own use expecting the bridge would never fail. His company paid off the premium and the agent was sentenced to 15 years in prison.

After the bridge's collapse, the insurance companies appointed a board of engineers to study the damage. The members were:

1. Hardy Cross, professor of civil engineering, Yale University
2. Shortridge Hardesty of Waddell and Hardesty, New York
3. Holton D. Robinson of Robinson & Steiman, New York
4. Wilber M. Wilson, research professor of structural engineering, University of Illinois
5. Clifford E. Paine of Chicago, Chairman

Another board of engineers was appointed by the WSTBA to assess the damage to the Tacoma Narrows Bridge. The board members were:

1. Francis Donaldson, Chief Engineer for Mason-Walsh-Atkinson-Kier Co. on the Grand Coulee Dam for one and a half years
2. Leif J. Sverdrup of the consulting firm Sverdrup & Parcel, St. Louis
3. Russel G. Cone, engineer for the Golden Gate Bridge and Highway District

The WSTBA filed a claim for $5,200,000, representing 80% of the $6,500,000 cost of the bridge. The 23 insurance companies offered $1.8 million, stating that the bridge could be restored to its original condition using the existing towers, cables, and piers. The final settlement offer of $4 million by the underwriters was accepted by the WSBTA in August of 1941. (Engineering News Record 1941)

9 LAWSUIT FILED BY THE CONTRACTORS

The three firms – Pacific Bridge Co. of San Francisco, General Construction Co. of Seattle, and Columbia Construction Co. of Portland, OR – who held the contract for the construction of the Tacoma Narrows Bridge, filed a lawsuit in the Washington State Supreme Court to compel the WSTBA to pay $619,915 for work done on the bridge. The WSTBA stated that it had not received the final grant from the PWA and, as a result, final payments had not been made to the Contractors.

On April 12, 1941, the Washington State Supreme Court ordered the WSTBA to pay the contractors the $619,915 plus 6% interest. In May of 1941, the WSTBA petitioned the Washington State Supreme Court for a rehearing to escape from paying the interest. In July of that year, the Supreme Court refused to reverse its previous decision and upheld the bridge contractors' claims. (Engineering News Record 1941)

10 TACOMA NARROWS REPLACEMENT BRIDGE

The new Tacoma Narrows Bridge was made wider with four lanes to cover the growth in population following World War II. It was built in about 30 months and cost about $14 million plus the lives of four workers who were killed during its construction. The superstructure was redesigned with great care and the design was validated by the results of the wind tunnel testing of a 100-foot (30.5 m) scale model of the new bridge.

The stiffening trusses were made 33 feet (10.1 m) deep, were open to allow the wind to pass through, and were braced horizontally. The new trusses were 58 times stiffer than the old system and the new bridge was stable even in winds of over 70 miles per hour. (Waugh 1993)

11 PEOPLE CONNECTED WITH THE FIRST TACOMA NARROWS BRIDGE

11.1 *Elmer Hayden (1868–1938)*

Mr. Hayden was senior partner in the law firm Hayden, Metzger, and Blair which provided the legal counsel necessary for the bridge to become a reality. Providing legal support to county commissioners, authoring legal documents, appearing before the state's Supreme Court, serving on committees and on the Chamber of Commerce's Board of Trustees, Mr. Hayden used every venue available to him to promote and support the bridge concept.

He was born in Indiana on October 25, 1868 and died in Tacoma at the age of 69 on August 16, 1938.

11.2 *Clark H. Eldridge (1896-1990)*

Mr. Eldridge completed engineering studies at Washington State College in 1920. Over the next 15 years, he distinguished himself in the Seattle City Bridge Engineer's office. Then, in 1936, Eldridge joined the State Highway Department. He was assigned to design two of the State's most colossal bridges, namely, the Tacoma Narrows Bridge and the Lake Washington Floating Bridge.

His boss, State Highway Director Lacey V. Murrow, took Eldridge's design and cost estimates to the PWA in Washington, D.C.in the spring of 1938. The PWA decided Eldridge's design was too expensive and required the WSTBA to hire Leon Moisseiff of New York as consultant for the superstructure and Moran, Proctor, and Freeman as consultants for the substructure. Eldridge supervised the construction.

After the collapse of the Tacoma Narrows Bridge, Eldridge joined the U.S. Navy and was sent to Guam in April 1941. During World War II, he was captured by the Japanese and spent the duration of the war in a prisoner-of-war camp.

The 1950 Narrows Bridge, with four lanes of traffic, used the same piers that Eldridge had designed, and with its 33-foot (10.1 m) deep Warren stiffening truss, it closely resembles the original Tacoma Narrows Bridge with the 22-foot (6.7 m) stiffening truss Eldridge designed for two lanes of traffic.

11.3 *Frederick B. Farquharson (1895–1970)*

Farquharson was in the right place at the right time as a professor of civil engineering at the University of Washington for most of his career, from 1925 to 1963. He pioneered aerodynamic studies of the ill-fated 1940 and the new 1950 Tacoma Narrows Bridges.

Farquharson stood on the 1940 Tacoma Narrows Bridge the day it collapsed. He intently monitored its behavior, snapped photos, and took a motion picture film of the disaster. He died at home on Jun 17, 1970 at the age of 75.

11.4 *Lacey V. Murrow (1904–1966)*

Mr. Murrow attended Washington State University and graduated with a B.S. in Military Science in 1926 and later in Engineering in 1935. In 1933, Murrow began an eight-year term as Director of the State Highway Department. In 1937, Murrow also served as Chief Engineer of the WSTBA and oversaw the completion of the first, ill-fated Tacoma Narrows Bridge.

From 1940 to 1946, he served as a command pilot in the U.S. Air Force, winning military honors including a Presidential citation with four cluster decorations, the Legion of Merit, the Order of British Empire, and the Croix de Guerre. On April 27, 1948, Murrow was promoted to Brigadier General in the Air Force. He served in Korea, Japan, and the U.S. before retirement.

11.5 *Leon Solomon Moisseiff (1872–1943)*

Mr. Moisseiff was born in Riga, Latvia on November 10, 1872 and he was educated there. He moved to New York in 1891 and graduated from Columbia University in 1895 with a degree in Civil Engineering.

He joined the City of New York in 1897 as a Draftsman in the Borough of the Bronx. He was transferred to the Department of the Bridges for Manhattan and became Chief Draftsman in 1900. Within a short time, he was promoted to the position of Assistant Engineer on the computation design of the Manhattan and Blackwell's Island Bridges. In 1902, when Lindenthal became the Commissioner of the Department of Bridges, he appointed Moisseiff as his personal assistant. After the departure of Lindenthal on December 31, 1903, Moisseiff returned to his old job title.

He was the first to recognize the significance of the Deflection Theory which had been outlined by Josef Melan in 1888. Moisseiff proved that the deflection theory allows important savings in material. His development and application of the Deflection Theory became the standard work of reference for designers of suspension bridges.

1910, the City of New York established a Division of Design of which he was the head with the title Engineer of Design. In 1915, he left the City of New York and opened an office as a consulting engineer. From that time until his death, he was actively engaged in the design of the superstructures for the Golden Gate Bridge in San Francisco, the Bronx-Whitestone and George Washington Bridges in New York City, the Ambassador Bridge in Detroit, the Ben Franklin Bridge in Philadelphia, and the Tacoma Narrows Bridge.

The Tacoma Narrows Bridge disaster effectively ended his career. He died at age 70 on September 3, 1943 at his summer home in Belmar, New Jersey, three years after the failure of the Tacoma Narrows Bridge, the bridge he described as his "most beautiful" bridge. (ASCE Transactions 1946)

12 CONCLUSIONS

In the second half of the 19th century, the design of a suspension bridge was based on intuition, prior experience, and involvement of the bridge engineer from conceptual design to actual construction.

The emphasis was on safety of the bridge. Examples include bridges designed by John A. Roebling – such as the Niagara Railway Suspension Bridge, Cincinnati Covington Bridge, and Brooklyn Bridge – and Leffert L. Buck and his projects such as the replacement of the stone towers by iron towers on the Niagara Railway Suspension Bridge and Williamsburg Bridge in New York. The stiffening truss was a major element of all bridges designed during this period.

However, during the first 40 years of the 20th century, the emphasis changed from safety to the initial cost of a bridge. This was brought about by promulgation of deflection theory by Moisseiff and its acceptance by two other prominent suspension bridge designers of that period, David B. Steinman and Othmar H. Ammann. The importance given to the stiffening truss for the stability of a suspension bridge was disregarded, resulting in its substitution by an 8-foot (2.4 m) deep plate girder by Moisseiff for the Tacoma Narrows Bridge. Most of the savings claims were as a result of the downsizing of the stiffening trusses.

Knowledgeable bridge engineers in New York, who were concerned about the underdesign of the Manhattan and Blackwell's Island Bridges, did express their concerns, but they wanted to remain annonymus for the fear of retaliation by the Department of Bridges. Their half-hearted attempt did result in the appointment of Ralph Modjeski of Chicago to check the calculations performed by or under the direction of Moisseiff for the Manhattan Bridge cables and towers. However, Modjeski's staff, like most engineers of that period (or even now), did not understand higher mathematics involved in the deflection theory. Also, Modjeski's firm was not known for the design of major suspension bridges. They did not have any rules of thumb by which to guess approximate sizes of various elements of cables and towers and compare them with those sizes determined by Moisseiff. They became dependent on Moisseiff to explain his calculation. In reality, this was not an independent check, but a mere rubber-stamping of the calculations performed by Moisseiff.

It was a shock to lose the third longest suspension bridge in the world within four months of its opening. However, it did start a worldwide study of aerodynamics as applied to long-span bridges using wind tunnels and other theoretical means, which continues unabetted. The real causes of the Tacoma Narrows Bridge's failure have been studied and debated without reaching any agreement among the scholars in this field. Because of this tremendous gain in our knowledge, we are fortunate not to have had a similar bridge failure since the original Tacoma Narrows Bridge.

ACKNOWLEDGEMENTS

The author gratefully acknowledges the assistance received from Mr. Michel Wendt, reference librarian of the Washington State DOT, Ms. Tammy Gobert of the Rensselaer Polytechnic Institute (RPI) library in Troy, NY, and Ms. Annie Sidou and Ms. Livia Bennett of Gandhi Engineering for providing reference materials and assistance during the preparation of this paper.

REFERENCES

Ammann, O.H., Karman, T. & Woodruff, G.B. (March 28, 1941). The Failure of the Tacoma Narrows Bridge. *A Report to the Honorable John M. Carmody, Administrator, Federal Works Agency, Washington, D.C.*

Engineering-Contracting 1908a. Reports on the Safety and Sufficiency of the Blackwell's Island Bridge, New York City: *30*(20), 344–346.

Engineering-Contracting 1908b. Discussions Before the Technical League on the Blackwell's Island Bridge: *30*(24), 405–406.

Engineering-Contracting 1908c. An Investigation Should Be Made of the Blackwell's Island and Manhattan Bridges: *30*(26), 425.

Engineering-Contracting 1908d. Discussion Before the Technical League on the Blackwell's Island Bridge: *30*(26), 440.

Engineering News Record 1941a. Tacoma Bridge contractors upheld by supreme court: *127*(2), 9.

Engineering News Record 1941b. Tacoma Bridge Authority settles insurance case: *127*(8), 7.

Finch, J.K. 1941. Wind Failures of Suspension Bridges or Evolution and Decay of the Stiffening Truss. *Engineering News Record 30*(26): 74–79.
Gandhi, K. 2006. Roebling's Railway Suspension Bridge over Niagara Gorge. *5th International Cable-Supported Bridge Operator's Conference*, New York, NY.
Gandhi, K. 2013. Lindenthal and the Manhattan Bridge Eyebar Chain Controversy. *7th New York City Bridge Conference*, New York, NY.
Gandhi, K. 2017. The Failure and Reconstruction of the Quebec Bridge. *9th New York City Bridge Conference*, New York, NY.
Gandhi, K. 2015. Charles Ellet, Jr., the Pioneer American Suspension Bridge Builder. *8th New York City Bridge Conference,* New York, NY.
Gimsing, N.J. 1997. *Cable-Supported Bridges – Concept and Design, Second Edition*. Hoboken, NJ: John Wiley & Sons.
Johnson, A. 2010. The Manhattan Bridge. *The Municipal Engineers of the City of New York, Paper No. 55*, New York, NY.
Petroski, H. 2009. Tacoma Narrows Bridges. *American Scientist 97*(15): 103–107.
Scientific American 1867. The Suspension Bridge Across the Niagara: *17*(23), 353.
Scott, R. 2001. *In the Wake of Tacoma – Suspension Bridges and the Quest for Aerodynamic Stability.* Reston, VA: ASCE Press.
Steinman, D.B. 1929. *A Practical Treatise on Suspension Bridges – Their Design, Construction, and Erection, Second Edition*. Hoboken, NJ: John Wiley & Sons.
Steinman, D.B. 1913. *Theory of Arches and Suspension Bridges*. Chicago: Myron C. Clark Publishing Co.
Waugh, K. 1993. *Galloping Gertie, Guide to the Records of the Tacoma Narrows Bridge 1932–1969.*

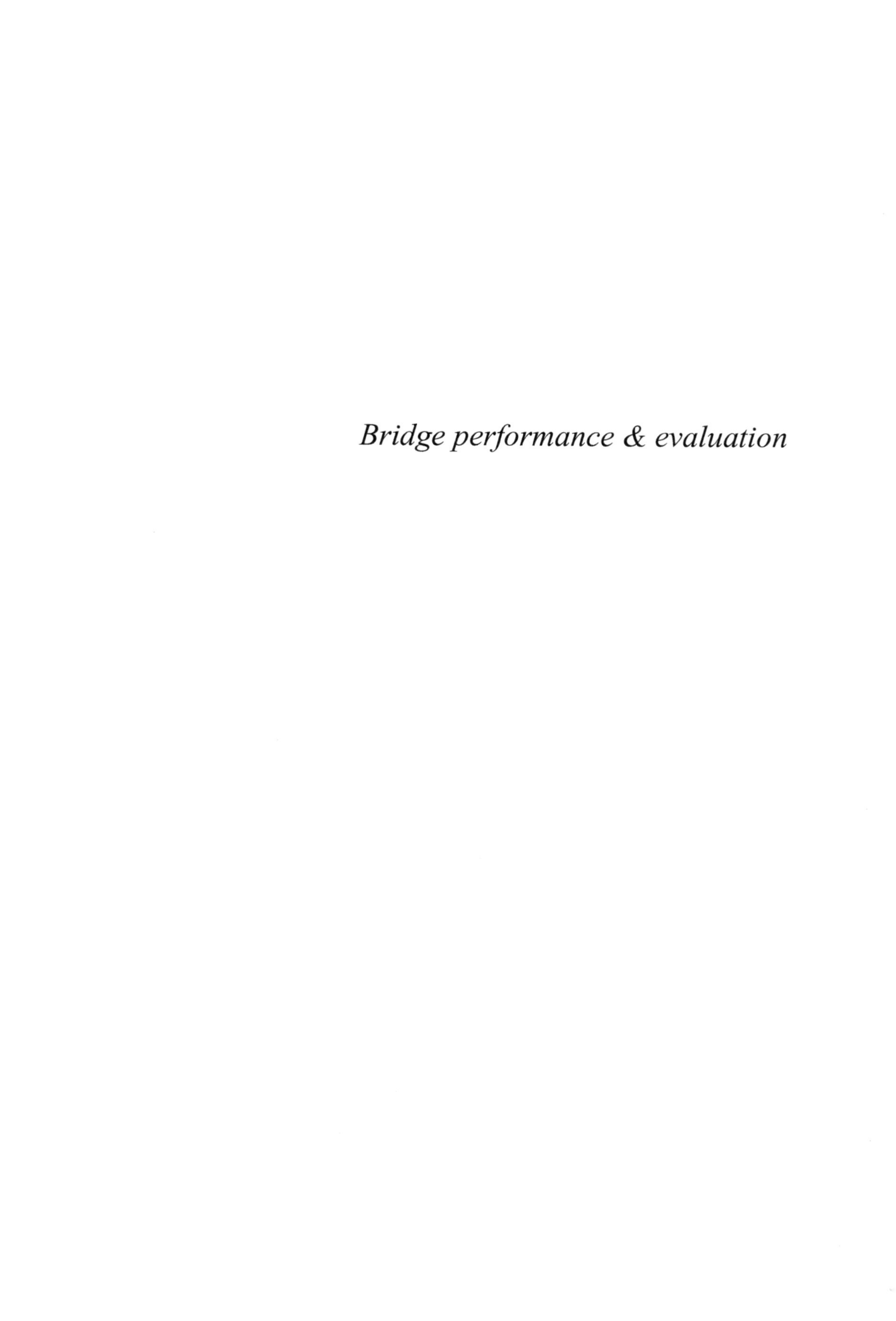

Bridge performance & evaluation

Chapter 14

Strength and deformation based performance evaluation of existing bridges

E. Namlı, D.H. Yıldız & N. Çilingir
Emay International Engineering and Consultancy Inc., Istanbul, Turkey

ABSTRACT: In Turkey which has a high seismic risk, it is required to design bridge structural systems so as to ensure a sufficient level of safety, as similar to other types of structural systems. In addition, it is of paramount importance to evaluate the seismic safety of existing bridge structural systems. In this context, currently it is often required to resort to deformation based performance evaluation methods in addition to the conventional strength based performance methods. In this paper, with reference to a particular project commissioned by İstanbul Metropolitan Municipality (İBB) which involved the inspection, evaluation and seismic retrofit of various existing bridges, the seismic performance of bridges have been studied and employed calculation methods explained. In this study, principles of "DLH Seismic Design Guidelines, General Directorate for Construction for Railways Harbours and Airports of (RHA) Ministry of Transportation (DLH 2008)" have been taken into consideration. Initially the existing bridge site inspections (cores, soil boring and structural in place measurements, etc) were carried out, followed by the establishment of structural calculation models. In the process which involved strength based evaluation response spectrum method, and on the other hand in the case of deformation based evaluation pushover analysis and nonlinear time history analysis have been performed. In the case of nonlinear analysis frame and finite element models have been used. By taking into account the particular bridge class and the expected performance level corresponding to this class, the seismic performance of bridges studied have been evaluated and those structures which exhibited insufficient level of seismic safety have been retrofitted.

1 INTRODUCTION

Istanbul is located in a zone with a high degree of seismicity and at the same time a high density of population. The provision the transportation system needed for the search, rescue and evaluation activities and the transportation of vital material, equipment and supplies is of paramount importance considering the case of a major earthquake. It thus follows the whole transportation system and its critical sections such as bridges constitute one of the vital systems which should be given due attention in the process of taking the required anti seismic measures.

Within the scope of the Project implemented by İstanbul Metropolitan Municipality (İBB) some bridges in the İstanbul metropolitan area have been inspected and evaluated by Emay International Engineering and Consultancy Inc., and it has been aimed to provide a sufficient degree of structural safety by preparing detailed and final seismic retrofit design for some of those bridges which required to be strengthened.

Besides the ground motion effect, the importance of the bridge with respect to transportation and structural risk are factors which determine the importance of a bridge from the point of an earthquake. The importance of a bridge with respect to transportation in town is determined by the importance and density of traffic on the particular highway, whether or not the road alignment is of emergency line and whether or not there are alternative alignments. In addition, whether or not the bridge structure is large, old and/or damaged is another factor which contribute to structural risk, such circumstances do exist in the bridges inspected and those for which seismic retrofit design has been prepared. Bridges in general proved to have insufficient degree of structural seismic safety due to the fact that existing specifications and codes used in the design of existing bridges include low seismic coefficients, do not attribute the necessary importance to details which would enhance ductility (lap lengths, confinement and other details) and that they are based upon the principles of linear elastic (working stress) design method rather than the principles of performance based method. In addition, unsatisfactory supervision of material, construction quality and workmanship, the use of low standard material class, vehicular impact damage and insufficiency due to general wear or external effects are structural weaknesses observed in the bridges inspected.

In the process of evaluation and seismic retrofit of bridges, the seismic technical code with regard to the DLH Seismic Design Guidelines, General Directorate for Construction for Railways Harbours and Airports of (RHA) Ministry of Transportation (DLH 2008) which include the strength and deformation based calculation approach has been used. Since all projects in this Project were classified as "special bridge", multi mode spectral analysis has been carried out and minimum damage (MD) performance level was checked for D2 seismic level by using the strength based calculation method. In addition, nonlinear methods (such as push-over or time history analysis) were used for D3 seismic level and limited damage (LD) performance level was checked by using the deformation based evaluation method.

2 PRINCIPLES OF BRIDGE SEISMIC EVALUATION AND CALCULATION METHODS

2.1 *Principles of initial evaluation*

Initial evaluation reports which include the result of studies concerning the geometrical properties of bridges, structural systems, soil conditions and extent of damage accompanied by relevant recommendations have been prepared and submitted. The material test results (concrete core specimen and/or steel material specimen), soil boring results, foundation base investigation by georadar, three-dimensional mapping and damage inspection have all been used as an input for the subsequent detailed analysis conducted so as to evaluate the structure.

2.2 *Principles of detailed evaluation*

In accordance with the information and input data acquired during the initial evaluation stage. An analytical model of the bridge has been established and by adopting the DLH (DLH 2008) seismic technical code the load carrying capacities of bridges under seismic loading have been checked. In accordance with the results of the detailed examination, the seismic safety of the structures has been checked by evaluation, in terms of both strength and deformation. For those structures observed to have adequate degree of seismic strength, drawings for bridge repair were prepared and for others observed to be insufficient various alternative retrofit methods were considered. By using such alternatives trial checks were repeated until the structure proved to be seismically safe.

In cases where such retrofit measures technically included great difficulties are proved to be uneconomical, the replacement of the bridge have been proposed as an alternative in a comparative study. Flow chart for the evaluation of bridges under seismic loading is given in Figure 1.

2.2.1 *Bridge classification*

The Code (DLH 2008) has been adopted for the evaluation of existing bridges. Hence following the classification of each bridge as "special bridge", "ordinary bridge", or "simple bridge" the evaluation of the bridges was completed by using the DLH 2008 earthquake levels (such as D1, D2 or D3).

Special Bridges:

- Bridges located on strategic highway sections
- Critical bridges expected to be used immediately after an earthquake

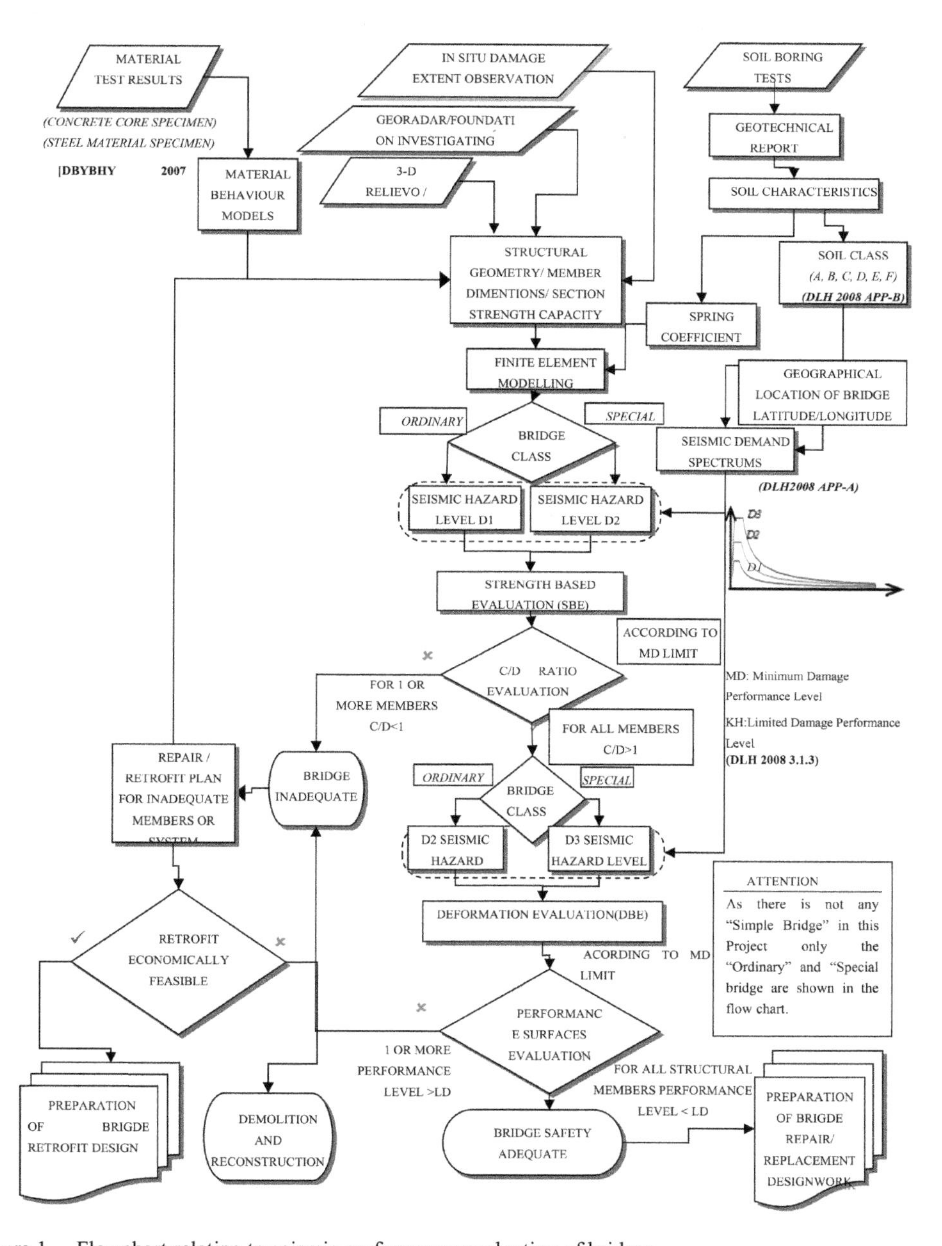

Figure 1. Flowchart relating to seismic performance evaluation of bridges.

Ordinary bridges:

- Bridges which are neither "special" nor "simple"

Simple bridges:

- Single span bridges which are not "special" with span length less than 10 m
- Bridges which are not "special" located on small areas with effective ground acceleration less than 0.1 g

Table 1. Seismic hazard levels (DLH 2008,1.2.1.).

Seismic hazard level	D1	D2	D3
DLH code article	1.2.1.1	1.2.1.2	1.2.1.3
Excedence probability within 50 years	50%	10%	2%
Return period	72 years	475 years	2475 years

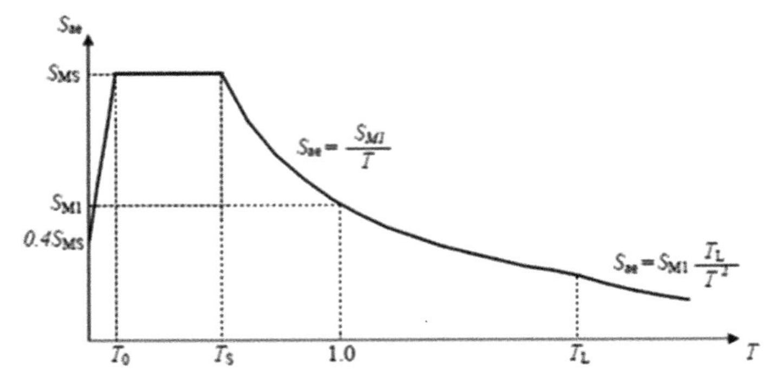

Figure 2. Seismic design response spectrum (DLH 2008 1.2.2.1 & 1.2.2.2).

2.2.2 *Seismic performance spectrum*

The seismic performance spectrum has been prepared by the method depicted in the Code (DLH 2008) Appendix A by using the bridge coordinates and adjusted in accordance with the soil class concerned. In the performance based bridge calculations 3 separate seismic levels were taken into account, namely D1, D2 and D3 levels (DLH 2008 1.2.1) in Table 1 D1 seismic level has the lowest intensity but the highest probability of occurrence, whereas D3 seismic level has the highest intensity with the lowest probability of occurrence. The return periods of D1, D2 and D3 earthquakes are 72 years, 475 years and 2475 years, respectively.

The spectral acceleration values depicted in the Code (DLH 2008) Appendix A have been given for both S_s (short period spectral acceleration) and S_1 (spectral acceleration for 1 second period), soil class B.

The S_s and S_1 values derived from coordinates are multiplied by adjustment coefficients in accordance with DLH 2008 Code by using the soil class and spectral acceleration values. The seismic spectrum has been established from equations given in the Code (DLH 2008). The general form of this spectrum is shown in Figure 2.

2.3 *Methods of detailed evaluation*

Two methods were used in the seismic evaluation of existing bridges: Strength based evaluation (SBE) and deformation based evaluation (DBE) (DLH 2008, 3.1.5). The seismic safety level of the structure was determined by checking the seismic performance of bridge and bridge members derived from analyses performed previously against performance limits given in the Code (DLH 2008). Minimal damage (MD) and limited damage (LD) performance level limits were used in the evaluation. Minimal damage performance level in such that either no damage at all is caused or very limited amount of damage occurs due to seismic action. In this case either traffic flow continues uninterrupted or problems which might arise can easily be removed within a few days (DLH 2008 3.1.3.1). A limited (or controlled) damage performance level (LD) however, is defined as a damage level where damages occurring due to seismic actions are permitted provided that such damages are structurally not very serious and can be repaired. In this case, it is reasonable

Table 2. DLH 2008 evaluation methods with respect to bridge class.

	Essential bridges		Ordinary bridges		Simple bridges
Seismic hazard level	D2	D3	D1	D2	D2
Evaluation method	SBE	DBE	SBE	DBE	SBE
Performance level	MD	LD	MD	LD	LD

Table 3. Pier/abutment behaviour coefficients (DLH 2008, Table 3.3).

	Performance Level	
Load-bearing system of pier	MD	LD
Single column or slender wall in transverse direction (flexure wall – $H/L_w > 3$)	1.5	2.5
Single span or multi span reinforced concrete or steel frame in transverse direction	2.5	5.0
Single span or multi span steel frame with bracing in transverse direction	2.0	3.5
Short wall in transverse direction (shear wall – $H/L_w \leq 3$)	1.5	2.0
Piers behaving as a cantilever in bending in the longitudinal direction	1.5	2.5
Piers behaving integrally with the bridge deck in bending in the longitudinal direction	2.5	4.0

to expect short period interruptions in bridge operations lasting a few days or weeks. (DLH 2008, 3.1.3.2).

Table 2 shows the required evaluation method and the seismic performance level to be used for a given bridge class (DLH 2008).

2.3.1 *Methods of detailed evaluation*

In the strength based evaluation method (SBE), starting with the linear elastic behaviour of the structure, seismic effects are determined and these effects are evaluated by capacity/demand ratio (C/D). The capacity/effect ratio method is used in accordance with the definitions made in FHWA Seismic Retrofit Manual 2006 section 5.4 –method C (FHWA-HRT-06-032 2006).

The surplus capacity is to be taken into account as follows:

$$r = \frac{C - D_g}{D_{EQ}} \tag{1}$$

where, r = capacity ratio; C = section capacity; D_g = effect of vertical loading (except that due to seismic loading); D_{EQ} = effect due to seismic loading.

For all bridge members, the capacity is considered adequate where the capacity/effect ratio(r) is greater than 1, is considered inadequate where this ratio is less than 1.

In the strength based evaluation process, the structural analysis was performed by linear- elastic analysis method by a mathematical model developed from the finite element program SAP2000. The seismic effects thus obtained were divided by seismic load reduction factor, R in order to take into account ductile structural behavior. The seismic load reduction factor (R) was determined by virtue of horizontal load bearing system characteristics (Table 3, DLH 2008 Table 3.3). However in the case of shear force evaluation, the seismic load reduction factor (R) is taken as 1 as this points to a brittle mode of failure; in another words a nonlinear ductile behavior is not applicable in this case. A flowchart for strength based evaluation is shown in Figure 3 in which the seat width is assessed in accordance with FHWA-HRT-06-032, and equation 7-3B of (AASHTO 2002).

2.3.1.1 *Column/wall load carrying capacity due to flexure and shear*

In columns and walls, capacity/demand ratio in two dimensional flexure is determined from a two dimensional bending moment interaction diagram accompanied by a vertical load. The bending

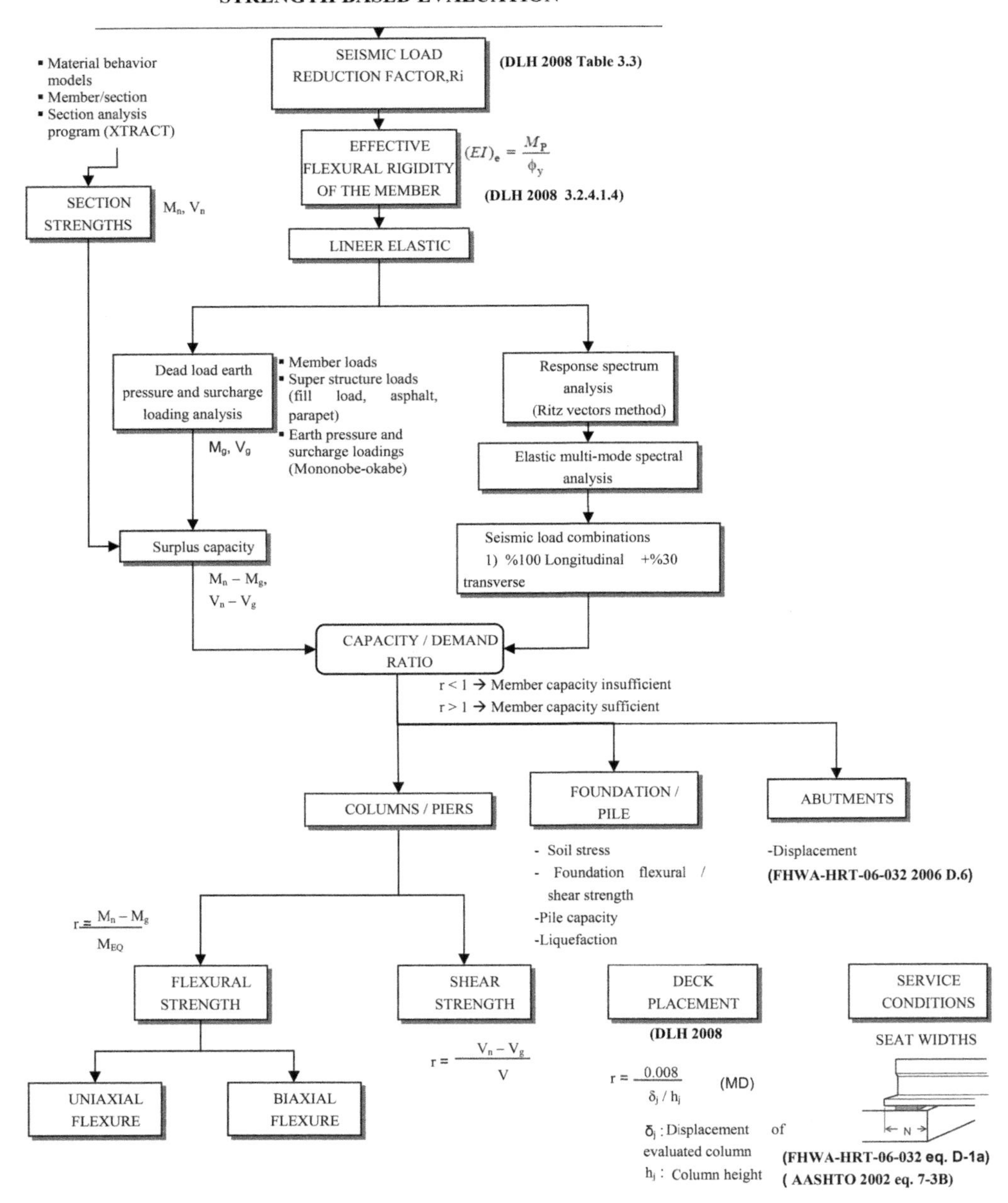

Figure 3. Flowchart for strength based evaluation.

moment effects in the diagram in both directions due to vertical loads are reduced in order to deduce the required reduced capacity. The capacity/demand ratio can be found by dividing the reduced flexural capacity by the resultant of seismic moments. This procedure is explained in Figure 4.

The capacity/demand ratio of bridge columns in shear is determined from the column surplus shear capacity of shear forces derived from the analysis.

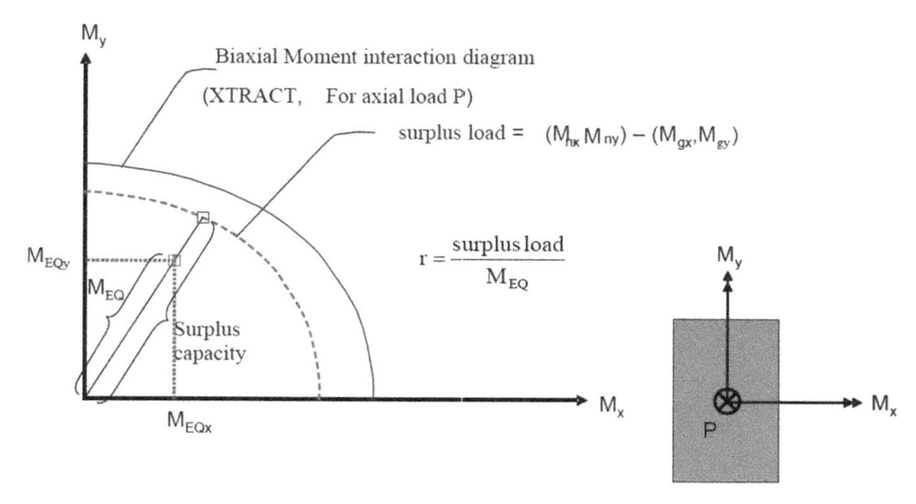

Figure 4. Calculation of column capacity/demand ratio.

2.3.1.2 *Column/wall relative displacement limit*

For columns, relative displacement capacity/demand ratio is evaluated, in addition. The displacement limits at the column deck joint zone is given in the Code (DLH 2008,3.2.3.6).

$$\delta_j = \beta R \Delta_j \tag{2}$$

where, δ_j = displacement which the evaluation be based upon; Δ_j = pier-deck joint displacement; R= behavior performance coefficient (R will be taken as equal to 1 provided that the displacements calculated from the analysis model is not reduced and coefficient $\beta = 1$for $R = 1.5$, otherwise $\beta = 2/3$).

The displacement subject to evaluation so calculated were divided by column heights in order to obtain relative displacements (δ_j/h_j). The relative displacement limits are given as 0.008 for minimal damage and 0.015 for limited (controlled) damage situation (DLH 2008, 3.2.3.6).

The displacement capacity/demand ratio(r) was found as follows;

$$r = \text{limited relative displacement / calculated relative displacement} \tag{3}$$

2.3.1.3 *Seismic load reduction coefficient*

In the course of the evaluation of bridge piers and abutments by the strength evaluation method (SBE) the influence of a possible nonlinear behavior which may be observed at minimum damage level (MD) upon the energy damping is taken into account by reducing seismic forces by certain amounts depending on the type and ductility of the structural member.

For bridges with different types of piers, the bridge load bearing system performance coefficient is calculated from the weighted average of the piers concerned. These coefficients are given in the preceding 3 (DLH 2008, 3.2.3.2.1).

Where $R \leq 1.5$, seismic load reduction factor is taken as $R_a(T) = R$. For cases where $R > 1.5$ however, $R_a(T)$ is determined based on the natural vibration period T as $1.5 + (R - 1.5) \times (T/T_S)$ (DLH 2008 3.2.3.2.2), where T is the first natural vibration period of the bridge load bearing system and T_s is the spectrum corner period.

2.3.2 *Deformation based evaluation*

The deformation based evaluation method, although much more complicated and tedious than the linear- elastic calculation method, is regarded as more realistic since it takes into account the nonlinear behaviour arising when the structural members and their connections approach, reach or exceed their strength capacity. In this method, materials and connections are defined by nonlinear stress-deformation relationships by defining the nonlinear material behaviour models taken from the related code (DBYBHY 2007 App.-7) and the relevant calculations are performed in accordance with the structural displacements. This approach is also known as "performance based evaluation". The calculated values of material deformation in the members derived from seismic analysis are compared against performance limits. The performance limits to be used were taken from the appropriate sections of the code (DLH 2008). It is envisaged to use limited damage (LD) performance limit for the deformation based evaluation method.

The nonlinear behavior of the members is defined as frame members concentrated in plastic hinge formation points, whereas for the wall (plate) members it is defined by sections of nonlinear layers. Plastic hinge formation point is a concept widely used in nonlinear analysis and is frequently named as plastic hinge. The plastic behaviour which is expected to occur at the end sections of frame members where the maximum forces may be observed, is determined by the section yield surface at the particular point. The yield surface defines the normal force P representing the start of hinging and bending moments M_1 and M_2 at both perpendicular axes. This relationship can be observed by various methods including section moment-curvature analysis or section fiber analysis. In the two-dimensional analysis, the yield plane changes to simply to yield curve related to normal force P and bending moment M. In this study the yield plane and hence the plastic hinge definition has been automatically defined in accordance with FEMA-356 (FEMA356 2000) hinges defined in SAP2000 program and section characteristics. The sections where plastic hinges would occur are determined to be at the ends of members expected to have plastic hinges taken to be at the midpoint of the calculated plastic hinge length. The plastic hinge lengths were calculated from the equation given below (FHWA-HRT-06-032 2006 7.8.1.1 & DLH 2008 3.2.4.1.1):

Column/wall plastic hinge length:

$$L_p = 0.08H + 0.0022 f_{yk} d_b \geq 0.044 f_{yk} d_b \tag{4}$$

Beam and pile plastic hinge length:

$$L_p = 0.5H \tag{5}$$

where, H(mm) = pier height; f_{yk} (Mpa) = characteristic yield stress for the reinforcing bar; d_b = diameter of bar and h (mm) = cross-section dimension in the direction considered.

2.3.2.1 *Analysis methods for deformation evaluation*

Incremental pushover analysis can be performed for deformation evaluation, provided that the first (prevalent) vibration mode effective mass participation ratio is greater than 70 percent (DLH 2008 4.4.4). In cases where this condition is not satisfied or for complicated and/or curved bridges requiring a more comprehensive study, a nonlinear time history analysis method has been employed. The incremental pushover analysis method has been used in calculations in accordance with DLH 2008 Section 3.2.4. A minimum of 7 sets of seismic records were employed which represent design earthquake spectrum for the time history analysis. The average values of internal forces, displacements and deformation effect, derived from the time history analysis result based on the 7 earthquake records were used in the evaluation.

2.3.2.2 *The evaluation of the member deformations due to seismic load*

The unit deformation limits for the respective performance limit can be taken from Table 4 given below (DLH 2008, Table 3.4).

Table 4. Unit deformation limits defined for pier plastic section (DLH 2008 Table 3.4).

Unit deformation	Performance level	
	MD	LD
Unit deformation of concrete in compression, ε_c	0,004	0,020
Unit deformation of reinforcing steel, ε_s	0,010	0,040

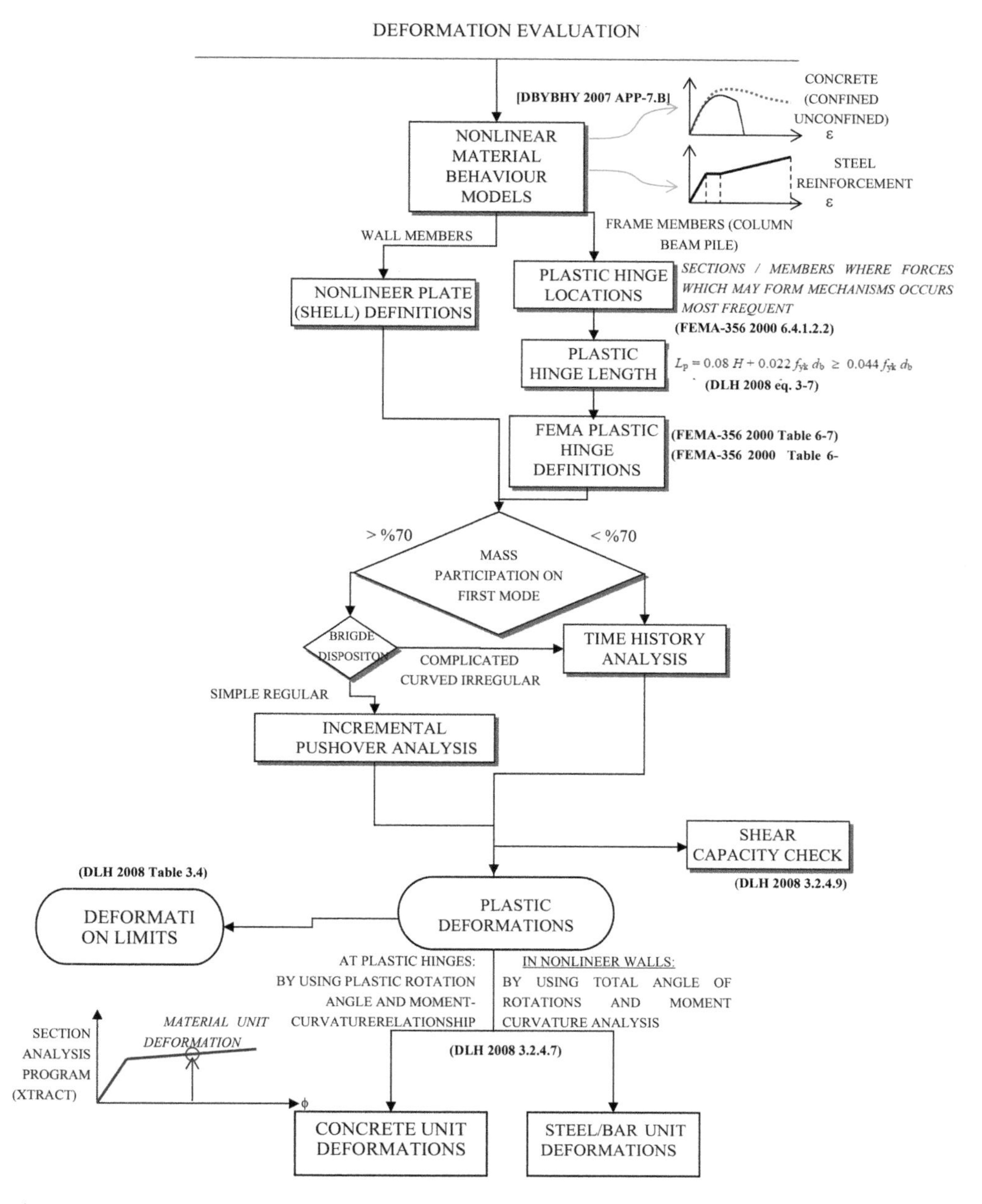

Figure 5. Flow chart for deformation based evaluation.

Figure 6. General view of Üsküdar-Haydarpaşa Overpass Bridge.

Figure 7. Extraction of concrete core specimens for material test (Üsküdar-Haydarpaşa Overpass Bridge).

The unit deformations given in the table have been transformed into the equivalent curvature and/or notation values by using the section moment-curvature (M-K) analysis.

$$LD = 0.004H + 0.016\frac{\rho_s}{\rho_{sm}} \tag{6}$$

where, ρ_s = transverse reinforcement ratio of the column; ρ_{sm} = minimum transverse reinforcement ratio.

Deformation based evaluation flow chart depicted in Figure 5. As an example in Figures 6, 7, 8 and 9 general view of the bridge, photos from initial in situ works and the detailed evaluation results is given of Üsküdar Haydarpaşa Overpass Bridge.

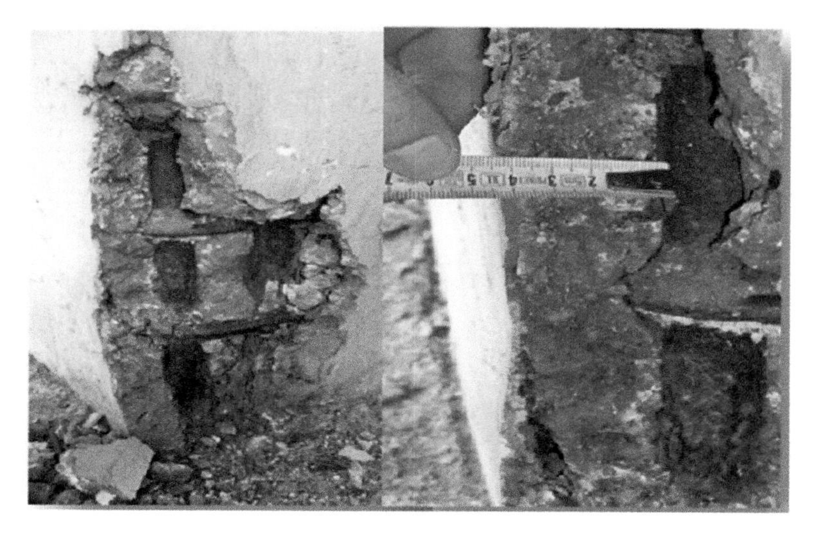

Figure 8. Measurement of existing steel rebar diameters (Üsküdar-Haydarpaşa Overpass Bridge).

- STRENGTH BASED EVALUATION
  Linear Elastic Spectral Analysis
  - Flexural moment
    - *Pier column bottom end: INSUFFICIENT*
    - *Pier column top end: INSUFFICIENT*
  - SHEAR FORCE
    - *Pier column bottom end: INSUFFICIENT*
    - *Pier column top end: INSUFFICIENT*
  - DECK DISPLACEMENT
    - *INSUFFICIENT*
  - FOUNDATION CHECKS
    - *SUFFICIENT*
- DEFORMATION BASED ANALYSIS
  NONLINEAR ANALYSIS
  - MATERIAL DEFORMATION LIMITS(COLUMNS)
    - *concrete: INSUFFICIENT*
    - *reinforcement: INSUFFICIENT*

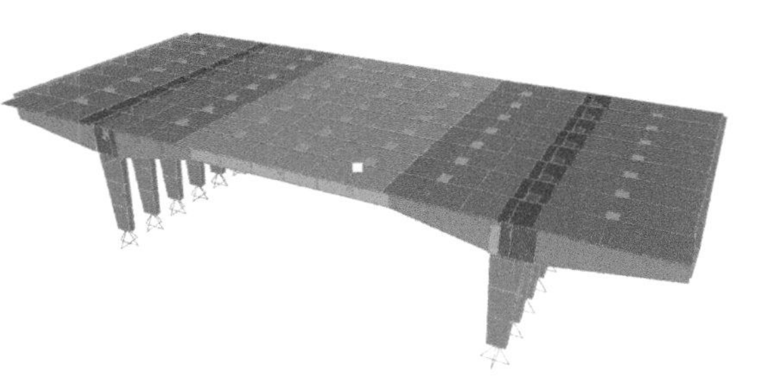

Figure 9. SAP2000 mathematical model and detailed evaluation results (Üsküdar-Haydarpaşa Overpass Bridge).

3 CONCLUSIONS

In this paper, with reference to a particular project commissioned by İstanbul Metropolitan Municipality (İBB) which involved the inspection, evaluation and seismic retrofit design of various existing bridges, the seismic performance of bridges have been studied and the employed calculations methods are explained. In this study, principles of "DLH Seismic Design Guidelines, General Directorate for Construction for Railways Harbours and Airports of (RHA) Ministry of Transportation (DLH 2008)" have been taken into consideration. Within the scope of this Project, the studies involved strength based evaluation response spectrum method and on the other hand in the case of deformation based evaluation pushover analysis and nonlinear time history analysis have been performed. In the case of nonlinear analysis, frame and finite element models have been used. By taking into account the particular bridge class and the expected performance level corresponding to this class, the seismic performance of bridges studied have been evaluated and those structures which exhibited insufficient level of seismic safety have been retrofitted. In conclusion, considering the high seismicity of our country such evaluations and studies should be generalized in order to minimize the unenviable results of the earthquake effects.

REFERENCES

AASHTO (2002) Standard Specifications for Highway Bridges. American Association of State Highway and Transportation Officials, 17th Edition.

DBYBHY (2007) Specification for Buildings to be Built in Seismic Zones, (in Turkish) Ministry of Public Works and Settlement, Ankara.

DLH (2008) Seismic Design Guidelines, (in Turkish) General Directorate for Construction Railways Harbors and Airports of (RHA) Ministry of Transportation, Republic of Turkey.

FEMA356 (2000) Prestandard and Commentary for the Seismic Rehabilitation of Buildings, American Society of Civil Engineers.

FHWA-HRT-06-032 (2006) Seismic Retrofit Manual for Highway Structures: Part 1 – Bridges, U.S. Department of Transportation Federal Highway Administration.

Chapter 15

Effect of thermal loading on the performance of horizontally curved I-girder bridges

G.W. William
AECOM, Morgantown, West Virginia, USA

S.N. Shoukry & K.C. McBride
West Virginia University, Morgantown, West Virginia, USA

ABSTRACT: This paper focuses on the response of horizontally curved steel I-girder bridges to changing thermal conditions. Bridge curvature complicates the structure's response to thermal loading as the bearing configuration must be able to handle expansion and contraction in the transverse, or radial, direction. Failure to properly design bridge bearings to accommodate thermal loads will lead to unaccounted for deformations and stresses in the superstructure. This paper presents a case study of the Buffalo Creek bridge structure subjected to changing thermal conditions prior to any in-service loading. Two detailed 3D finite element models of the bridge were created, one modeling the piers as rigid members and one modeling the piers as flexible members, and both were subjected to uniform temperature increase and decrease. Results indicate that uniform thermal loading leads to lateral displacement along the I-girder web centerlines, lateral distortional buckling in the web cross section, and thermal stresses in the I-girder webs. Although pier flexibility is shown to reduce the magnitude of thermally induced local and lateral distortional buckling and thermal stresses, I-girders experience larger lateral displacements when the piers are flexible. Additionally, even the introduction of pier flexibility does not relieve all thermal stresses in the I-girder webs. At some locations, when the piers are rigid, the I-girder stresses exceed the AASHTO web bend-buckling capacity as well as the overall stress capacity of the section. In both the cases of flexible or rigid piers, this study shows that uniform thermal loading will lead to increase out-of-plane web deformations and increased web stress levels, which will both combine to decrease the load carrying capacity of the bridge when subject to subsequent live-loading conditions.

1 INTRODUCTION

The number of curved bridge structures constructed in the United States has steadily risen over the past several decades. As of 2004, over one-third of all steel superstructure bridges constructed were curved (Davidson et al., 2004). Curved bridge popularity experienced a boom since a curved bridge can offer the designer solutions to complicated geometrical limitations or site irregularities as compared to traditional straight bridges. Additionally, as the use of high-performance steel has become more prevalent, engineers have become able to design more complicated structures as the girder can handle greater loads (Linzell et al. 2004). A previous alternative to constructing a bridge using a curved girder section was to use a chorded structure composed of a series of straight girder sections oriented in a curve to produce a curved bridge. However, using curved girder sections provides aesthetic as well as cost benefits over these traditional chorded structures.

Studies have been conducted for quite some time on the behavior of curved beams, but research on the analysis and design of horizontally curved bridges in the United States began in 1969 when the FHWA formed the Consortium of University Research Teams (CURT) whose work resulted in the initial development of working stress design criteria and tentative design specifications. This work, along with most of the research conducted prior to 1976, was gathered by The American Society of Civil Engineers and the American Association of State Highway and Transportation Officials and compiled into recommendations for the design of curved I-girder bridges (Armstrong 1977). Later, Load Factor Design criteria was developed out of the work of Stegmann and Galambos (1976) and Galambos (1978) as well as the working stress design criteria in the first set of Guide Specifications for Horizontally Curved Highway Bridges (AASHTO, 1980). In 1992, the Federal Highway Administration (FHWA) initiated the Curved Steel Bridge Research Project (CSBRP) as a large scale experimental and analytical program aimed at developing new, rational guidelines for horizontally curved steel bridges. This work resulted in the Guide Specifications for Horizontally Curved Highway Bridge (AASHTO, 2003).

The use of curved girders adds complexities in the bridge design, construction, and analysis that are not present when straight members are used (DeSantiago *et al.* 2005). I-beams are designed to primarily carry vertical bending loads and do not perform well when lateral loading or torsion is placed on the member. However, curved bridges experience torsion and lateral forces under normal loading conditions which will affect the stability of the I-girders (Zhang *et al.* 2005, Kim *et al.* 2007, Fasl et al. 2015). Additionally, much more care must be taken in designing the erection procedures for a curved I-girder bridge because curved steel members will experience lateral deflections in addition to vertical deflections under gravity loading. (Most problems that have occurred with curved girder bridges have been related to fabrication and assembly procedures or unanticipated or unaccounted for deformations that occur during construction (Grubb et al., 1996). In the curved I-girder bridge system, nonuniform torsion results in warping normal stresses in the flanges. Also, because of torsion, the diaphragms or cross frames, or both, become primary load-carrying members (Davidson and Yoo, 2003). While cross frames are secondary members in straight girder systems, they are designed as main members on curved bridges because they stabilize the girders and redistribute the loads. In general, the simple addition of curvature to a bridge system leads to structural intricacies that do not exist in straight bridges.

Presently, curved I-girder bridge design procedures treat thermal loading as a secondary loading. AASHTO LRFD Guide Specifications (2017) specify that for metal structures a range of temperatures from 0 to 120°F (−17.8 to 48.9°C) should be considered. In addition, AASHTO Guide Specifications (2017) state that the load effects due to a temperature gradient across the superstructure depth shall be added to the uniform temperature effects. AASHTO Specifications (2017) acknowledge that although temperature changes in a bridge do not occur uniformly, bridges are usually designed for an assumed uniform temperature change. An assumption that is often made is that the bearing orientation on a curved bridge is such that as thermal expansion and contraction occurs, the bridge is allowed to move freely along rays emanating from a fixed point, causing the thermal forces to be zero. This presumes that the conditions at the bearings act precisely as designed, that the temperature change is in fact uniform, and the constraints of the concrete deck have no effect on the expansion and contraction of the girders.

The concept of thermal loading on horizontally curved I-girder bridges is a topic which has received very little attention. In contrast to straight bridges, thermal effects will be greater on curved structures because the thermal expansion and contraction will invoke both longitudinal and transverse responses, as compared to the primarily longitudinal response for straight bridges. Thus, curvature will likely cause temperature conditions to have an impact on cross member forces, cross section buckling, girder load carrying capacity, and cross member fitting, just to name a few. It is evident investigations must be performed to study the impact, if any, changing environmental conditions will have on the behavior and performance of curved I-girder bridges.

The main focus of this research is to develop a 3D finite element model of an existing curved I-girder bridge which accurately replicates the behavior of the structure and employ it to study the influence of uniform thermal loading on I-girder web distortions, both longitudinally and through the web depth, at the stage just after the completion of construction but before any live loading is placed on the bridge and how these thermal deformations might impact the performance of the curved I-girder bridge.

2 BUFFALO CREEK BRIDGE

Buffalo Creek Bridge is in Logan County, West Virginia and carries WV Route 10 over Old WV10, Buffalo Creek, and CSX Railroad as shown in Figure 1. The bridge was originally constructed in 2001. This 4-span bridge is a consists of eight steel curved steel girder and one stub girder to accommodate a turning lane at the north end of the bridge topped with a composite concrete deck. The bridge has a 7.5% cross slope as shown in Figure 2. The bridges is supported over three piers, an integral abutment at south end, and semi-integral abutment at the north end. The structure was designed for HS-25 loading using Load Factor Method in accordance with the AASHTO Standard Specifications for Roads and Bridges (1996) including all Interim Specifications through 1999, and Guide Specifications for Horizontally Curved Bridges (1993). Only the deck was designed utilizing an empirical deck with a specialized concrete overlay, designed in accordance with the Load and Resistance Factor Design method. Originally, the bridge was designed sandwich concrete deck consisting of 165-mm thick reinforced concrete substrate overlaid with a 50-mm thick micro-silica modified concrete. In October 2004, full-depth longitudinal cracks along the edges of the interior main girders were observed in the deck surface throughout the structure along with heavy map cracking in spans No. 1, 2 and 4. As a result, the deck was replaced in 2007 with a 215-mm (8.5-inch) full-depth monolithic Class H concrete deck with closely spaced steel reinforcement (Shoukry et al. 2010).

Figure 1. Buffalo Creek Bridge.

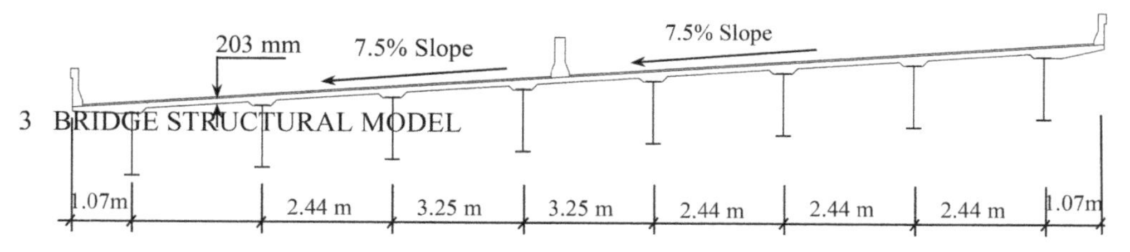

Figure 2. Bridge cross section.

3 BRIDGE STRUCTURAL MODEL

3 BRIDGE STRUCTURAL MODEL

A detailed 3D finite element model was developed for the Buffalo Creek Bridge using ADINA finite element software that accounts for large deformations and geometric nonlinearity. Owing to the curvature and varying elevations of the bridge, an FE model of the full structure was created to reproduce the exact geometry of the bridge. The initial challenge in creating the Buffalo Creek bridge model was reproducing the complicated girder geometry as shown in Figure 3. Each of the 8 girders is steel, curved, I-beam girder with stiffeners along their length, with three girders (girders 3, 4, and 5) having a uniform radius of curvature and five girders (girders 1, 2, 6, 7, and 8) having varying radii of curvature. In addition to the bridge curvature, the structure has varying elevation changes (cross slopes) along both the radial and transverse directions.

This study consists of two major categories of FE models: those without the full piers modeled and those with the piers modeled. Figure 3(a) shows a full model of the Buffalo Creek bridge, complete with all three piers modeled. The piers were modeled using eight-node shell elements with the bearings represented using a series of spring elements.

Because of the large size of the model, every effort was made to minimize the computing cost associated with solving the model while maintaining the accuracy and usefulness of the results. Therefore, eight-node shell elements were used to model most of the bridge components including the girders, cross diaphragms at piers, abutment walls, piers, and deck. Most of the area of the bridge girders is modeled using elements with lengths of approximately 0.90 m. However, the element mesh is refined to a maximum length of 0.15 m. on the girders at the abutment, pier, and mid-span locations allowing a more thorough investigation of behavior in these areas. The sizes of the elements modeling the abutment wall vary and are chosen so that the ends of the girders and the abutment share nodal points, creating a rigid connection between the girders and the abutment. The main structure of the bridge piers is discretized as elements no larger than 0.30 m on any edge. However, element subdivisions of the girder haunches are adjusted to allow a one-to-one node connection between the piers and the girders.

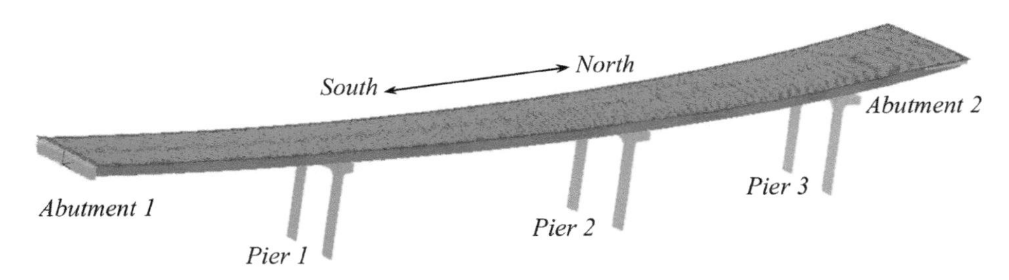

(a) Full FE model of Buffalo Creek Bridge

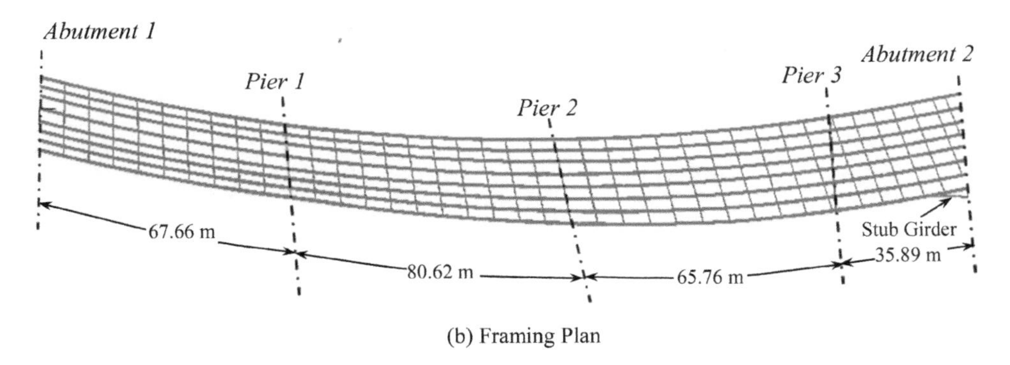

(b) Framing Plan

Figure 3. FE model of Bufffalo Creek Briddge.

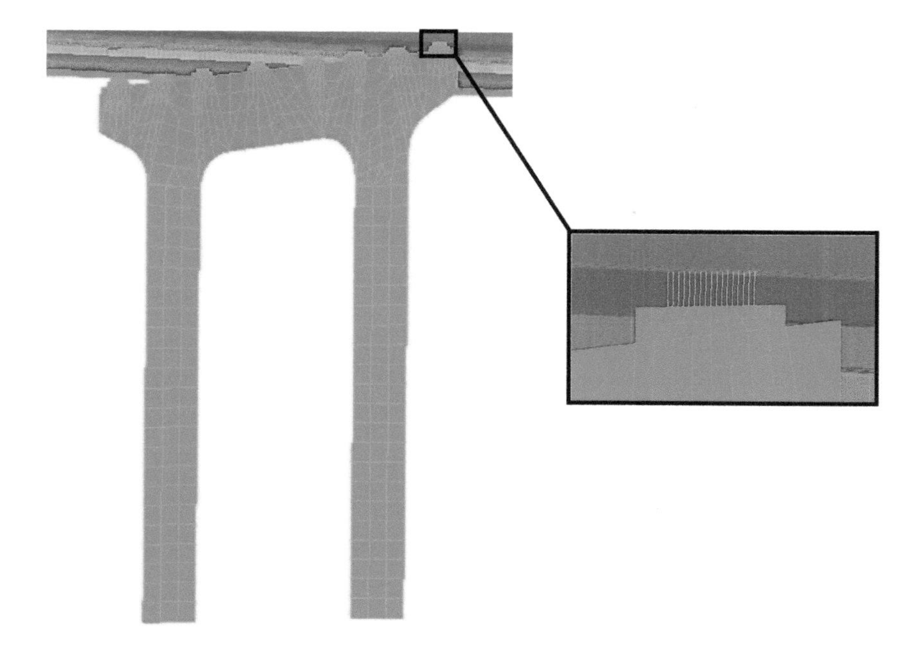

Figure 4. FE model of Buffalo Creek Bridge Pier 1.

The bridge deck is idealized as a multilayered shell element to accurately represent the separate layers of reinforcement. Use of the multilayered shell element allows specifying an arbitrary number of individual layers to make up the thickness of the shell with each layer assigned different material properties. The multilayered shell element is a useful and efficient way to model a composite section without having to use a bulky 3D element. Being as the reinforced deck is a composite section, the properties of each layer of the deck are computed and used to form the multilayer shell. Shell elements making up the bridge deck are specified as having a maximum length of 0.90 m.

The cross frames and diaphragms consist of two horizontal members (top and bottom) and two diagonal members. For the intermediate cross members, the top and diagonal members are of type L4 × 3 × 3/8 and the bottom horizontal members are L4 × 3 × 1/2. The top and diagonal members at each pier are L6 × 4 × 1/2 sections and the bottom member is an I-beam of type W24 × 162 with two stiffeners on each side of the web. The stresses within the cross members are not of paramount concern for this study, so most of the cross members are discretized as Hermitian beam elements. All members except the bottom members at the pier are represented using beam elements by simply specifying the beam cross-sectional properties and orientation. Bottom cross member sections at the piers cannot be represented in this way because of the presence of the bearing stiffeners. Consequently, the geometry of these members is replicated using eight-node shell elements.

To further understand the behavior of the Buffalo Creek Bridge, it was deemed necessary to investigate how introducing pier flexibility will affect the behavior of the bridge superstructure. The concrete pier columns were modeled using 8-node shell elements with every effort made to exactly replicate the pier geometry. Figure 4 has a depiction of the model of Pier 1. The columns are assumed to be fixed at their bottom. To accurately simulate the bearings over the piers, the top surface of the pier cap and the girder bottom flange (at the locations of the piers) have their surfaces subdivided in such a way that there is one node on the top surface of the pier cap exactly 0.20 m, representing the elastomeric bearing height, directly below each node on the girder bottom surface. These corresponding girder and pier nodes are tied together with a series of springs intended to mimic the behavior of the bridge bearings. To simulate fixed bearings, six springs are modeled

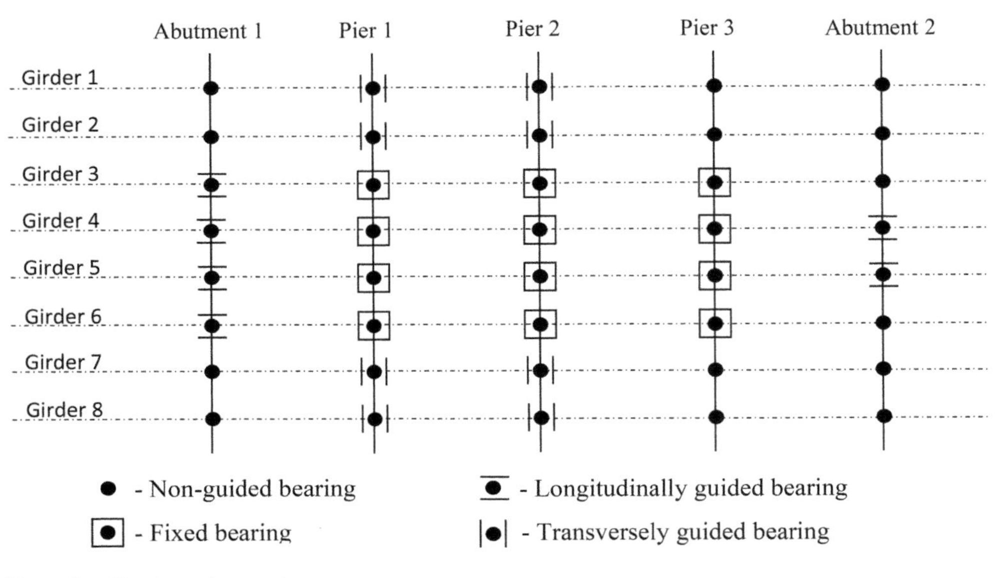

Figure 5. Bearings at supports.

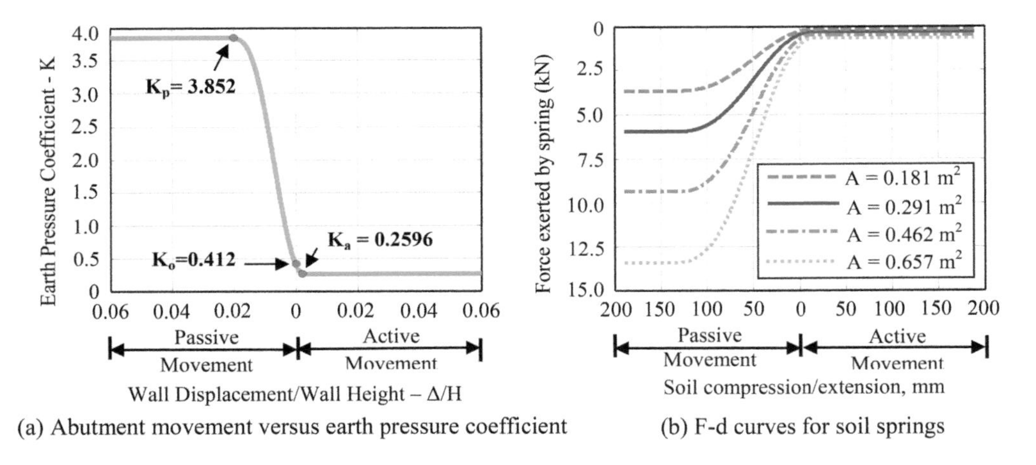

(a) Abutment movement versus earth pressure coefficient (b) F-d curves for soil springs

Figure 6. Force-displacement relations for soil springs.

between each node, connecting the x-, y-, and z-translations and rotations. The stiffness of the springs is set at an extremely high value ($k = 1 \times 10^{21}$) which allows the spring to model a tied connection between the specified degrees of freedom. For the other types of bearings, springs are modeled between the pier haunch and girder nodes connecting the appropriate degrees of freedom to simulate each desired bearing type shown in Figure 5.

In bridge design, it is assumed that the shear studs attached to the top flange of bridge girders will provide full composite action between the top flange of the bridge girders and the bottom surface of the concrete deck. Full composite action implies that under any given loading condition the displacement at locations where the deck and girder top flange come in contact are equal. For this study, full composite action is assumed between the girders and deck.

The south abutment (Abutment 1) is constructed as a semi-integral abutment as shown in Figure 7. Thus, the abutment wall is not continuous from the bridge deck to the support piles as in integral

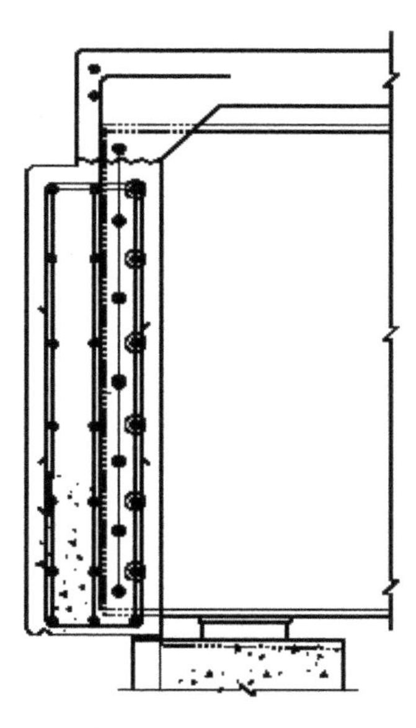

Figure 7. Typical semi-integral abutment detail.

abutment design. Top portion of the wall, end diaphragm, incases the girder ends, and is supported on bearings on abutment stem. The wall of Abutment 1 is modeled as rigidly connected to the girder ends with the appropriate bearings modeled at each girder. The backfill behind abutment wall and the soil surrounding the piles are represented by a series of nonlinear spring elements. The stiffness of such springs is defined by the National Cooperative Highway Research Program design curves (Barker et al. 1991) as shown in Figure 6(a). The force in each spring varies from active to passive earth pressure depending on the wall displacement at its location. In this study, the unit weight of the backfill was found to be 18 kN/m^3 and the angle of internal friction $\phi = 39°$.

4 MODEL RESULTS

The steel superstructure (girders and cross members) of the Buffalo Creek Bridge were subjected to gravity loading. The resulting vertical girder deflections were compared with the deflection values in the bridge design sheets. Figure 8 illustrates such a comparison for Girder 2 as an example. The deflected shape under steel superstructure gravity loading predicted by the FE analysis is in a good agreement with the deflection values given in the girder camber tables. Figure 9 illustrates a comparison of the deflections of Girder 2 due to the weight of the steel superstructure and concrete deck. The greatest differences in the sets of deflection values occur roughly at the middle of each span where the vertical span deflections are the largest. The percent error between the two sets of deflection values defined as the percentage of the absolute error to the camber value was used to determine how the two sets of deflection are close to each other and shown in Figures 8 and 9. The percent errors are less than 5% in Spans 1 and 2, less than 10% in Span 3 and increase to higher values in Span 4. This can mainly be attributed to the deflections at these locations being very small, but all the absolute error values at mid-span 4 are less than 2.5 mm. As a result, even though the percent error values at midspan 4 are large, the vertical girder deflection values due to deck and steel superstructure loading predicted by the FE analysis are acceptable.

The webs of I-girder bridges are designed such that their main function is to resist shear forces while maintaining the relative distance between the top and bottom flanges under all anticipated

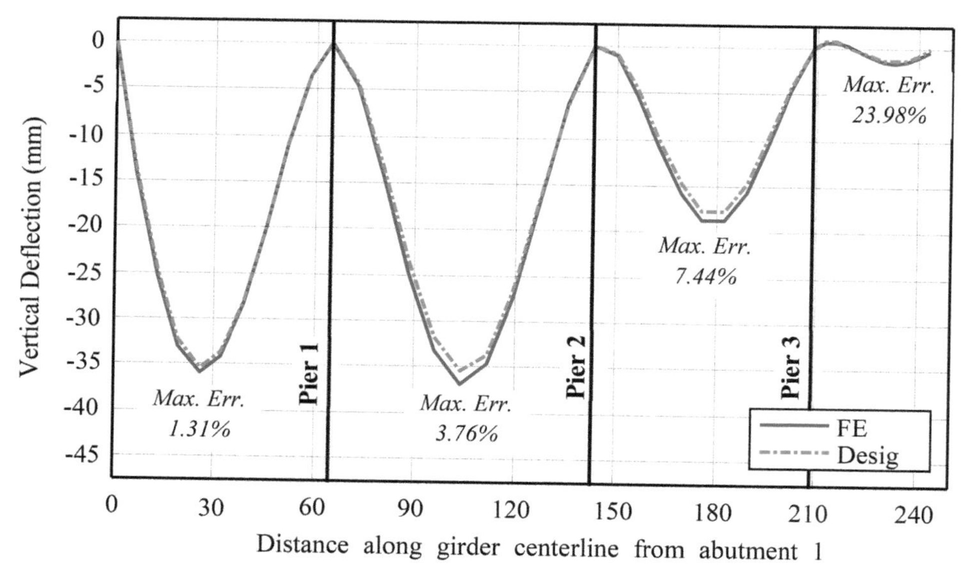

Figure 8. FE and design deflections of girder 2 due to steel self-weight.

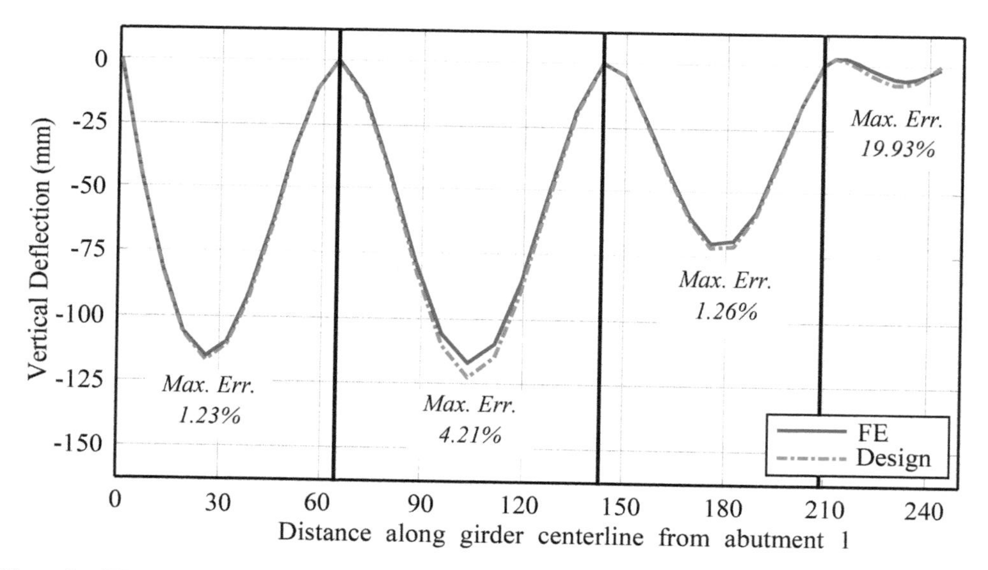

Figure 9. FE and desig deflection of girder 2 due to steel superstructure and deck.

loading conditions. Top and bottom flange sections of the I-girders are designed to carry most of the loading placed on the structure. Out-of-plane web deformations occur because of deflection and have a negative effect on the loading carrying capacity of bridges constructed using I-beams. I-girder out-of-plane distortions most often come in the form of lateral distortional buckling which is the combination of the local buckling and lateral buckling modes as depicted in Figure 10.

In curved structures, the curvature of the girder introduces a torsional component to the structural response, even under simple self-weight loads, that is not present in straight girder bridges. Figure 11

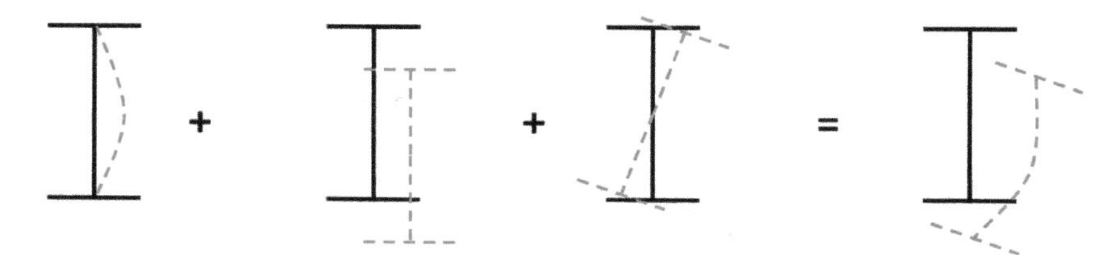

Figure 10. Buckliing modes of I-girders.

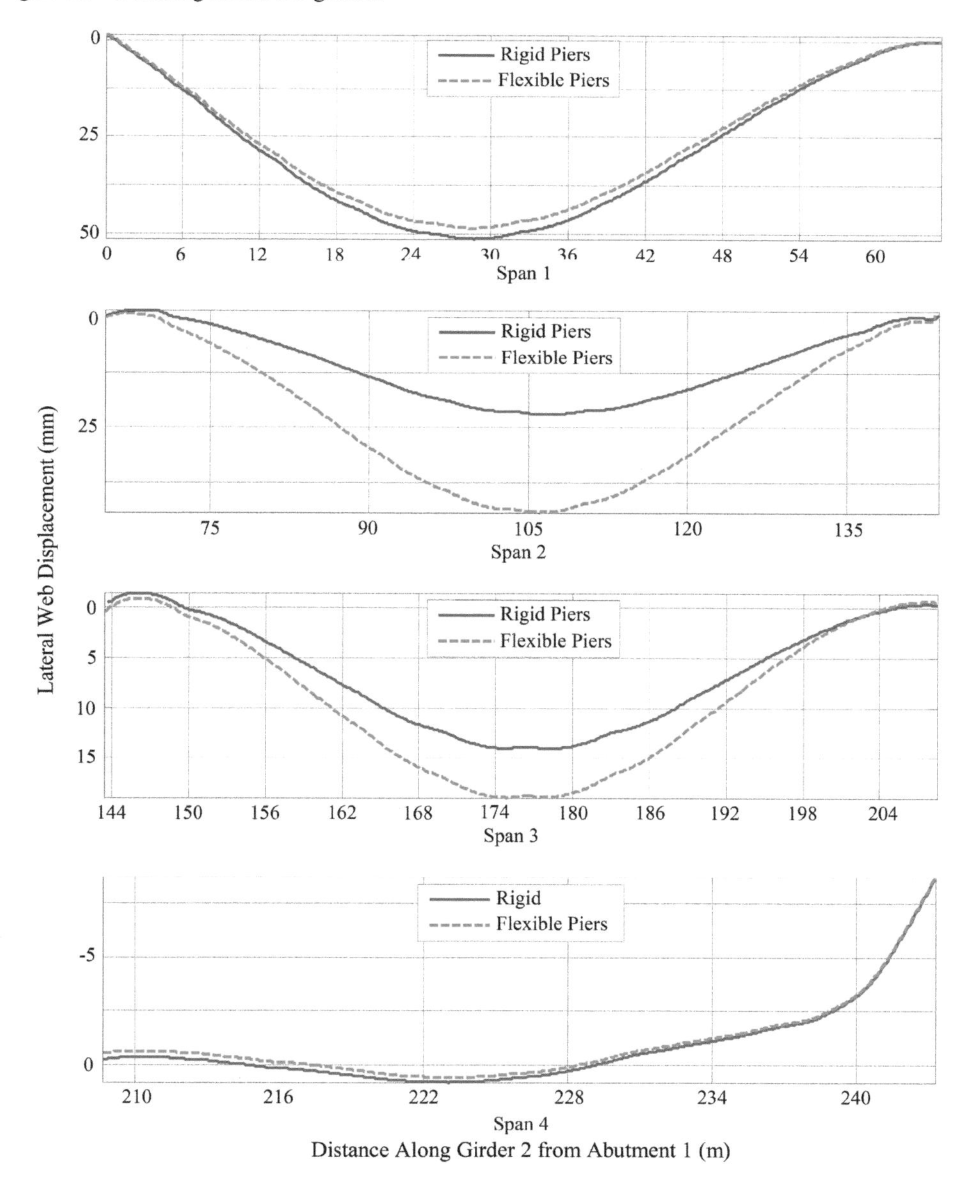

Figure 11. Girder 2 lateral web centerline displacement due to gravity loads.

presents the lateral displacement at the web centerline of Girder 2 subjected to gravity loading from steel superstructure and concrete deck, which indicates that I-girder webs experience global lateral displacements. The difference in magnitude of gravity induced global lateral web displacement between cases with rigid and flexible piers is evident in Figure 11. The largest disparity between displacements occurs in Spans 2 and 3, where one or both of ends of the span are supported by bearings designed to restrict longitudinal movement along the girder centerline. Although the bearing itself is designed to eliminate girder displacement at the pier, the flexibility of the piers will allow the girders to displace at these locations. This additional movement at the supports when modeling the piers as flexible members leads to an increase in the overall lateral deformation of the I-girders in these spans compared to when the piers are modeled as rigid members.

The presence of lateral displacements in the curved I-girder cross section prior to placing the bridge in-service is likely to lead to increased levels of girder buckling once in-service loads are introduced. The out-of-plane displacement will lead to subsequent in-service loading not being applied through the centroid of the I-girder cross section; thus, further increasing the out-of-plane distortion of the cross section. Although these initial out-of-plane distortions may not correspond to concerning high levels of stresses in the cross section and it is well known that steel girders maintain a certain level of post buckling strength after some initial buckling, the increase in out-of-plane distortion of the cross section will decrease the load carrying capacity of the I-girder (Kala *et. al.* 2005 and White and Jung 2007). According to the Bridge Welding Code (2007), sweep deviations are horizontal displacements from a perfectly straight (in this case curved) alignment. Section C-3.5.1.4 also states that most bridge members are flexible and allow some lateral adjustment during erection without damage. However, the finite element results plotted in Figure 11 show that gravity loading on the structure results in sweep deviations in the bridge girders of over 2 in. in span 1 of the Buffalo Creek Bridge.

Out-of-plane web deformation is defined as the I-girder web deforming laterally from the plane created between the top flange and bottom flange centerline and is a form of local buckling as shown in Figure 10. Figure 12 shows that even during the early stages of construction, out-of-plane web deformation is occurring about the web centerline of the curved I-girders. Although there is out-of-plane deformation present, the magnitudes of these deformations are small, with the maximum value of 1.1 mm in Span 2, which is only 8% of the web thickness. Deflections of this magnitude likely will not have any effect on the capacity of the structure. However, it should be noted that the deflections shown in Figures 11 and 12 are occurring without consideration of any initial imperfections due to fabrication, lateral forces, or thermal forces. The consideration of any small initial imperfections that are sure to arise during fabrication would almost certainly increase the magnitude of the lateral web deflections. Although these displacement levels do not appear to be concerning, the presence of out-of-plane displacement and associated reduction in initial web stiffness under this initial loading could lead to reductions in I-girder load carrying capacity in later stages of the structure's life.

The out-of-plane deformation profiles plotted in Figure 12 show that modeling the bridge piers as rigid or flexible members doesn't have a significant impact on the out-of-plane web displacement when loaded with gravity. The plots also indicate that there are regions along the length of each span, especially spans 1, 2, and 3, where the magnitude of the out-of-plane displacement is significantly less than the rest of the span. Cross referencing with the design sheets reveals that these areas of lesser out-of-plane displacement correspond to locations where intermediatestiffeners are present on the girder webs in between the cross members. Intermediate stiffeners on the girders obviously minimize the magnitude of out-of-plane displacement caused by gravity loading. Next, the plots show that the out-of-plane web displacement behavior of each span is different near the piers compared to in the center of the span. Under gravity loading, the girders will be subjected to a negative bending moment at the pier locations and a positive bending moment at the midspan. It appears from Figure 12 that I-girder sections under positive and negative bending moment experience out-of-plane web displacement in opposite directions. Finally, these out-of-plane web displacements in opposite directions at the piers and mid-spans indicate that the girders are experiencing a degree of longitudinal buckling under self-weight loading.

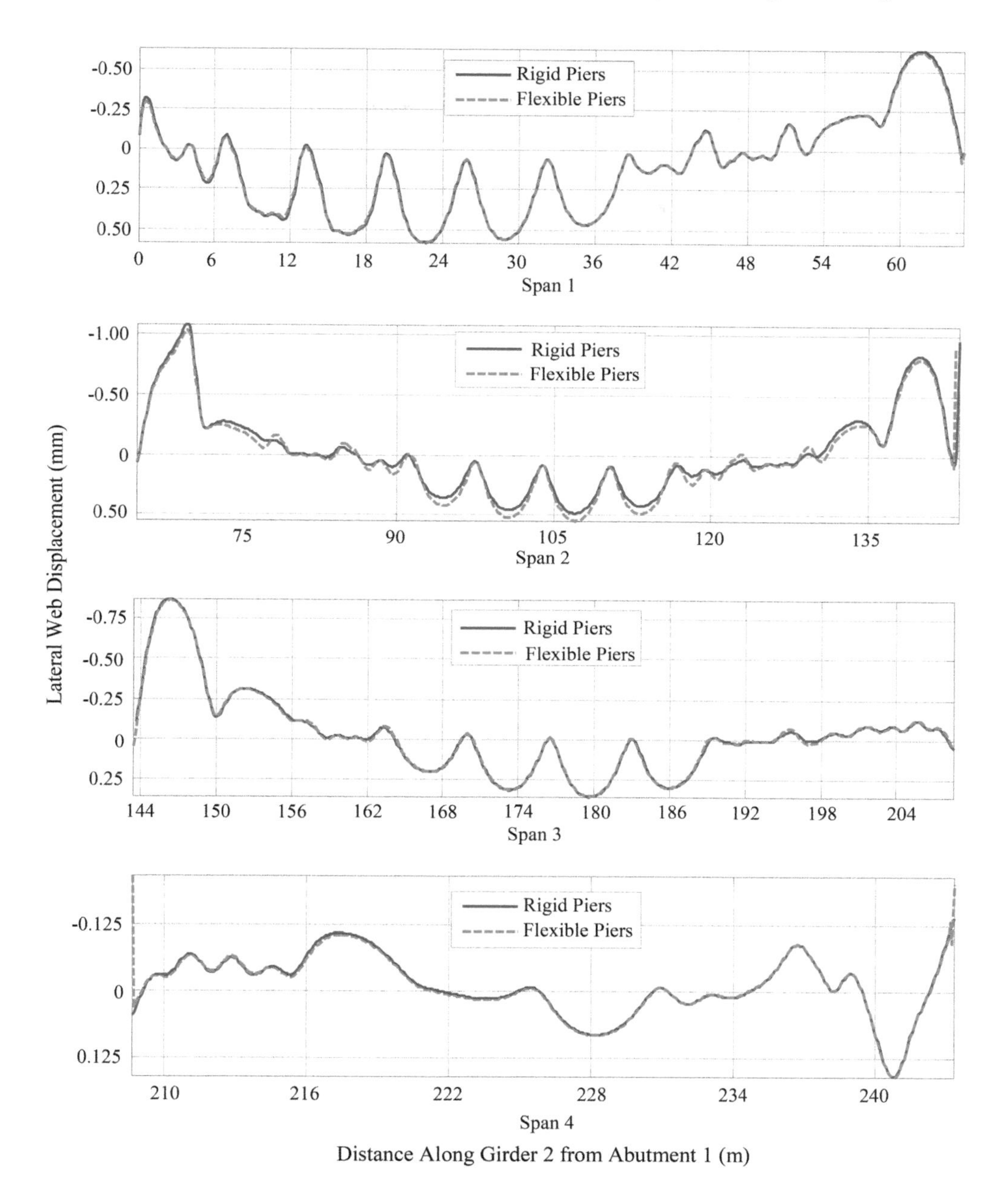

Figure 12. Girder 2 out-of-plane web deformation due to gravity loads.

5 EFFECT OF TEMPERATURE VARIATIONS

To study the effect of thermal loading, the behavior of the bridge once −25°C and +25°C uniform temperature variations are introduced. Thermal loading is added to the model after the structure has displaced under self-weight and the deck has fully cured, allowing it to contribute its full stiffness to the system. The lateral displacement profiles of girder web centerlines when subject to gravity

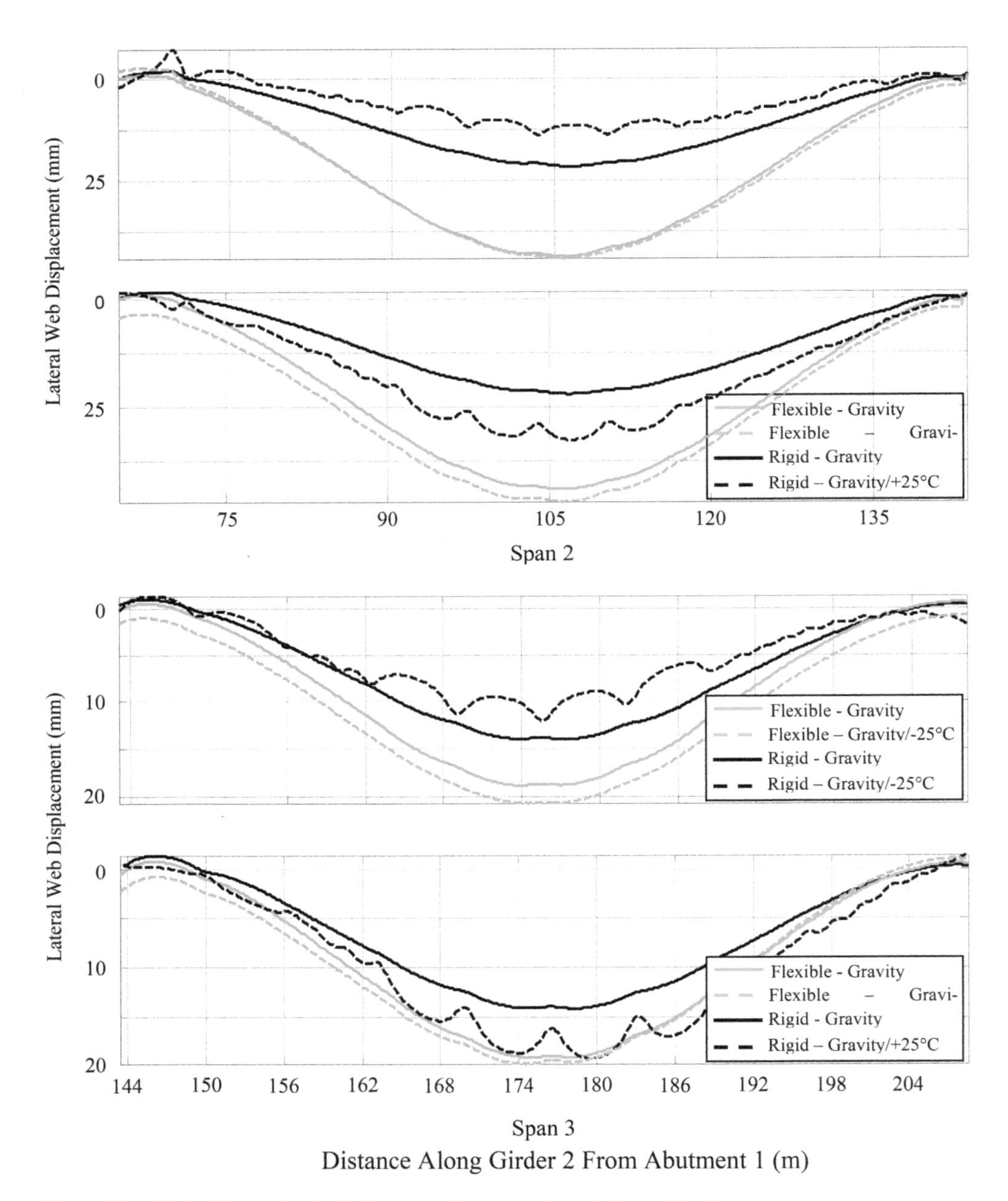

Figure 13. Girder 2 lateral web displacement due to gravity and thermal loads.

loading followed by +/−25°C thermal loading. Figure 13 shows lateral web displacement profiles for Spans 2 and 3 of Girder 2 when loaded with self-weight followed by a uniform +25°C and −25°C thermal loading.

Although Figure 11 shows that modeling the bridge piers as flexible members yields higher lateral web displacements in spans 2 and 3 under self-weight loading than modeling the piers as rigid members, results in Figure 13 show that lateral web displacements in span 2 are more significantly impacted by thermal loading when the bridge piers are modeled as rigid. Temperature

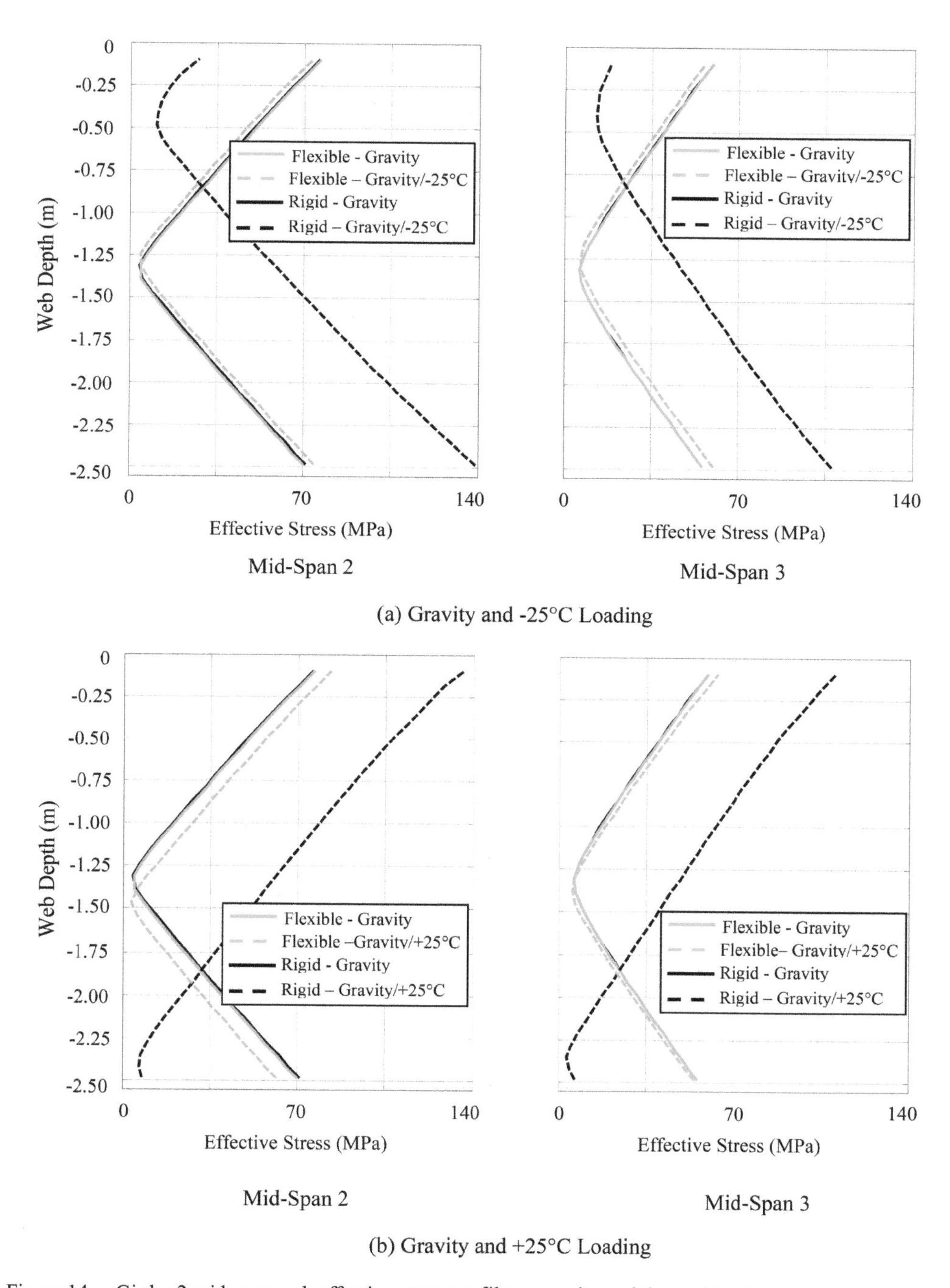

Figure 14. Girder 2 mid-span web effective stress profiles – gravity and thermal loading.

increase leads to up to 18.7% greater lateral displacements in span 2 when the piers are modeled as flexible members and up to 51.1% greater lateral displacements when piers are modeled as rigid. Conversely, temperature decrease causes up to a 13.1% decrease in lateral displacement in span 2 when the piers are flexible and up to a 50.9% decrease when the piers are rigid. Profile plots also show that when the girders in span 2 are supported by rigid piers and loaded with gravity and thermal loading, the I-girders begin to show signs of lateral displacement along the web centerline at mid-span.

As bridge girders expand and contract under changing thermal conditions, this expansion and contraction will create forces on the bridge bearings according to the type of bearing at each location and the curvature of the superstructure. When the piers are modeled with flexibility, the forces imposed on the bearings from girder expansion and contraction will lead to the piers displacing according to the magnitude of the forces and the orientation of the bearings. The displacement of the piers serves to dissipate some of the thermally induced forces in the I-girders. On the other hand, the only girder movement at the bridge supports when piers are modeled as rigid members is movement allowed by bearing design. Piers do not move in response to thermal expansion and contraction, thereby preventing the I-girders from transferring any thermal forces to the bridge piers.

Results in Figures 11 and 13 show that uniform thermal loads increase both global and local bucking of the steel I-girders, and studies have previously shown that increase in the early age bucking of steel I-girders will be detrimental to their load carrying capacity. The degree to which these unanticipated thermal loads impact structural capacity can be investigated through the study of the effective stress in the I-girder sections. Figure 13 shows the effective web stress profiles at mid spans 2 and 3 of Girder 2. The plots in Figure 13 reveal that modeling the bridge piers as rigid or flexible members has little to no effect on the effective stress profiles under gravity loading. Because effective stress is a measure of a sections overall state of stress, the symmetry of these effective stress profiles indicates the bending moment of the section being mostly symmetric about the web centerline. The addition of uniform thermal loading has minimal impact on the effective stress in the cross sections under investigation when the bridge piers have flexibility. However, web effective stress profiles are impacted by uniform temperature loading when the bridge piers are assumed rigid and profiles show that temperature loading, both −25°C and +25°C, cause the effective stress profiles to no longer exhibit symmetry about the web centerline. This is an indication that thermal loading is introducing an axial stress component to the web stress profile. The magnitude of such stresses could likely be higher for in-service applications as relatively mild temperature increases and decreases of 25°C were chosen for this study. Stresses presented here that are caused by thermal loads are additional stresses not considered by designers, and therefore consume I-girder capacity that was designed to accommodate the design loads. As I-girders only have a finite capacity available to handle all load combinations that can arise, results here show that thermal loading on the Buffalo Creek Bridge in the early stages of construction before the introduction of any live loading could decrease the bridge capacity, and in some cases, may lead to premature buckling of the I-girder webs.

6 CONCLUSIONS

The main focal point of this research study has been to investigate the impact changing thermal conditions will have on the I-girders of curved steel I-girder bridges. Based on the detailed FE modeling of the Buffalo Creek Bridge, a curved I-girder bridge, conclusions can be drawn as follows:

1. I-girders experience global lateral displacement, in the form of lateral web displacement, due to only self-weight loading on the superstructure. Overall, pier flexibility yields greater global I-girder lateral bucking under self-weight loading, but lateral web deformation is present for both the flexible and rigid pier cases.
2. Design sheets consider the camber deviation of the I-girders due to self-weight loading, but no calculations are made for the lateral deformation from gravity. Results indicate that lateral deformations reach up to 40% of the magnitude of camber values. For curved steel I-girder bridges, lateral camber values should be considered as the curvature of the structure will typically

always lead to lateral deformations being present and lateral deformations, although small at this stage, are typically more detrimental to I-girder capacity.
3. At the stage of gravity loading, web out-of-plane displacement, or local buckling, is observed in the I-girder webs. These deformations can be thought of as initial imperfections comparable to imperfections from fabrication because they are not considered during design. The initial local buckling of the webs reduces the initial web stiffness, will lead to larger displacements at lower load levels, and reduces the load carrying capacity of the I-girder
4. For the case of gravity loading, modeling the piers as either rigid or flexible members does not significantly impact the local web buckling.
5. Compared to the rigid pier case, pier flexibility greatly reduces the magnitude of thermally induced effective stresses in the I-girder webs as a result of the piers allowing expansion and contraction of the superstructure. On the other hand, assuming the piers as rigid members leads to measurable increases in the overall state of stress in the webs. The larger the number of degrees of freedom constrained by the bearings on a span, the larger the magnitude of stress thermal loading induces in the web.

ACKNOWLEDGEMENTS

The research work presented in this paper was funded by West Virginia Department of Transportation. The authors acknowledge the support and encouragement of Mr. Jimmy Wriston, Deputy Secretary of Transportation and Acting Commissioner. Acknowledgements are also extended to the support and cooperation of the Engineers of the Engineering Division and District 2.

REFERENCES

American Association of State Highway and Transportation Officials 1980. *Guide Specifications for Horizontally Curved Highway Bridges, 1st Edition*, Washington, D. C.
American Association of State Highway and Transportation Officials 1993. *Guide Specifications for Horizontally Curved Highway Bridges, 2nd Edition*, Washington, D. C.
American Association of State Highway and Transportation Officials 1996. *Standard Specifications for Highway Bridges, 16th Edition*, Washington, D. C.
American Association of State Highway and Transportation Officials 2003. *Guide Specifications for Horizontally Curved Highway Bridges, 3rd Edition*, Washington, D.C.
American Association of State Highway and Transportation Officials 2017. *AASHTO LRFD Bridge Design Specifications, 8th Edition*, Washington, D. C.
Armstrong, W.L. 1977. Curved I-Girder Bridge Design Recommendations, *Journal of the Structural Division* 103(5): 1137–1168.
Barker, R.M., Duncan, M., Rojiani, K.B., Ooi, P.S.K., Tan, C.K., and Kim, S.G. 1991. *Manuals for the Design of Bridge Foundations*. National Cooperative Highway Research Program No. 343, Transportation Research Board, Washington, DC.
Davidson, J. S. and Yoo, C.H. 2003. Effects of Distortion on Strength of Curved I-Shaped Bridge Girders, *Transportation Research Record* 1845: 48–56.
Davidson, J.S., Abdalla, R.S. and Madhavan, M. 2004. *Stability of Curved Bridges during Construction*. UTCA Report 03228, University Transport Center for Alabama, University of Alabama, Tuscaloosa, Alabama.
DeSantiago, E, Mohammadi, J., and Albaijat, H.M.O. 2005. Analysis of Horizontally Curved Bridges Using Simple Finite-Element Models. *Practice Periodical on Structural Design and Construction* 10(1): 18–21.
Fasl, J.D., Stith, J.C., Helwig, T.A., Schuh, A., Farris, J., Engelhardt, M.D., Williamson, E.B., and Frank, K.H. 2015. Instrumentation of a Horizontally Curved Steel I-Girder Bridge during Construction. *J. Structural Engineering* 141(1):
Galambos, T.V. 1978. *Tentative Load Factor Design Criteria for Curved Steel Bridges*. Research Report No. 50, School of Engineering and Applied Science, Civil Engineering Department, Washington University, St. Louis, Missouri, USA.
Grubb, M.A., Yadlosky, J.M., and Duwadi, S.R. 1996. Construction Issues in Steel Curved-Girder Bridges, *Transportation Research Record* 1544: 64–70.

Howell, T. D. and Earls, C.J. 2007. Curved Steel I-Girder Bridge Response during Construction Loading: Effects of Web Plumbness, *J. Bridge Engineering* 12(4): 485–493.

Kala, Z., Kala, J., Skaloud, M. and Teply, B. 2005. Sensitivity Analysis of the Effect of Initial Imperfections on the i) Ultimate Load and ii) Fatigue Behavior of Steel Plate Girders. *J. Civil Engineering and Management*, 11(2): 99–107.

Kim, W. S., J. A. Laman and Linzell, D.G. 2007. Live Load Radial Moment Distribution for Horizontally Curved Bridges, *J. Bridge Engineering* 12(6): 727–736.

Linzell, D., Hall, D. and White, D. 2004. Historical Perspective on Horizontally Curved I Girder Bridge Design in the United States, *J. Bridge Engineering* 9(3): 218–229.

Shoukry, S.N., William, G.W., McBride, K.C., Riad, M.Y., Wriston, J.D. 2011. Buffalo Creek Bridge: A Case Study of Empirical Versus Traditional Deck Design, *J. Bridge Structures* 6(3–4): 139–153.

Stegmann, T.H., and Galambos, T.V. 1976. *Load Factor Design Criteria for Curved Steel Girders of Open Cross Section*. Washington University Research Report No. 43, Washington University, St. Louis, MO.

White, W.D. and Jung, S. 2007. Effect of Web Distortion on the Buckling Strength of Noncomposite Discretely-Braced Steel I-Section Members. *J. Engineering Structures* 29: 1872–1888.

Zhang, H., Huang, D., and Wang, T.L. 2005. Lateral Load Distribution in Curved Steel I-Girder Bridges, J. Bridge Engineering 10(3): 281–290.

Chapter 16

Structural performance of continuous slab-on-steel girders bridge subjected to extreme climate loads

A. Mohammed, B. Kadhom & H. Almansour
National Research Council Canada (NRC), Ottawa, Ontario, Canada

ABSTRACT: The expected continuation or acceleration of climate change could induce additional stresses that increase the risk of failure of critical bridge components. Bridge structures are designed based on historical weather records, where the climatic patterns reflect the local climate in which the bridge is located. This study investigates the impact of extreme ambient temperatures and very high thermal gradients on the safety and serviceability of multi-span slab-on-steel girder bridges. A three-span slab-on-girder bridge designed according to the Canadian Highway Bridge Design Code and simulated by a 3D nonlinear finite element model is investigated for the effect of average ambient and differential temperatures as predicted by climate change models. The results show increases in the moments in critical sections of the bridge superstructure and very large longitudinal and lateral displacements. More investigations are required to further investigate the safety of the bridge elements and connections.

1 INTRODUCTION

Climate change is one of the biggest challenges facing the earth, due to the associated global changes in temperature, precipitation and wind patterns, and increased frequency and severity of extreme weather events, such as heat waves, storms and floods. The expected continuation or acceleration of climate change threaten the integrity and robustness of transportation infrastructure.

In this context, bridges are the most critical links of surface transportation infrastructure as they are exposed to exterior conditions, in which climate loads could dramatically vary over short and long terms. Bridge structures are designed based on historical weather records, where the climatic patterns reflect the local climate in which the bridge is located. It has always been observed that bridge performance is highly affected by the weather (Nemry and Demirel 2012) and (Wang et al. 2010). Tong et al. (2000) investigated the temperature distribution in steel bridges; they highlighted that steel bridges might undergo significant temperature changes under the combined influence of solar radiation, daily air temperature variation and wind speed. Liu and DeWolf (2007) reported that the global temperature changes the effect of natural frequencies by affecting the stiffness of bridges. In their study, specific mechanisms were proposed to explain the reasons behind this phenomenon, such as 1) Young's modulus of concrete decreases with increasing temperatures, 2) boundary conditions change due to temperature variation, and 3) changes in the bridge deck asphalt layer when the temperature changes. They recommended additional research to gain a further understanding of how structural stiffness changes are related to temperature variations so that this information can be incorporated into the bridge elements damage models.

Bridges could be significantly impacted by climate changes, mainly through the increase in average temperatures, and increases in the different types of climate extremes such as hot and cold days. Climate change could also be correlated to the observed increase in frequency and magnitude of extreme climatic events, and hence, resulting in a rise of extreme climatic loads on bridges (Wang et al. 2010). The changes in ambient and differential temperatures may lead to significant changes in the structural response of the bridge components and structural systems in terms of large deformations and high-stress levels that could exceed those allowed by existing design codes. Such effects could impact the structural integrity of the bridge system, for example, substantial differential temperature could result in damaging the composite action between the slab and girders in slab-on-girder bridges. The overall bridge deformations due to significant changes in the thermal stresses are induced due to the variations either in the temperature gradient over the cross-section or high changes in the overall extension due to uniform temperature (ambient temperature). Such thermal loading influences the design of the bridge structural elements and joints. Failure to allow effects like repeated cycles of heating and cooling may magnify the distress in various parts of the bridge (Tong et al. 2000). For instance, the elevated ambient temperature could severely damage the expansion joints and affects its functionality (Chang and Lee 2001).

The response of bridges to the changes in temperature caused by climate change could lead to a significant change in deformations and stresses; therefore, it is essential to incorporate these changes in the bridge design. Typically, in the bridge design code, climate loads such as temperatures, precipitation level, and wind speed are derived from historical climate data. The current studies indicated that a non-stationary nature of climate parameters is observed, where historical climate data is not reliable to predict the future trend of climate loads. Based on assuming a specified rise in the average global temperature (e.g., 2°C or 4°C), climate change models can predict with acceptable uncertainty the increases and fluctuations of average regional or local ambient temperatures. This would include the prediction of the increases in length of hot and cold waves and their frequency. The predictions of the precipitation, ice accretion, and wind loads are more complicated and have very large uncertainties. Hence, the focus of the current study is on the temperature loads only.

The hourly temperature changes throughout the day and as the sun rises and goes down in a continuous cycle, the temperature fluctuates leading to contraction and expansion of the superstructure. The seasonal temperature change corresponds to the maximum mean temperature change expected to occur during the year. The greatest superstructure expansion will generally occur on summer days, while the most significant contraction will occur on winter nights. These extreme temperatures cause substantial thermal displacements. A nonlinear temperature distribution (thermal gradients) over the bridge superstructure depth is caused by the difference between the extreme daily temperature on the top surface of the bridge superstructure affected by the direct solar radiation and the bottom surface that is under the ambient air temperature.

Slab-on-girder bridge is the most common bridge type in North America, and generally, it is designed for a life span of 75 years. It is observed that the high-temperature variation and extreme temperatures can seriously affect the structural performance of this type of bridges, especially when the bridge is continuous over two or more spans. The objective of this paper is to investigate the impact of extreme ambient temperatures and very high thermal gradients on the safety and serviceability of multi-span slab-on-steel girder bridges. The current research aims at investigating the response of multi-span continuous slab-on-girder bridge designed according to the Canadian Highway Bridge Design Code (CHBDC S6-14 2014) under different temperature load scenarios using 3D non-linear finite element model. The emphases are on (i) investigating the structural performance of the bridge superstructure when subjected to different levels of temperatures combined with service loads, and (ii) exploring whether the CHBDC design requirements could accommodate the deformations and stresses resulting from the extreme temperatures predicted by climate change models.

2 BRIDGE MODELLING

2.1 *Bridge characteristia*

A 3-span, continuous, slab on steel I-girder bridge with spans length equal to 50 m, 65 m and 50 m is considered in this FE analysis and design. The platform width is 12.88 m which accommodates three lanes for a CL-625 Truck, and each lane is 4.0 m wide. The wearing surface is 15 mm thick

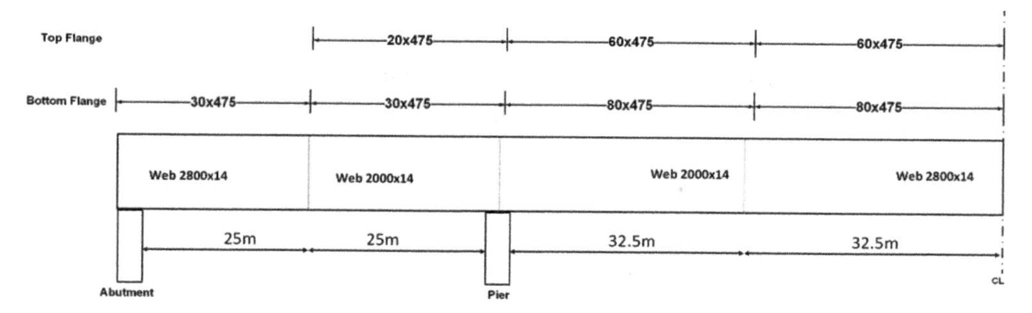

Figure 1. Typical girder elevation (interior and exterior).

Table 1. Material properties of slab concrete and steel girders.

Property	Value
Slab concrete modulus of elasticity: E_c	24.8 GPa
Compressive strength of slab concrete: f_c'	30 MPa
Poisson ratio of concrete slab: ν	0.2
Shear modulus of slab concrete, G	10669.58 MPa
Weight per unit volume for both slab and girder concrete	24.0 kN/m^3
Coefficient of thermal expansion, A	9.9E-6°C^{-1}
Modulus of elasticity of steel girders: E_s	200 GPa
Yield strength: f_y	350 MPa
Ultimate strength: f_u	480 MPa

bituminous overlay. The haunch is 75 mm, and shoulder width on the bridge is 2.0 m. The bridge is assumed to have no skew or curvature in the plan. The superstructure consists of four steel I-girders integrated with a 0.25 m thick reinforced concrete slab. The girders have variable section heights, and they are spaced at 3.6 m c/c. A typical girder elevation is shown in Figure 1. Material properties for the concrete slab and steel girders are given in Table 1.

2.2 *FE model of the bridge*

CHBDC S6-14 permits the use of finite element method for structural analysis of short and medium span bridges. The finite element model (FEM) for the bridge of this study is created using CSiBridge 20 (v20.2.0 2019), which is an integrated tool for modelling, analysis, and design of bridge structures following different codes and specifications such as AASHTO, CHBDC, etc. In this investigation, CSiBridge analyzes the superstructure on a girder-by-girder basis while ignoring the effect of torsion. The substructure is assumed to be very rigid, and the boundary conditions of the bridge superstructure are simplified as a hinge in the left abutment and as rollers on the two interior piers and the right abutment as shown in Figure 2. Dead loads are comprised of the weight of the structure, the pavement surface and sidewalks. The live load on a bridge is the result of vehicular traffic. The bridge is analyzed to evaluate its structural response when subjected to Canadian Highway Bridge Design truck loading (CHBDC truck CL-625) and lane loading.

A 3-D FEM is generated for the bridge using shell elements. The shell element is three or four-nodes three-dimensional with six degrees of freedom; at each node, there are three displacements (U1, U2, U3) and three rotations (φ1, φ2, φ3). In the 3D model, the vehicular traffic direction is along the X-axis of the bridge, while the Y-axis and Z-axis represent the bridge transverse direction and vertical orientation, respectively. In the model, the material is assumed to be homogenous and

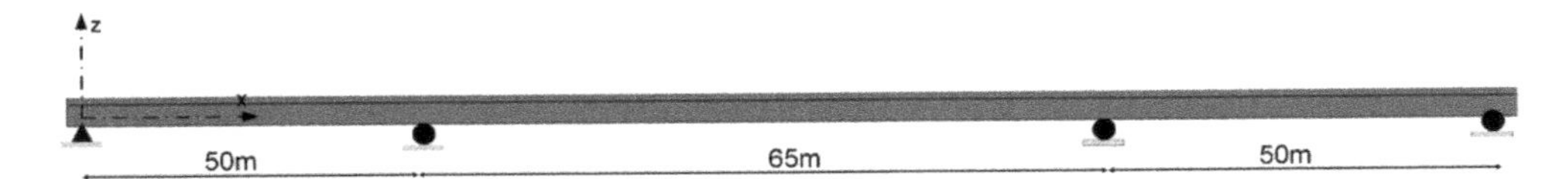

Figure 2. The bridge with boundary conditions.

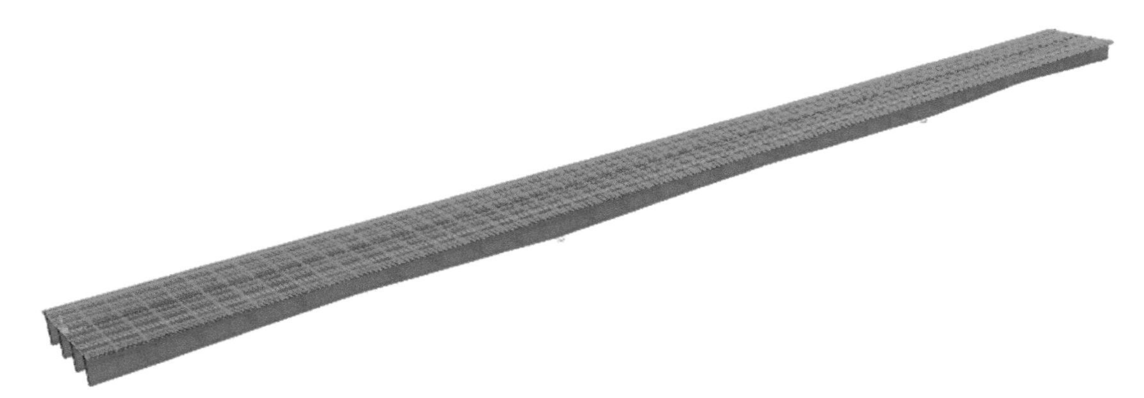

Figure 3. View of 3D model of steel girder bridges of 165 m.

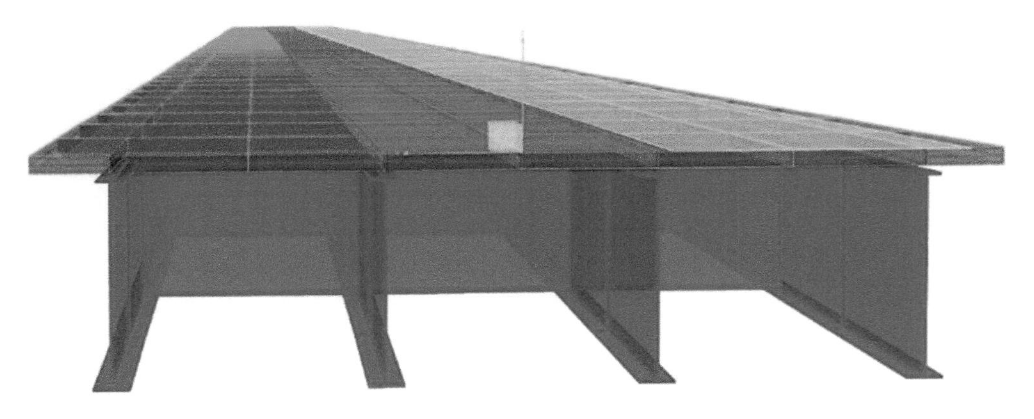

Figure 4. View of 3D model of steel girder bridges: 4 Girders, diaphragms and lanes.

isotropic. Cracking of the bridge's deck is ignored, and the girders and slab are assumed to act compositely. Figures 3 and 4 show the 3D view model of the bridge as created by CSiBridge.

3 PARAMETRIC STUDY

In order to perform a parametric study on the temperature loads using linear or nonlinear static and dynamic analysis for the bridge structural systems, the CSiBridge's 3-D model described above is employed assuming that the bridge is located in critical Canadian cities. These cities are expected to experience ample changes in climate loads. The climate change prediction models show very high variation in the temperatures for the selected cities. The temperatures range between +36°C (maximum recorded temperature in hot days) and −50°C (coldest recorded temperature). This

Figure 5. Deformed shape of the bridge superstructure under dead load only.

Figure 6. Deformed shape of the bridge superstructure under temperature load only (+36°C).

Figure 7. Deformed shape of the bridge superstructure under temperature load only (−50°C).

parametric investigation focuses on the comparison of the bridge performance when different load combinations are applied including climate loads. The objective is to see whether the extreme climate loads could result in any safety or serviceability issues.

There is no data available for bridge analysis under thermal loads to compare with. Thus, in order to verify the effectiveness of the FEM, and to ensure the model suitability in performing the thermal loading, the bridge is considered under dead load only, and then thermal loading is applied. Dead and live loads are defined in CSiBridge. Dead loads consist of the self-weight of bridge components (deck and girders) as well as superimposed loads (asphalt and barriers). The live load consists of CL625 Truck and lane loads. The thermal loading is applied as a load case and is considered in two different cases: (i) ambient temperature and (ii) thermal gradients.

In the design stage, loads are combined according to CHBDC S6-14 load combinations for serviceability, and ultimate limit states SLS and ULS:

SLS Combination = 1.0 Dead Load + 0.9 Live Load
ULS Combination 1 = 1.2 Dead Load + 1.7 Live Load
ULS Combination 2 = 1.2 Dead Load + 1.4 Live Load + 1.0 Thermal Load

3.1 *Performance of the bridge under ambient temperature*

Based on the temperatures predicted by the climate change models, six critical cities are selected. The cities are Montreal, St John's, Iqaluit, Yellowknife, Winnipeg and Saskatoon. The predicted extreme temperatures are +36°C, +30°C, +20°C, −50°C, −43°C, and −40°C for Montreal, St John's, Iqaluit, Yellowknife, Winnipeg and Saskatoon, respectively. The predicted temperature is applied on the bridge as uniform or ambient temperature and is considered as a load case by itself or a component of a load combination as per CHBDC S6-14.

Figure 5 shows the deformed shape of the bridge superstructure due to the dead load with maximum central span deflection of 55 mm. Figure 6 shows the deformed shape of the bridge in the City of Montreal (+36°C) with a maximum deflection equal to 9.8 mm at the central span and longitudinal expansion of 50 mm at right end of the bridge. For the bridge in Yellowknife (−50°C), the superstructure gradually bends upward by 14 mm, and the right end moved longitudinally from its original position by 70 mm as shown in Figure 7.

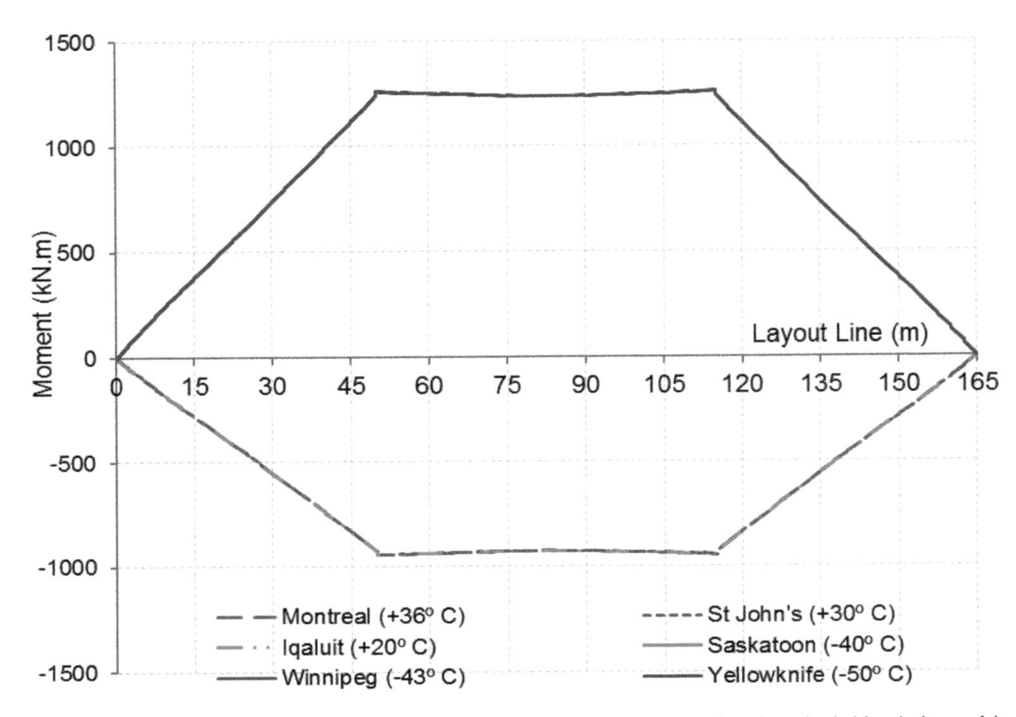

Figure 8. Moment along the bridge length due to ambient temperature only when the bridge is located in different cities.

The previous figures show the behaviour of the bridge under dead load and ambient temperature (positive and negative). Moreover, the bridge is investigated in the other selected locations with different ambient temperatures. Figure 8 shows the moment distribution over the bridge length for different cities. It can be seen that in a relatively warm city like Iqaluit (+20°C), the maximum positive moment is 945.0 kN.m, which is the same moment when the same bridge is located in warmer cities like St John's (+30°C) and Montreal (+36°C). The corresponding moment reversed to negative with an absolute value of 1260.0 kN.m in extremely cold cities, Yellowknife city for instance with an ambient temperature of −50°C.

Figure 9 shows the longitudinal displacement distribution of the bottom flange of any of the bridge girders along the bridge length when the bridge is located in different cities. The absolute value of the longitudinal displacements of the bridge increased compared to the correlated displacements when the bridge is subjected to the dead load only (which are below or equal to 14 mm). The absolute value of the longitudinal displacements are increased up to 72 mm in the coldest city (Yellowknife: −50°C) and decreased to 54 mm in the hottest city (Montreal: 36°C). Figure 10 shows the vertical displacement distributions of the bottom flange of any of the bridge girders along the bridge length when the bridge is subjected to the dead load on, which are much higher than that displacement when the bridge is under the temperature loads. The warmer cities like St John's and Montreal show vertical displacement distributions opposite to the gravity deflection; while for the extremely cold cities, the vertical displacement distributions are in the direction of the gravity.

The variations of stresses over an interior girder along the longitudinal axis due to CHBDC S6-14 load combination of the dead and thermal loads are shown in Table 2. When only the dead load is considered, the stress values range from −84 to 98 MPa. However, when the ambient temperature is +36°C (Montreal location), the maximum stress values decrease up to 24.5 MPa. For Yellowknife, the maximum stress is (84 MPa). On the other hand, the results show that there are no significant changes in the force distributions over the interior and exterior girders.

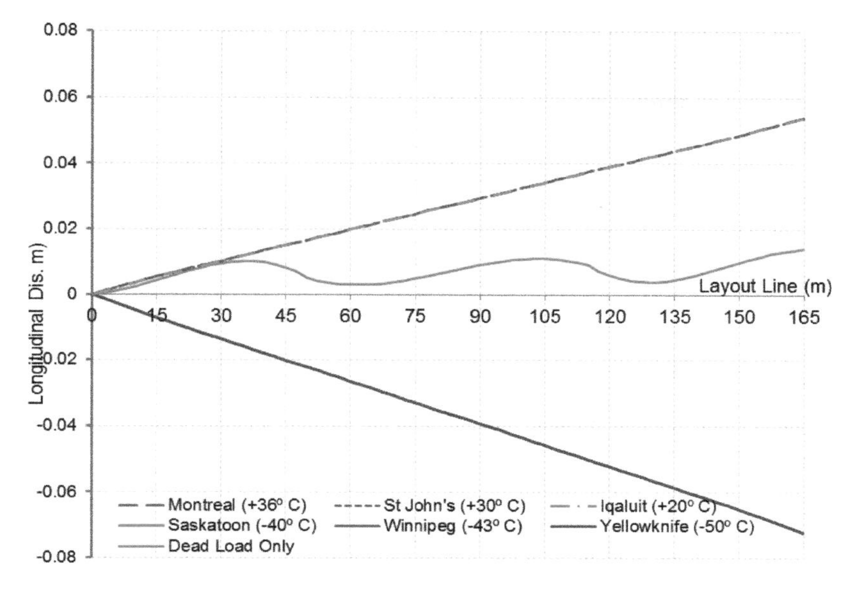

Figure 9. Longitudinal displacement of a girder along the bridge length due to ambient temperature when the bridge located in different cities.

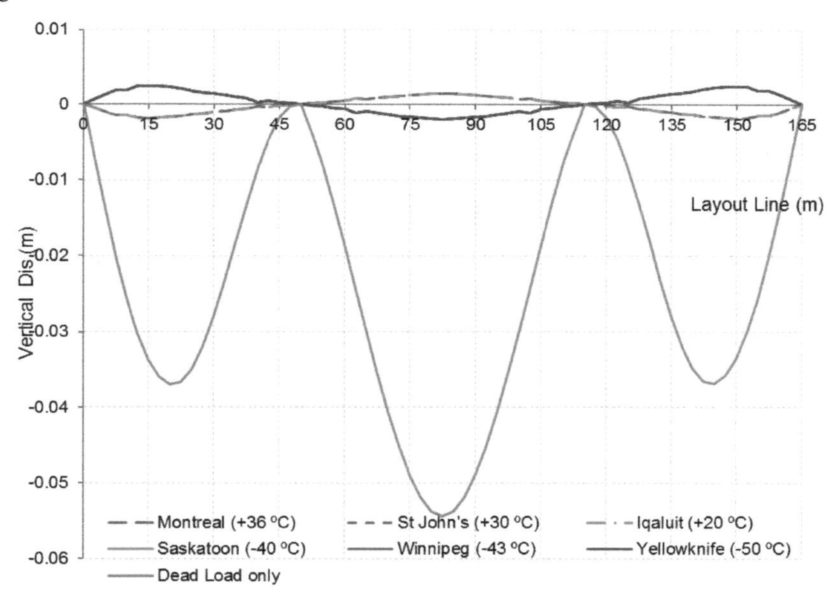

Figure 10. Vertical displacement of a girder along the bridge length due to ambient temperature when the bridge located in different cities.

Table 2. Maximum stress variation over the exterior girders

Maximum stresses due to dead load only	Maximum stresses due to temperature load	
+98 MPa	Montreal City (+36°C) +24.5 MPa	Yellowknife City(−50°C) +84 MPa

Figure 11. Deformed shape of the bridge superstructure under thermal gradients only.

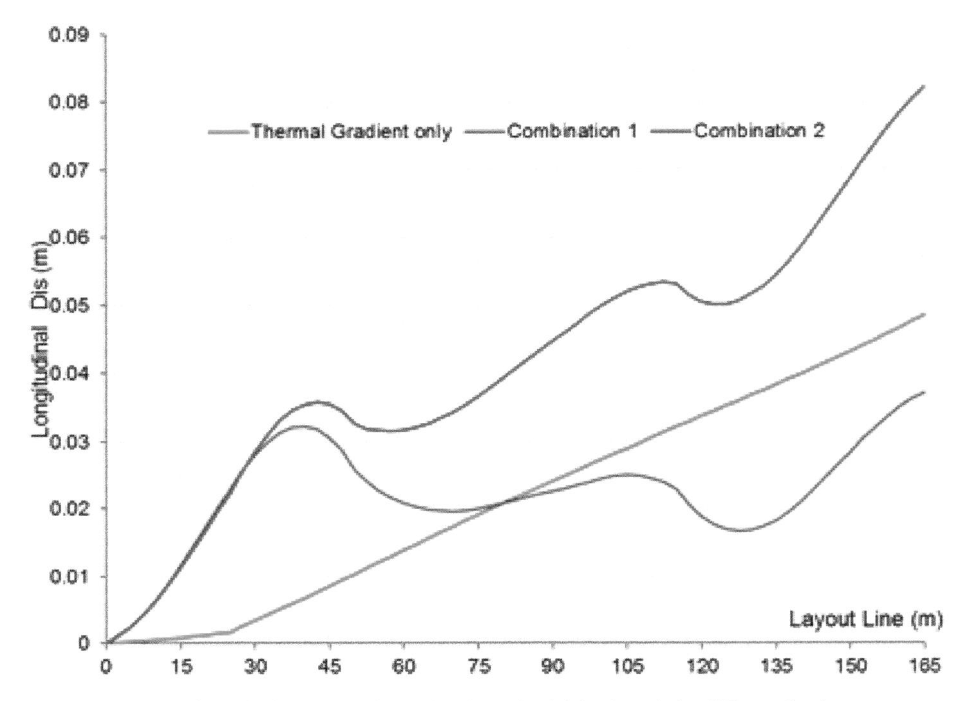

Figure 12. Longitudinal displacement of a girder along the bridge length for different load cases.

3.2 *Performance of the bridge under nonlinear thermal gradient*

In this study, the thermal gradient is modelled as a piecewise linear gradient with up to five linear segments (CSiBridge Manual). The following three load cases are considered:

(i) Bridge under thermal gradient load only,
(ii) Bridge under ULS combination1 (which named combination1), and
(iii) Bridge under ULS combination 2 (which named combination 2).

In each load combination, the thermal gradient is applied as five linear segments ranged from 50°C, 40°C, 30°C, 20°C, and 5°C. ULS combination 1 and ULS combination 2 are specified in CHBDC S6-14 as mentioned above.

Figure 11 shows the deformed shape of the bridge superstructure due to the thermal gradient with a maximum vertical displacement equal to 17 mm, and the right end of the bridge is shifted horizontally by 47.5 mm. Figures 12 and 13 show the longitudinal and vertical displacements along the entire length of the bridge for different load cases.

The maximum moment caused by the thermal gradient is relatively low (4000 kN.m) compared to the maximum moment resulting from the load combination of the dead and live loads. However, load combination 2 shows a high increase of up to 20% in the negative moment over the interior supports (Figure 14).

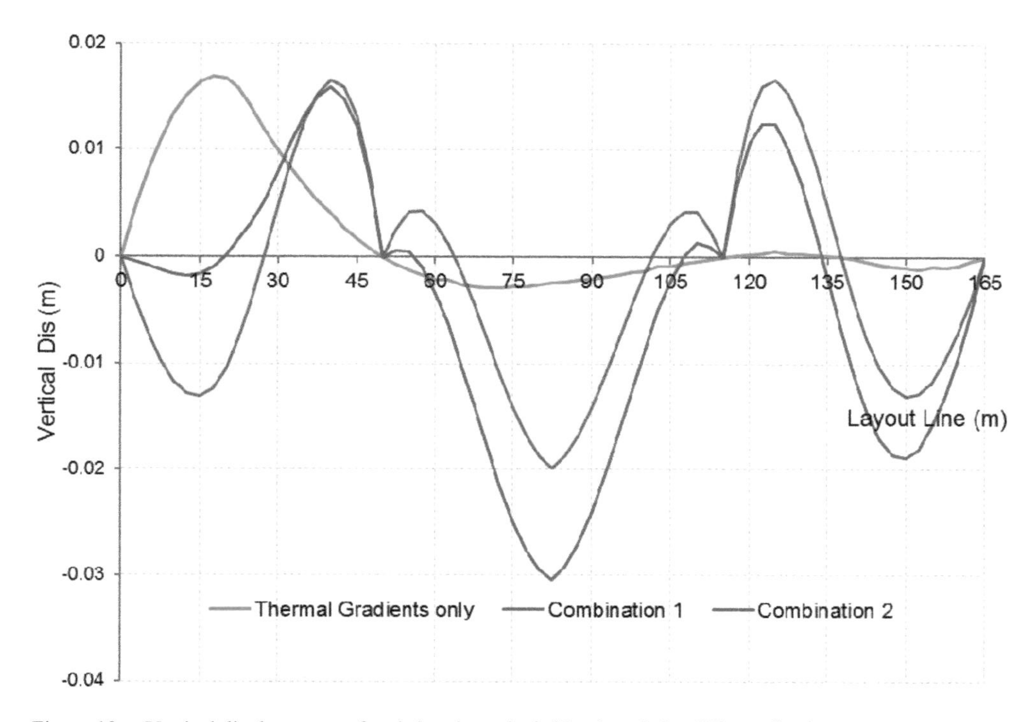

Figure 13. Vertical displacement of a girder along the bridge length for different load cases.

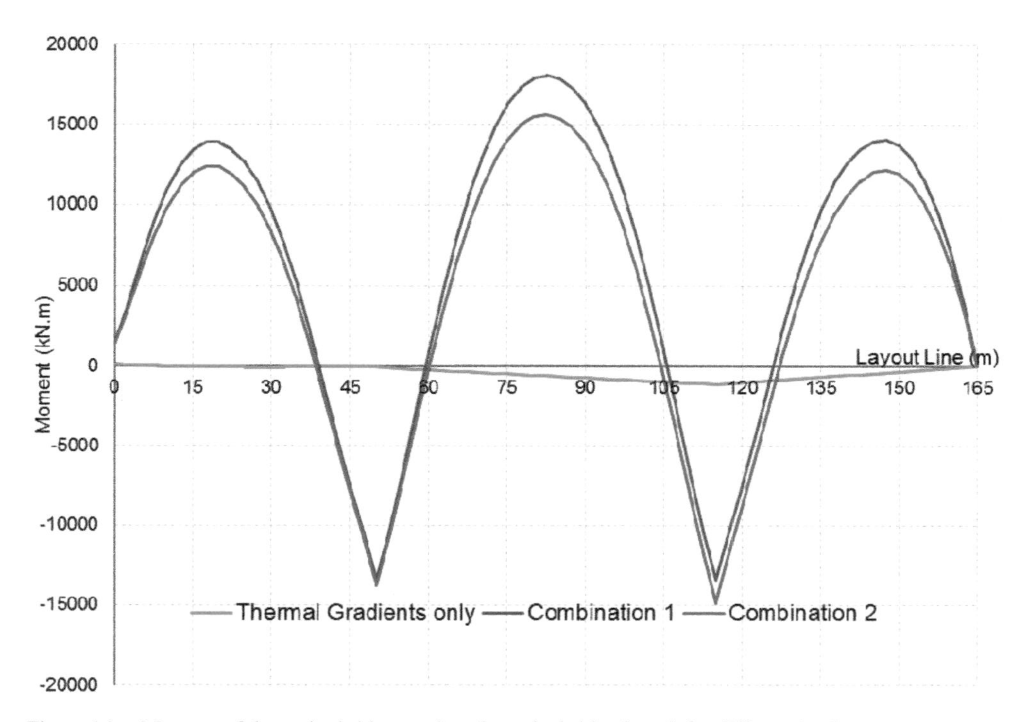

Figure 14. Moment of the entire bridge section along the bridge length for different load cases.

4 BRIDGE RESPONSE TO EXTREME TEMPERATURES PREDICTED BY CLIMATE CHANGE MODELS COMPARED TO CHBDC S6-14

As shown in previous sections, the exposure of multi-span continuous concrete slab-on-steel girders bridge to extreme temperature loads would lead to large deformations and high-stress levels. These high changes in the structural response of the bridge components and the structural system could exceed the safety and serviceability limits in the current CHBDC S6-14. The bridge considered in the previous sections is assumed to be located in different cities and exposed to various levels of ambient temperatures and thermal gradients. The load combinations in CHBDC S6-14 related to temperature loads and gravity loads are used in this comparison.

The results in the previous sections show amplification of up to 25% in the moment magnitude along the bridge length. The model results show large longitudinal and lateral displacements that reach the CHBDC S6-14 allowable limits. On the other hand, the model results show a slight increase in the stresses in the deck slab which proves the adequacy of CHBDC S6-14 slab design requirements. Further investigations on the safety of connection between the deck-slab and the steel girders are part of an ongoing research project.

5 CONCLUSIONS

Among the many effects of the climate changes on existing bridges, the increase in the average ambient temperature, increase in the frequency, and length of hot waves are predicted with acceptable uncertainty. The expected continuation or acceleration of climate change threaten the integrity and robustness of bridge infrastructure. This study investigated the impact of extreme ambient temperatures and very high thermal gradients on the safety and serviceability of a multi-span slab-on-steel girder bridge. The bridge is designed according to CHBDC S6-14 and simulated using a three-dimensional nonlinear finite element model. The bridge is assumed to be located in critical Canadian cities that expected to experience large changes in climate loads where the temperatures range between +36°C (as a maximum predicted the average ambient temperature in a hot day) and −50°C (as a coldest predicted average ambient temperature). The structural performance of the bridge superstructure is considered when subjected to different levels of temperatures combined with different loads. The aim is to investigate whether the CHBDC S6-14 design requirements can accommodate the deformations and stresses resulting from the temperatures predicted by climate change models. From the case study performed in this investigation, it is found that the extreme temperature loads would lead to large deformations and high-stress levels where the structural response of the bridge components could exceed the safety and serviceability limits in the current CHBDC S6-14. The results of this study show large increases in bending moments in the critical sections of the interior and exterior girders. Furthermore, there are substantially longitudinal and lateral displacements that reach the CHBDC S6-14 allowable limits. However, the safety of the connection between the deck-slab and the steel girders should be further investigated in future studies.

This study shows only preliminary results of ongoing research at the NRC Canada that focus on evaluating the impact of climate change on bridges. The potential aim is to propose modifications to the CHBDC adopting the climate change effects and maintaining the same levels of structural safety and serviceability when bridges are subjected to climate loads as predicted by environmental climate models.

REFERENCES

Canadian Highway Bridge Design Code, S6-14. Canadian Standard Association (CSA), Toronto, Canada, 2014.

Chang. L. & Lee. Y. Evaluation and policy for bridge deck expansion joints; Final report: FHWA/IN/JTRP-2000/1, 2001.

CSiBridge. v. 20.2.0, Computers and Structures, Inc., Walnut Creek, CA, 2019.
Liu, C., & DeWolf, J. Effect of Temperature on Modal Variability of a Curved Concrete Bridge under Ambient Loads; Journal of Structural Engineering. 12(133); PP 1742–1751, 2003.
Nemry. F. & Demirel. H. Impacts of Climate Change on Transport: A focus on road and rail transport infrastructures; European Commission, Joint Research Centre (JRC), Institute for Prospective Technological Studies (IPTS), 2012.
Tong, M., Tham. L., Au. F. & Lee, P. Numerical modelling for temperature distribution in steel; Computer & structures; 6(30). PP 583–593, 2001.
Wang, X., Nguyen, M., Stewart, M. G., Syme, M. & Leitch, A. Analysis of Climate Change Impacts on the Deterioration of Concrete Infrastructure Part 3: Case Studies of Concrete Deterioration and Adaptation; CSIRO, Canberra. ISBN 9780643103672, 2010.

Chapter 17

Application of heat straightening repair of impacted highway steel bridge girders

W. Zatar & H. Nguyen
Marshall University, Huntington, West Virginia, USA

ABSTRACT: Damage to highway steel bridge members may occur when an excessively high, over-sized truck dynamically impacts a bridge. The collision could result in significant plastic deformations in localized areas of the steel girders. The research team investigated the effect of heat straightening on the behavior of impacted highway steel bridge girders. An extensive list of highway structures in the state of West Virginia, which have been struck and damaged by over-height vehicles, has been developed. Detailed conditions of each struck bridge and times of damage/heat-straightening repairs are documented. Current West Virginia Department of Transportation (WVDOT) process for recording data in damage inspection reports of struck and misaligned member is reviewed and reported. WVDOT decision-making process for whether, or not, to repair damaged members is based on evaluating the remaining load-carrying capacity of the damaged members combined with viewpoints of district engineers and the agency consultants. The research team demonstrated how to determine if the damaged girder should be repaired to its original condition, or if it has sufficient remaining moment capacity to safely resist the expected load in its distorted condition. Flowcharts for assessing members' structural damage and designing of a retrofit process following FHWA guide for heat straightening of damaged steel bridge members are developed. Important recommendations and precautions for heat-straightening repairs are established taking into account recommendations from FHWA "Guide for Heat-Straightening of Damaged Steel Bridge Members," NCHRP Report 604, as well as other research reports and references from a comprehensive literature review that was completed during the first phase of the study.

1 INTRODUCTION

The use of heat-straightening to repair damaged steel members dates back to the 1930s with low-grade steels. For many years, heat-straightening of damaged steel bridge girders has been more of an art than a science. Several safety concerns have historically limited its validity as a repair technique. When applied appropriately, heat-straightening, along with mechanical techniques including pressing or jacking, could offer an economical and a viable alternative to replacement of damaged steel girders (Zatar and Leftwich 2009; Zatar and Nguyen 2018). The West Virginia Department of Transportation, Division of Highways (WVDOH) neither has guidelines nor special provisions for completing single and multiple heat-straightening repairs of damaged steel bridge members. It is therefore necessary to develop procedures, guidelines and specifications to assist WVDOH making decisions on whether, or not, to heat straighten impacted steel members. This research covered:

1) Developing a list of structures in West Virginia that have been struck and misaligned, and numbers of damages and heat straightening
2) Presenting current WVDOH process for recording data in their bridge inspection reports of struck and misaligned members

Table 1. List of bridge structures struck by over-height vehicles in West Virginia.

#	Bridge Name	Bridge Number	County	Condition	Times of Damage	Times of HS	HS Documented
1	OYLER AVENUE BRIDGE	10-61/28-1.25	010 - Fayette	Damaged-HS	7	2	Y
2	BENEDICT RD OVERPASS WB	06-64-31.67	006 - Cabell	Damaged-HS	4	1	Y
3	US 52 OVERPASS EB	06-64-6.45	006 - Cabell	Damaged-HS	3	1	Y
4	KENNA OVERPASS BRIDGE North	18-77-124.89	018 - Jackson	Damaged-HS	3	2	Y
5	Veterans of All Wars Bridge	10-38-3.53	010 - Fayette	Damaged-NR	3	NA	NA
6	DUNBAR TOLL BRIDGE	20-25/47-0.1	020 - Kanawha	Damaged-NR	3	NA	NA
7	I-64 INSTITUTE INTERCHANGE WB	20-64-49.98	020 - Kanawha	Damaged-R	3	NA	NA
8	INSTITUTE RAMP B	20-64-50	020 - Kanawha	Damaged-R	3	NA	NA
9	MONTROSE DRIVE I-64 OVERPASS	20-60/64-0.25	020 - Kanawha	Damaged-HS	2	2	N
10	I-64 SOUTH CHARLESTON EB	20-64-53.8	020 - Kanawha	Damaged-HS	2	1	Y
11	SOUTH PARK BRIDGE	21-7-4.27	021 - Lewis	Damaged-HS	2	1	Y
12	Dallas Pike Bridge EB	35-70-11.06	035 - Ohio	Damaged-HS	2	1	Y
13	POWER PLANT BRIDGE (CSWB)	37-2-4.43	037 - Pleasants	Damaged-HS	2	1	Y
14	ROCKPORT I/C N (CSWB)	54-77-161.66	054 - Wood	Damaged-HS	2	1	Y
15	US 50 I/C N (CSPG)	54-77-176.42	054 - Wood	Damaged-HS	2	2	Y
16	GLENVILLE TRUSS	11-33-16.59	011 - Gilmer	Damaged-NR	2	NA	NA
17	I-64 ST. ALBANS RAMP BRIDGE	40-64-43.5	040 - Putnam	Damaged-NR	2	NA	NA
18	EARL M VICKERS MEMORIAL BRIDGE	10-6-0.12	010 - Fayette	Damaged-R	2	NA	NA
19	CR 11 OVERPASS N (CSWB)	18-77-143.32	018 - Jackson	Damaged-R	2	NA	NA
20	DUPONT INTERCHANGE	02-901-5.32	002 - Berkeley	Damaged-HS	1	1	Y
21	FLATWOODS ICHG S	04-79-67.01	004 - Braxton	Damaged-HS	1	1	Y
22	Ramp A Jennings Randolph	15-30-0	015 - Hancock	Damaged-HS	1	1	N
23	I-77 EDENS FORK I/C NB	20-77-106.11	020 - Kanawha	Damaged-HS	1	1	N
24	KINGMONT ROAD OVERPASS	25-64/01-0.04	025 - Marion	Damaged-HS	1	1	Y
25	ROSS BOOTH MEMORIAL BRIDGE	40-34-21.34	040 - Putnam	Damaged-HS	1	1	N
26	I-64 WINFIELD INTERCHANGE BRIDGE	40-64-15	040 - Putnam	Damaged-HS	1	1	Y
27	KNOB HILL OVERPASS	41-3/21-1.3	041 - Raleigh	Damaged-HS	1	1	Y
28	Richard Snyder Memorial Bridge	52-7-0.07	052 - Wetzel	Damaged-HS	1	1	N
29	WILLIAMSTOWN-MARIETTA BR (CSTT)	54-31-19.01	054 - Wood	Damaged-HS	1	1	N
30	West Virginia 95 Overpass (CSWB)	54-77-173.22	054 - Wood	Damaged-HS	1	1	N
31	WASHINGTON AVE OVERPASS	06-52-0.45	006 - Cabell	Damaged-NR	1	NA	NA
32	HICO BRIDGE	10-60-26.17	010 - Fayette	Damaged-NR	1	NA	NA
33	DUCK CREEK ROAD OP	17-79-107.55	017 - Harrison	Damaged-NR	1	NA	NA
34	KANAWHA TURNPIKE I-64 OVERPASS	20-61/13-0.03	020 - Kanawha	Damaged-NR	1	NA	NA
35	HARTS RUN BRIDGE	13-64-175.31	013 - Greenbrier	Damaged-R	1	NA	NA
36	PRICKETTS CREEK ROAD OP	25-79-138.77	025 - Marion	Damaged-R	1	NA	NA

Note: WB = West Bound; EB = East Bound; CSWB = Continuous-rolled Steel Wide-flange Beam; N = North; S = South; CSPG = Continuous Span Plate Girder; I/C = Interchange; NB = North Bound; CSTT = Continuous Steel Thru Truss; OP = Overpass; HS = Heat-Straightened or Heat Straightening; R = Repaired; NR = Not Repaired; Y = Yes; N = No; NA = Not Available.

3) Providing an example of an actual damage/special inspection report
4) Developing flowcharts to assess damaged structures and designing of a retrofit process following FHWA guide for heat straightening of damaged steel bridge members
5) Providing recommendations and precautions for heat-straightening repairs

2 LIST OF BRIDGE STRUCTURES STRUCK BY VEHICLE IMPACT IN WEST VIRGINIA

According to 2017 National Bridge Inventory (NBI) database (FHWA 2017), West Virginia has 7,228 highway bridges. Of these bridges, large populations of the inventory are steel-girder bridges. Many overpass highway bridges within the state have been struck by over-height vehicles as shown in Table 1. Some bridge members have been hit multiple times, and damaged steel girders were heat-straightened, replaced, or repaired/retrofitted. Below are details of the bridges that have been impacted more than once and heat-straightened at least once.

3 CURRENT WVDOH PROCESS FOR RECORDING DATA IN BRIDGE INSPECTION REPORTS OF STRUCK AND MISALIGNED MEMBERS

According to WVDOH, inspection reports are categorized into different types such as periodic, in-depth, inventory, interim-condition, and damage/special inspection reports. Damage inspection report are required by WVDOH when over-height vehicles hit a bridge. A damage inspection report is an important source of information for engineers to make a decision on whether, or not, to repair or replace damaged members. Detailed information on the current process for recording data in the damage inspection report are given in this section. The report consists of five major sections including purpose, procedure, findings, drawings, and photographs.

- Purpose: This section should incorporate collision information and general facts such as accident date and truck company (if applicable).
- Procedure: The following information should be documented in the procedure section: names of all inspectors; time in hours and minutes that inspection team arrived; detailed bridge inspection procedures and equipment; how long were a lane, road, or shoulders closed during the inspection.
- Findings: Only damage related to the crash should be documented in this section. The following information should be reported: bridge railing height at the crash site (measured if truck jumped over it); Non-Destructive Testing (NDT) on diaphragm connection cracks (Dye Penetrant method is preferred); NDT at impact points to examine flange cracks (Dye Penetrant method is preferred); temporary measures taken to ensure the safety of traveling public; lane, road, or shoulders needing closed after the inspection; cracks, concrete spalling, scrapes, broken strands, bulges, distortions, and other defects.
- Drawings: Required drawings/sketches include: distorted girder plan and cross section (drawing for each distorted beam, horizontal/vertical distortions, impact points, locations of nicks, web bulges, and scrapes, etc.); damaged diaphragm connection details (drawing for each diaphragm connection, dimensions, crack and bolt-hole locations, bolt shank and diaphragm sizes, diaphragm connection plates, weld details, etc.); girder layout plan for cracks (new cracks and changes in existing fatigue cracks); and elevation views showing clearance at both road edges and centerline (drawing for each damaged beam).
- Photographs: Good photographs are important to document damages of impacted structures. Inspectors should capture: big picture view of the accident; close up of all impact points, damaged connections/members, nicks, bulges, distortions, scrapes, and cracks; photos under each distorted beam showing the entire deflected zone; current elevation/deck views in both directions; truck photos (pictures of entire truck stuck under the bridge, impact points on truck, identification numbers on truck door, truck/trailer license, etc.); photos of all clearance signs and parapets if truck went over the edge of the bridge.

Do's and don'ts regarding damage/special inspection report are listed in Table 2. It is important to note that if inspectors find serious damage during a routine inspection, they should prepare a separate damage report instead of combining with any other reports. Damage/special inspection reports aim at gathering legal data that could also be used for preparing plans for repair. Therefore, inspectors should collect as much detailed information as possible so that they will not waste time and money for another visit.

4 AN EXAMPLE OF AN ACTUAL DAMAGE/SPECIAL INSPECTION REPORT

An example of a damage/special inspection report for an impacted bridge (US 50 Interchange Northbound bridge) is given in this section. The bridge is located in Parkersburg, West Virginia. It has been hit and heat-straightened twice (bridge #15 in Table 1). Details of this bridge are shown in Figure 1.

Table 2. Do's and don'ts for damage/special inspection report.

Do's	Don'ts
– Prepare a separate damage report if you find serious damage during a routine inspection	– Combine the special damage report with any other reports
– Cooperate with the police officer investigating the accident	– Initiate conversation with the driver or police officer
– Take lots of pictures, draw lots of drawings, take lots of measurements, and record lots of data	– Worry about server space
– Remember this is a report to gather data for court and to prepare repair plans	– Think this report is just like a normal routine or in-depth inspection report
– Think about what you would want to see, measure or know if you were on a jury or fixing the bridge	– Miss the chance to get information that will require another visit
– Describe in detail the actual dimensional damage	– Just say the beam is "bent" or "severely damaged"
– Think like a lawyer or a designer	– Just think like a bridge inspector
– Get to the crash site as soon as possible	– Miss the chance to get as much information as possible

4.1 *Purpose of the damage/special inspection report*

The primary purpose of the damage/special inspection report of this bridge was to determine if the damaged beam on the US 50 Interchange Northbound Bridge in Parkersburg, West Virginia should be repaired to its original condition or if it has enough remaining moment capacity to resist the live load in its distorted condition. The Heat-Straightening Repairs of Damaged Steel Bridge guidelines (Avent & Mukai 1998) and the most recent Bridge Load Rating Modeling and Analysis Software (LARS bridge program) were used to justify the decision and to provide recommendation for repair.

4.2 *Existing conditions of the bridge*

A team from WVDOH district personnel inspected the US 50 Overpass Bridge (Figure 1). This bridge carries interstate I-77 over US 50 on September 24, 2014. The inspectors discovered that a westbound truck damaged the east exterior beam. The lower flange was distorted laterally out of plane west 3-3/16″ over 53′ and vertically bent 1″ over a few feet. One of the diaphragms was damaged, and the non-composite top flange was pulled away from the deck for 40′ on the east side of the beam. Unfortunately, the damage was from a "hit-and-run." Since there is no insurance company to bill for the repairs, the WVDOH had to repair this bridge with taxpayer's money. The damaged beam is a 64″ deep riveted built-up member with angles for flanges.

4.3 *Repair options*

There were four options for repairing the bridge:

1) Heat straightening entire damaged beam to its original condition
   a) Heat straighten entire beam
   b) Fill the 40′ gap between the deck and the top flange with epoxy
   c) Replace the damaged diaphragm cross members
   d) Paint damaged areas
2) Replace the lower flange and heat straighten the damaged web to its original condition supporting the Dead Load
   a) Remove the lower flange angles
   b) Heat straighten the web
   c) Replace the lower flange with new bolted angles

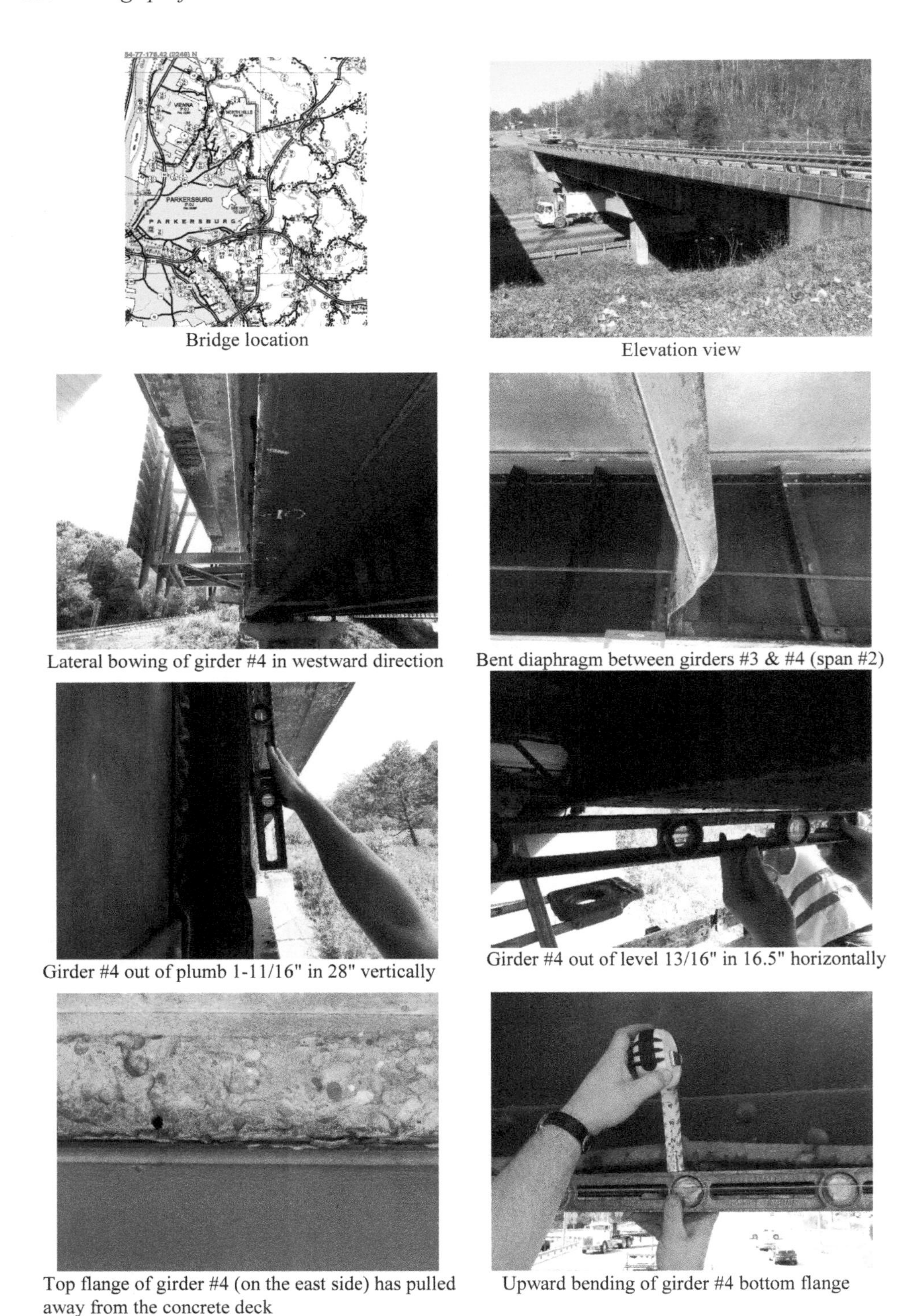

Bridge location

Elevation view

Lateral bowing of girder #4 in westward direction

Bent diaphragm between girders #3 & #4 (span #2)

Girder #4 out of plumb 1-11/16" in 28" vertically

Girder #4 out of level 13/16" in 16.5" horizontally

Top flange of girder #4 (on the east side) has pulled away from the concrete deck

Upward bending of girder #4 bottom flange

Figure 1. Details of US 50 interchange northbound bridge (source: WVDOH's inspection report).

   d) Fill the 40′ gap between the deck and the top flange with epoxy
   e) Replace the damaged diaphragm cross members
   f) Paint damaged areas
3) Do not straighten the damaged beam
   a) Make a judgment call on the amount of Live Load the distorted beam can carry
   b) Leave the horizontal and vertical distortion in the existing beam
   c) Fill the 40′ gap between the deck and the top flange with epoxy
   d) Replace the damaged diaphragm with shortened cross members
   e) Grind smooth discontinuities on the beam and paint damaged areas
4) Vertically heat straighten lower flange but leave lateral distortion in lower flange

Make a judgment call on the amount of Live Load a 3-3/16″ laterally distorted beam can carry.

a) Grind smooth discontinuities on the beam
b) Only heat straighten the lower flange angle to relieve the 1″ vertical displacement
c) Leave the beam horizontally distorted 3–3/16″
d) Fill the 40′ gap between the deck and the top flange with epoxy
e) Replace the damaged diaphragm with shortened cross members
f) Paint damaged areas

4.4 *Beam analysis*

This section of the report determined what degree of repair was needed to maintain the pre-damaged live load capacity of the entire structure. Only a finite element analysis or load testing could accurately determine this. However, research data, referenced in the Heat-Straightening Repairs of Damaged Steel Bridge (Avent & Mukai 1998), produced strength reduction guidelines. These guidelines and the most recent LARS bridge analysis were used to justify the decision.

5 ASSESSMENT PROCESS OF DAMAGED STRUCTURE

According to "Guide for Heat-Straightening of Damaged Steel Bridge Members," assessment process of the damaged structure includes the following steps (Figure 2):

1) Initial inspection and evaluation for safety and stability: this step involves visually inspecting, recording, and documenting the major aspects of damage with measurements and photographs. During this inspection, a preliminary list of repair requirements should be made. The safety and stability of the bridge should be evaluated by reviewing the design drawings and computations. The specific cause of damage (e.g. oversize, overweight, or overload vehicle impact, fire, blast, earthquake, wind, etc.) should be investigated as they may affect final decision on repair.
2) Detailed inspection for specific defects: this step consists of evaluating three critical aspects including signs of fracture, the degree of damage, and material degradation. The signs of fracture can be determined by one of the following methods: dye penetrant, magnetic particle, ultrasonic testing, or radiographic testing. The degree of damage can be evaluated using two different criteria namely the angle of damage (φ_d), which is a measure of the change in curvature, and the strain ratio (μ), which is a measure of the maximum strain occurring in the damaged zone.

The material degradation may influence the heat-straightening decision. Several visual signs (e.g. melted mill scale, distortion, black discoloration of steel, crack and spalling of adjacent concrete) may suggest steel materials exposure to high temperature. Tests can be then conducted at suspicious regions using the following metallurgical test methods: (1) A chemical analysis; (2) A grain size and microstructure analysis; (3) Brinell hardness tests; (4) Charpy V-notch tests; and (5) Tensile tests to determine yield, ultimate strength, and percent elongation.

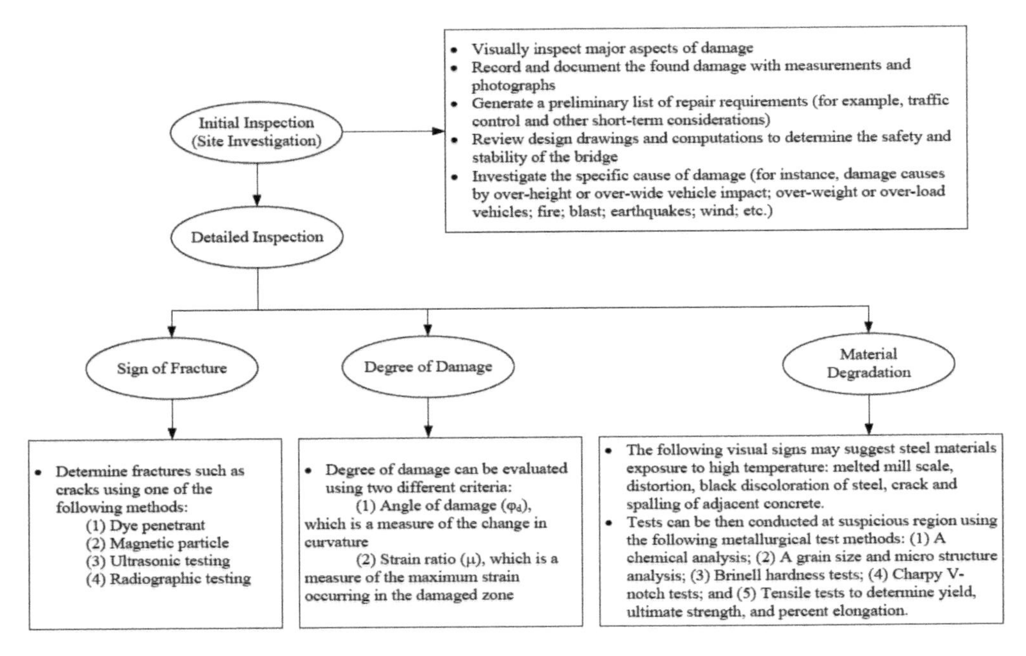

Figure 2. Flowchart of assessment process of damaged structure (FHWA 2013).

6 PLANNING AND DESIGN PROCESS

The repair(s) should be planned and designed after the completion of the assessment process of damaged structures. The planning and design process included the following steps (Figure 3):

1) Analyze the degree of damage and maximum strains induced: Heat-straightening repairs have been proven to be successful for strains up to 100 times yield strain (i.e. $\mu = 100$). Repair decisions should also be based on metallurgical analysis and expert opinion.
2) Conduct a structural analysis of the system in its damaged configuration: This step includes: (1) Determining the structure's capacity in its damaged configuration for safety purpose; and (2) Computing residual forces induced by the impact damage, which may affect safety and influence level of applied restraining force during heat straightening.
3) Select applicable regions for heat straightening repair: the major considerations for deciding a heat-straightening repair are the degree of damage and the presence of fractures and previously heat-straightened members.
4) Select heating patterns and parameters: Since typical damage is often a combination of the fundamental heating patterns (vee, edge, line, spot, and strip heats), the heat-straightening pattern combo is usually required. The key is to select the combination of patterns to fit the damage.
5) Develop a constraint plan and design the jacking restraint configuration: The maximum allowable jacking force for members may be calculated by a licensed Professional Engineer, by the methods outlined in US DOT Report No. FHWA-IF-99-004, "Heat-Straightening Repairs of Damaged Steel Bridges," (Avent & Mukai 1998).
6) Estimate heating cycles required to straighten members: The number of heating cycles (n) can be estimated as $n = \varphi_d / \varphi_p$, where φ_d is the degree of damage and φ_p is the predicted plastic rotation per heating cycle.

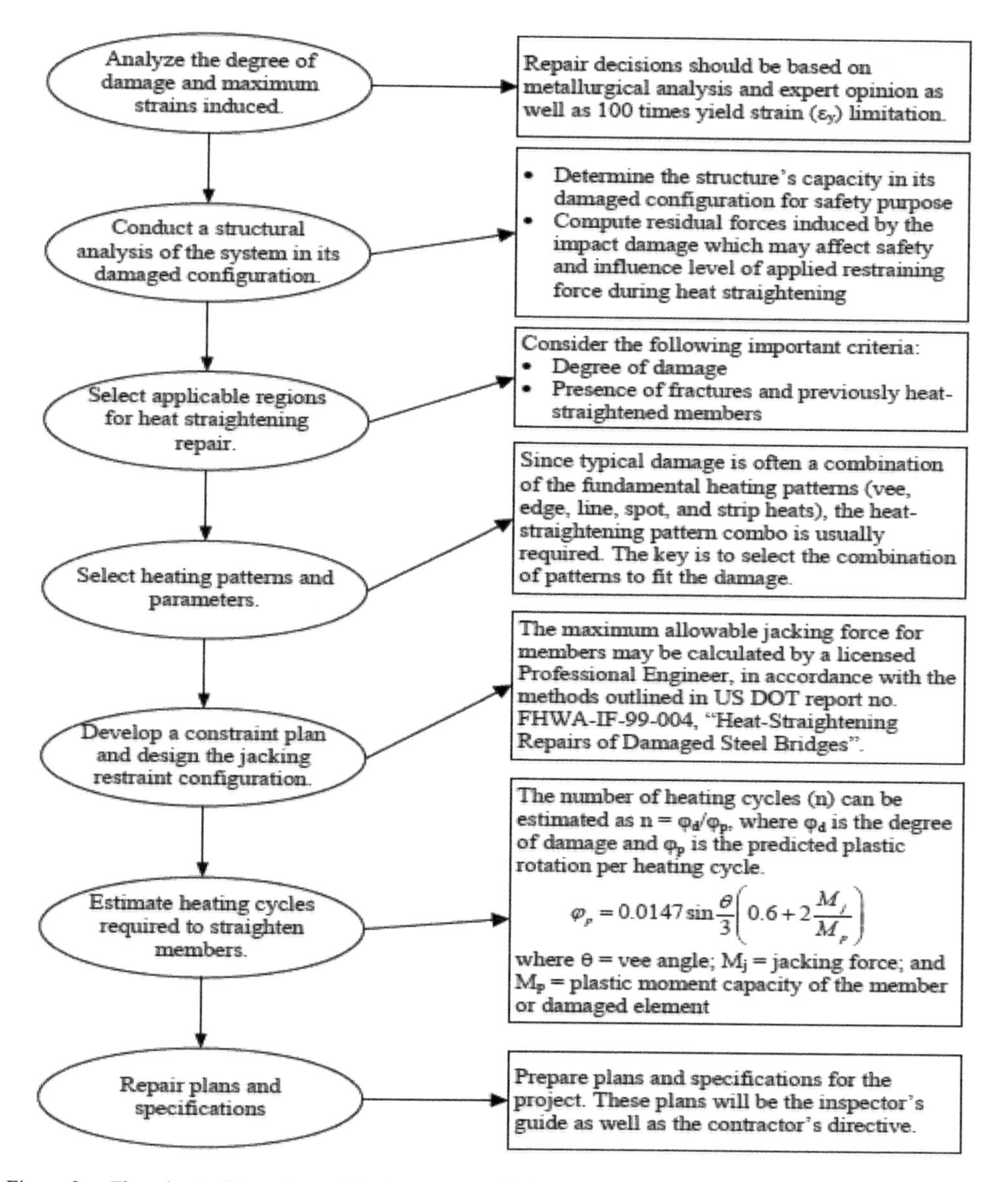

Figure 3. Flowchart of planning and design process (FHWA 2013).

7) Repair plans and specifications: This is the final step in the planning and design process. Plans and specifications will be prepared, and they will serve as the inspector's guide as well as the contractor's directive.

7 CONCLUSIONS AND RECOMMENDATIONS

The following recommendations and precautions for heat-straightening repairs were developed, taking into consideration the research findings, recommendations provided in NCHRP Report 604

(Connor et al. 2008), and other research reports/references from comprehensive literature review that was completed at the beginning of this study.

- It is recommended to limit the number of heat-straightening repairs at the same location to two as the third repair at the same location may result in a consistent decrease in base metal fatigue life (Connor et al. 2008).
- Test data have shown that there is no significant change in the upper shelf energy absorption before and after the heat straightening process for any grade of steel.
- Steel expands when heated and contracts during cooling. Heating temperatures should be limited to 620°C (1150°F) for non-quenched and tempered steels, 590°C (1100°F) for A514 and A709 Grade 100 and 100W quenched and tempered steels and 565°C (1050°F) for A709 Grade 70W quenched and tempered steel. Higher heats may adversely affect the material properties of the steel and lead to a weaker structure. For cooling, many practitioners allow the surface of the steel to cool below 315°C (600°F) before accelerating cooling. Rapid cooling is dangerous if the steel has been over-heated and may produce brittle "hot spots." However, once the steel has cooled below the lower phase transition temperature, rapid cooling is not harmful (FHWA 2013).
- A Federal Highway Administration Guide, published in 2013, has defined five fundamental heat-straightening patterns known as (a) the Vee- heat; (b) the edge heats; (c) the line heats; (d) the spot heats; and (e) the strip heats. The Vee-heat pattern is most effective in bending the damaged element about its major axis. The edge heat is often used to heat-curve rolled shapes in the fabricating shop. A line near the edge of the member is heated if a smooth, gentle bend is desired. The line heat is most effective in bending the damaged element about its minor axis. The spot heat is most effective in smoothing local bulging in the damaged element. The strip heat is most effective in shortening the damaged element. Typically, the heat-straightening patterns used to repair damaged steel bridge girders are combinations of the four heat-straightening patterns including Vee-, line-, spot-, and strip-heats. The selected heat-straightening patterns for a certain repair project shall be developed based on the structural damage caused by the collision and the need to keep portions of the bridge open to traffic while repairs are being conducted.
- Controlling the applied restraining forces is an important step during heat-straightening. The use of jacks to apply restraint can greatly shorten the number of heating cycles required as proven by research data. Over-jacking can, however, result in buckling or a brittle fracture during or shortly after heat straightening. Jacking forces should thus be limited to prevent a sudden fracture. The recommended procedure is to calculate the plastic moment capacity of the damaged member and limit the moment resulting from the combination of initial jacking forces and dead loads to one-half of this value. If practitioners do not take this precaution, brittle fractures or excessive deformation may occur. The jacking forces should always be applied in the direction tending to straighten the beam.
- The temperature of heated steel should be continuously monitored to make certain the material properties would not be adversely affected by overheating the steel. The heating temperature should be kept at 1150°F (620°C) for both Grade 36 and Grade 50 steels. One of the most accurate methods to monitor the heating temperature is to use temperature-sensing crayons. These crayons melt at a specified temperature and are available in increments as small as 14°C (25°F). By using two crayons that bracket the desired heating temperature, accurate control can be maintained. The crayons will burn if exposed directly to the flame of the torch. Therefore, the torch must be momentarily removed (one or two seconds) so that the crayons may be struck on the surface. An alternative is to strike the crayon on the backside at the point being heated. Another temperature monitoring method is to use a contact pyrometer. This device is basically a thermocouple connected to a readout device. The pyrometer should be calibrated with temperature crayons prior to using. Infrared devices are also available. These devices record the temperature and provide a digital readout. In addition to the crayons, pyrometer, or infrared devices, visually observe the color of the steel at the torch tip can be used. Under ordinary daylight conditions, a halo will form on the steel around the torch tip. At approximately 650°C (1200°F), this halo will have a satiny silver color in daylight or bright lighting at night. The observation of color is particularly

useful for the technician using the torch to maintain a constant temperature. However, this is the least accurate method of monitoring temperature.

- The amount of heat applied to a steel surface is a function of the type of fuel, the number and size of the orifices, the fuel pressure, and resulting heat output at the nozzle tip. Selecting the appropriate tip size is primarily a function of the thickness of the material. A tip that is too small for the thickness will result in insufficient heat input at the surface that does not penetrate effectively through the thickness. If the tip is too large, there will be a tendency to input heat into the region so quickly that it is difficult to control the temperature and distortion. The maximum torch tip size shall be limited to 1 inch (Kansas DOT 2007).
- According to a study on elevated temperature properties of A588 weathering steel by Princeton and Rutgers Universities (Garlock et al. 2014), at ambient temperature, the A588 steel has a larger ultimate stress (σ_u), similar yield stress (σ_y), smaller fracture toughness, and slightly larger hardness than other steel grades. At elevated temperatures, σ_u of the A588 steel decreases faster with increasing temperatures and the A588 steel is more ductile for temperatures less than 1000°F compared to other steel grades. The A588 steel seems to be affected more than other steel grades in terms of residual strength at 1500°F. At other temperatures, there is no significant effect on either material.

ACKNOWLEDGEMENTS

The research project was sponsored by the West Virginia Department of Transportation (WVDOT) Division of Highways (WVDOH), the Federal Highway Administration (FHWA), and the US Department of Transportation (USDOT). The authors would like to acknowledge the financial support provided by the WVDOH and USDOT. We would also like to acknowledge the great support provided by Glenn Lough of the West Virginia Department of Transportation.

REFERENCES

Avent, R. R., & Mukai, D. (1998). *Heat-Straightening Repairs of Damaged Steel Bridges: A Technical Guide and Manual of Practice* (Report No. FHWA-IF-99-004).

Connor, R. J., Urban, M. J., & Kaufmann, E. J. (2008). *Heat-straightening repair of damaged steel bridge girders: fatigue and fracture performance*. NCHRP 10-63, NCHRP Report 604, Transportation Research Board, Washington, D.C.

FHWA (2013). *Guide for Heat-Straightening of Damaged Steel Bridge Members*. Federal Highway Administration, Washington, D.C.

FHWA (2017). *National Bridge Inventory ASCII files*. Retrieved from https://www.fhwa.dot.gov/bridge/nbi/ascii.cfm

Garlock, M. E. M., Glassman, J. D., & Labbouz, S. (2014). *Elevated Temperature Properties of A588 Weathering Steel* (No. CAIT-UTC-021). Center for Advanced Infrastructure and Transportation, Rutgers, The State University of New Jersey.

Kansas Department of Transportation (2007). *Heat-Straightening (in-Place) of Damaged Structural Steel*. Division 700, July, Kansas DOT.

Zatar, W. and Leftwich, S. (2009). *Effect of Repeated Heat-Straightening on Behavior of Impacted Highway Bridge Steel Girders. Final Report (Research Project RP202),* West Virginia Department of Transportation, Division of Highways, Charleston, West Virginia.

Zatar, W. and Nguyen, H. (2018). *Effect of Heat-Straightening on Behavior of Impacted Highway Bridge Steel Beams – Phase II*. Draft Report (Research Project RP284), West Virginia Department of Transportation, Division of Highways, Charleston, West Virginia.

Chapter 18

Fracture detection in steel girder bridges using self-powered wireless sensors

M. Abedin, S. Farhangdoust & A. Mehrabi
Department of Civil and Environmental Engineering, Florida International University, Miami, FL, USA

ABSTRACT: Fracture Critical members are steel tension components whose failure is expected to result in collapse of the bridge. It is required to inspect fracture-critical bridges using "arms-length" approach, which is costly and time consuming. Structural health monitoring can be used as alternative approach for inspection providing both accuracy and economy. This paper investigates the feasibility of using a handful of self-powered wireless sensors for continuous monitoring and detection of fracture in steel plate girder bridges. A detailed finite element analysis was carried out on a multi-girder bridge using available traffic data. The time histories of displacement obtained for intact and fractured scenarios show that vibration amplitude was significantly increased for fractured girder, and strain variation was recorded especially in the vicinity of fracture, conditions that can be detected with relevant sensors. Moreover, the amplitude and frequency of the vibration was significant enough to provide the required power for typical sensor(s).

Keywords: Bridges, Steel Girder, Fracture Critical, Damage Detection, Health Monitoring, Self-powered Sensor

1 INTRODUCTION

According to the AASHTO LRFD Bridge Design Specifications (2017), "Fracture Critical members (FCMs) are steel tension members or steel tension components of members whose failure would be expected to result in collapse of the bridge." It is required that inspection of these bridges be carried out using "arms-length" approach, which is costly and is a drain on the total bridge budget. Structural health monitoring (SHM) could be used as an alternative approach to inspection providing both accuracy and economy for maintenance of this type of bridges. The approach discussed here will also be applicable to other bridges with fracture critical members.

SHM refers to a wide spectrum of activities and approaches to determining the changes in a structure and therefore determining its integrity and functional adequacy. This may range from routine visual inspection to sophisticated non-destructive evaluation techniques. It can be performed through periodic and on demand inspection, or can be carried out through continuous monitoring systems installed on the structure. One may consider SHM in two major categories; methods for which the structure is installed with sensors (sensor-based SHM), and methods for which health of the structure is evaluated without a sensor and using external devices. Non-destructive evaluation methods, visual inspection and other vision-based methods can be considered in the latter group. Sensor-based SHM system typically has three major subsystems: a sensor, a data acquisition, and a diagnosis subsystem. The diagnosis subsystem generally includes data processing, data mining, damage detection, and can go further to expand to model updating, and structural safety and reliability determination. The accuracy of results is largely dependent on not only the type and sophistication of the sensors and instruments, but also on the variety, quantity and quality of the measured data.

In a larger scale, optimal sensor placement has been a concern among researchers and has been studied widely. The goal is to improve the ability of sensor subsystem with the least number of sensors possible (Yi et al. 2011; Yuen and Kuok 2015). The quality of the collected data is also a matter of attention (Huang et al. 2016). Inaccurate results could lead to false alarms or to miss the recording of an event, therefore lead to unsafe structure and other consequences. A sensor commonly consists of different elements, including sensing component, transducer, signal-processing and communication interface module. Malfunction, harsh environment and normal wear and tear, as well as other factors such as electromagnetic interference, may lead to distortion of results and false data. For damage detection in large-scale and distributed systems, using large number of sensors is a common trend because it promises more coverage hence better chance of detection (Sohn 2014). This in turn introduces challenges for data collection and processing, and for meaningful interpretation of results. Developing a simple economical, flexible and at the same time accurate sensor is very much in demand. Sensor types developed for SHM purposes in recent decades include electric strain gages, piezoelectric sensors, cement-based strain gauge, corrosion sensors, Nano material-based sensors, wireless sensors, accelerometers, inclinometers, acoustic emission sensors, wave propagation devices and various fiber optic sensors (FOS). Additionally, advanced sensor and sensing technologies, such as fiber optics and Bragg grating sensors (Antunes et al. 2012; Ding et al. 2017) and those based on Global Positioning System (Moschas and Stiros 2013) have been recently developed for strain, displacement, and other response measurement.

Each of the sensor types has a certain application and works better in certain conditions. Nevertheless, actual use of these sensors presents some challenges in real environment. For instance, PZT (lead-zirconate-titanate)-based active and passive damage detection technologies can be used as an acoustic emission (AE) sensor, which receives the stress wave signal generated by damage occurred in a structure. Cement-based strain sensor is considered as one appropriate candidate to solve the incompatibly issue. Dispersed sensors have also been used for monitoring purposes. Incorporated with fibers, conductive Nano-particles, magnetized or magnetic metals, PZT, or a combination can give concrete and other medium a sensing ability.

1.1 *Statement of problem*

Bridges identified with fracture critical members are required detailed inspection at maximum 2-year interval which is costly and is a drain on the total bridge maintenance budget. In the positive bending moment region of steel girders, the bottom flanges are considered to be fracture-critical elements depending on the number of girders and bridge configuration (Connor et al. 2018; Hebdon et al. 2017). Inspection of girders' bottom flanges near mid-span over a busy roadway is costly, time consuming and causes traffic disruption and potential safety hazards (Rahimi et al. 2019; Samimi et al. 2019). To address cost, traffic interruption and safety issues associated with conventional inspection, Structural health monitoring (SHM) could be used as an alternative providing both accuracy and economy for maintenance of this type of bridges.

Traditional wired sensors may enable continuous monitoring of the bridge. However, the cost of installing wired sensors for longer span bridges, providing continuous power, and their maintenance may make them impractical in many cases. To address the shortcomings associated with wired sensors, the use of wireless sensors has offered a valuable alternative. Nevertheless, these sensors typically rely on battery energy source for operation and the cost of periodic battery replacement for large scale monitoring would constitute a major expense. New developments for "energy harvesting" to feed the sensors could be a potential solution for providing continuous power to permanent sensors. Energy harvesting techniques are mainly based on solar energy, thermal gradients and vibration energy. Among these techniques, vibration energy because of providing high level of energy and the ability for being embedded (as in concrete structures) could be used for a continuous large-scale monitoring (Elvin et al. 2006).

There are two main questions need to be addressed to verify the feasibility of using wireless self-powered sensors for detection of fracture in fracture critical members. One is whether the bridge vibration under live load would generate enough energy to power the sensors, and the other, if variation in stress/strain at predefined locations and dynamic characteristics of the bridge resulted from fracture of a girder could be exploited as a means for detection of the fracture.

1.2 *Objectives and approach*

The objective of this study is to explore new means for detecting fracture in fracture-critical bridge members that are both accurate and affordable at the same time. A timely detection of the onset of fracture will allow the maintenance crew to address the situation before the progress in damage threatens the public safety and requires major closures and costs.

To achieve the objective of this investigation, a two-fold investigation was performed. One was to verify that the changes in stress/strain state or dynamic characteristics of the bridge is adequate and clear to be detected by a set of optimally located sensors. And the other was to determine whether the frequency and amplitude of the vibrations of the bridge from traveling vehicles and that caused by the fracture is large enough to power or trigger the wireless self-powered sensors installed strategically to detect the occurrence of the fracture, as well as the feasibility of using self-powered wireless sensors for continuous monitoring of steel plate girder bridges. One means for facilitating the use of self-powered wireless sensors is by harvesting the vibration energy of the bridge. In addition to the bridge vibration providing power for the sensors, variation in dynamic characteristics of the bridge resulted from fracture of girder can also be exploited as a means for detection of the fracture. The bridge vibration and displacement amplitude depends on the bridge characteristics and traffic loading. These parameters could vary based on the bridge geometry, design, location, type of vehicles crossing the bridge, and Average Daily Truck Traffic (ADTT). A knowledge of the vibration characteristics of the bridge in its intact and fractured conditions can be developed by actual measurements or preferably by dynamic analysis of the finite element (FE) model of the bridge. This can lead to establishment of response thresholds for the bridge for indication of damage if certain threshold is surpassed. In this case, the sensor can be kept in waiting mode until the threshold is reached. The sensor can then begin emitting a signal as indication of the damage/fracture. Alternatively, the stress/strain variation in the vicinity of the fracture recorded using self-powered wireless sensors on a continuous basis can be used to determine the type, location and intensity of the fracture event.

Once one of the girders fractured, the bridge vibration characteristic and displacement amplitude could change significantly because of reduction in bending stiffness potentially resulting in greater vibration amplitudes that may be adequate for functioning of the sensors. Therefore, even if the amplitude and frequency of the intact bridge vibration would not be adequate to provide the necessary power for the sensors, the vibration of the fractured bridge will potentially activate the sensors. The novelty of the proposed approach is that the self-powered sensors that are dormant for the intact bridge can start monitoring once the bridge is fractured, and send a warning to the owner for taking action.

2 CHARACTERISTICS OF BRIDGE VIBRATION

A multiple steel plate girder bridge tested at the University of Nebraska–Lincoln (Kathol et al. 1995) was selected to investigate the bridge dynamic response for intact and fractured scenarios. This bridge is a full-scale simple span bridge with a span length of 21.3 m (70 ft) and is 7.9 m (26 ft) wide accommodating for two traffic lanes. The superstructure consists of three welded steel plate girders made composite with a 0.2 m (7 1/2 in.) reinforced concrete deck as shown in Figure 1. The girders are spaced 3 m (10 ft) on center and the reinforced concrete deck has a 0.9 m (3 ft) overhang and the railing system is a typical Nebraska Department of Road (NDOR) open concrete bridge rail, with 0.3 × 0.3 m (11 × 11 in.) posts spaced 2.4 m (8 ft) on center. Because of availability of reliable and extensive experimental results that included service and ultimate load testing and associate failure modes, modeling of this bridge was thought to provide a good background for validation of the FE modeling method adopted for this study.

Several tests were conducted on this bridge to evaluate the effect of diaphragms, elastic behavior and ultimate load carrying capacity of the bridge. The ultimate test, which consisted of loading the bridge to collapse, was selected for validating the capability of FE modeling adopted in this study for predicting the elastic behavior and ultimate capacity and failure modes. The bridge failure in

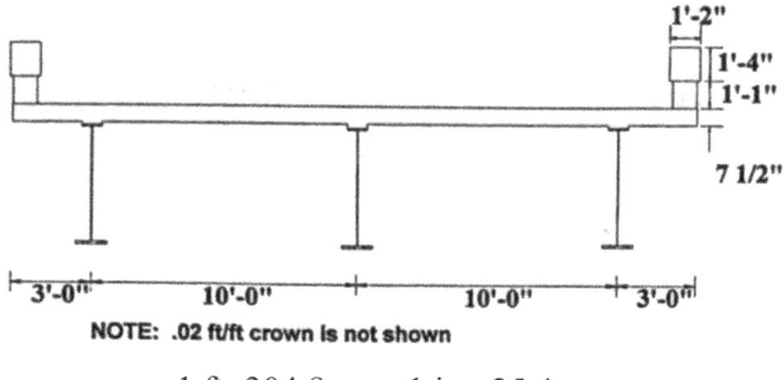

1 ft=304.8 mm, 1 in.=25.4 mm

Figure 1. The cross section of University of Nebraska–Lincoln Multiple Plate Girder Bridge (Kathol et al. 1995).

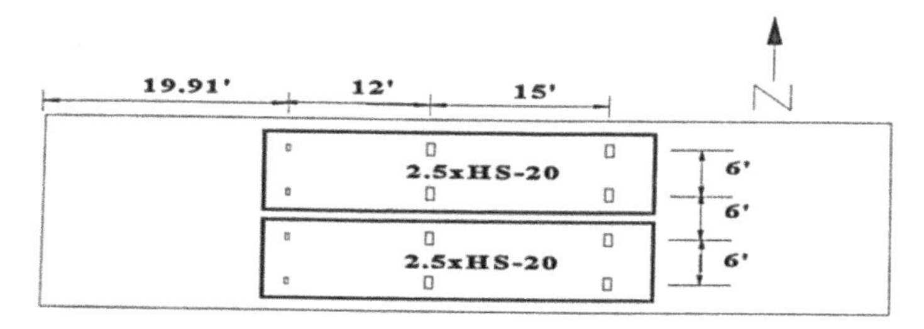

Figure 2. The loading configuration of the ultimate test (Kathol et al. 1995).

Figure 3. The University of Nebraska–Lincoln Multiple Plate Girder Bridge (Kathol et al. 1995).

the laboratory testing was governed by local punching shear failure in the deck under the loading plates. Figures 2 and 3 show the loading configuration of the ultimate test.

2.1 *Finite element method*

To be able to verify the adequacy of bridge vibration in providing the energy required for self-powered sensors, as well as understanding of the dynamic characteristics for two conditions of

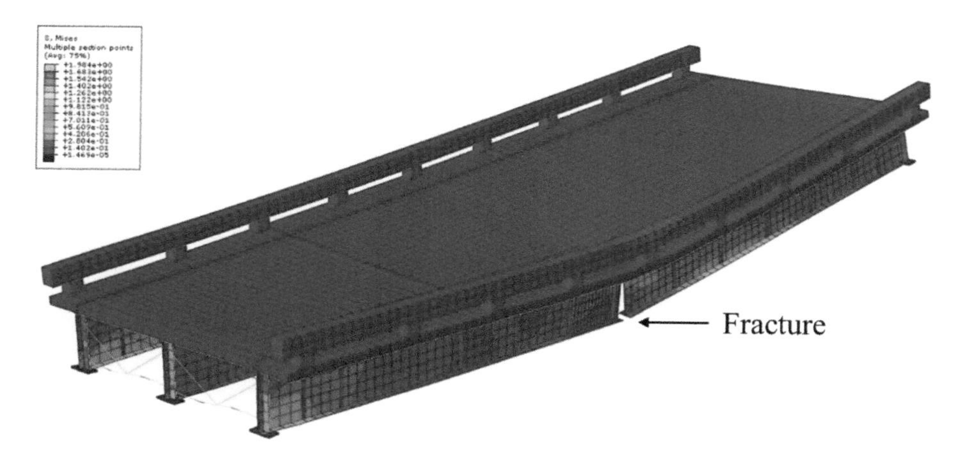

Figure 4. Simulation of damage (fracture of exterior girder) in the finite element model.

intact and fractured girder(s), a detailed finite element (FE) analysis was necessary. Finite element modeling has been recognized as a reliable means for detailed analysis of steel plate girder bridges to investigate their vibration and displacement amplitude in both intact and fractured girder conditions. This method offers an efficient alternative to differential equation and finite difference methods utilized by others (Farhangdoust et al. 2019). Construction of a detailed FE model of the bridge and analysis under loading of various configuration is a time-consuming and costly activity. Modeling of every detail in the bridge is neither economic nor always necessary. Additionally, solution methods available for numerical analysis of FE models are numerous and not always end to proper convergence and accurate results. Hence, application of FE modeling and analysis can be quite complex, and finding an optimum level of refinement and modeling details, as well as proper solution method require performing some experimentation and validation. Validation can be performed by modeling and analysis of bridges that are tested and for which adequate data on the behavior is available. The FE model for the steel plate girder bridge adopted for this study was created in the environment of ABAQUS (Dassault 2016) to simulate the response of the bridge under the intact and fractured girder scenarios. Figure 4 shows simulation of damage (fracture of exterior girder) in the finite element model. The proper modeling techniques, analysis procedure and material inputs were investigated thoroughly.

2.1.1 *Material*

Multi-linear inelastic material model with isotropic hardening is used for the behavior of steel plates, diaphragms and reinforcement in both tension and compression. The linear elastic behavior was defined by the specification of the modulus of elasticity and Poisson's ratio, which were 200,000 MPa (29,000 ksi) and 0.3, respectively. Yield and ultimate stress of steel material is considered as the typical value used in bridges, 345 MPa (50 ksi) for the steel plates and 414 MPa (60 ksi) for the concrete reinforcing bars is assumed as the yield strength of steel materials. An effective damping ratio of 3% is assumed for both steel and concrete materials. According to von Mises theory, the material yields when the equivalent stress exceeds the yield criterion. A linear elasticity with the concrete damage plasticity (Lubliner et al. 1989) is used for the concrete elements and for the initial elastic behavior, modulus of elasticity is calculated based on the ACI 318-14 (2014) (for normal-weight concrete) and a Poisson ratio of 0.2 was used.

2.1.2 *Finite element validation*

For simulating the bridge behavior during construction, finite element analysis was divided into two main steps: bridge construction and final analysis for live loading. For the first step, an initial

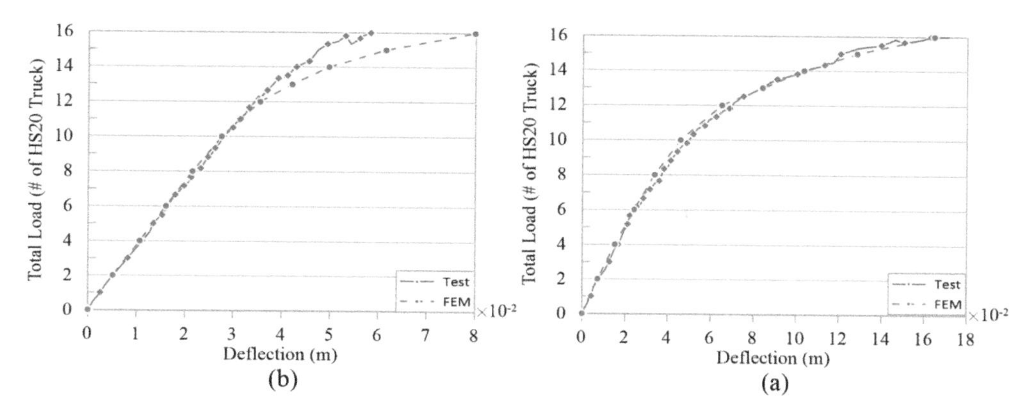

Figure 5. Comparison of load- deflection curves obtained from experiment and FE model. (a) Interior Girder and (b) Exterior Girder.

implicit static analysis was used to incorporate loading effect through the erection and construction phase when the concrete is not hardened yet and the section acts non-compositely with only the girders carrying the dead load. During the bridge construction, only the girders carry the deck, and the dead load deflections in the girders remain locked after the concrete deck hardens. For this reason, the stiffness and mass of the concrete and reinforcing rebar were reduced to a very low value during the construction phase and equivalent dead load of the deck was applied on the top flange of the girders based on the tributary area. Moreover, self-weight of the structural steel of the girder components was applied on the model at this stage. By reducing the stiffness of the deck to negligible, only girders carry the load and there will be no stress and strain on the concrete deck at the end of construction phase once the concrete deck has hardened.

The results of the first step was used as an initial predefined state for the final analysis step. In other words, initial states (stresses, strains, displacements and forces) for the final analysis step is the final state at completion of bridge construction. From this point on, the girder and slab sections act compositely together. Therefore, the initial equivalent uniform dead load of the concrete on the girders considered in the analysis for the previous step was removed and replaced by concrete with its actual stiffness and mass. The concrete damaged plasticity was also activated in this stage. Moreover, based on the construction procedure, railing elements were added at this step. At the final step, increasing loads were applied on the bridge by using the explicit dynamic solution method.

Figure 5 shows the comparison of load- deflection curves between experimental and FE results for exterior and interior girders. Deflection refers to vertical displacement of the bottom flange of each (exterior or interior) girder at the mid-span. Deflection was measured using displacement transducers in the laboratory tests and calculated through analysis using finite element method (FEM) for analytical simulation. As shown in these graphs, the FE model can predict the global behavior of the bridge during the elastic and plastic states. It was also able to simulate the local failure due to punching shear in the deck and cracking in the railing.

2.2 *Dynamic analyses*

Once the FE model of the bridge was validated by comparing the results with the experimental tests, dynamic analyses were performed for the bridge in intact and fractured scenarios. These analyses used simulation of actual traffic loading to investigate the dynamic response of the bridge in these two scenarios. In the fractured scenario, one of the exterior girders was fractured through the bottom flange and web at the middle of the span to investigate the worst fractured scenario that may occur for the single span three steel plate girder bridge. The two-step analysis of the bridge construction and final analysis for live loading was conducted for the steel plate girder bridge as explained

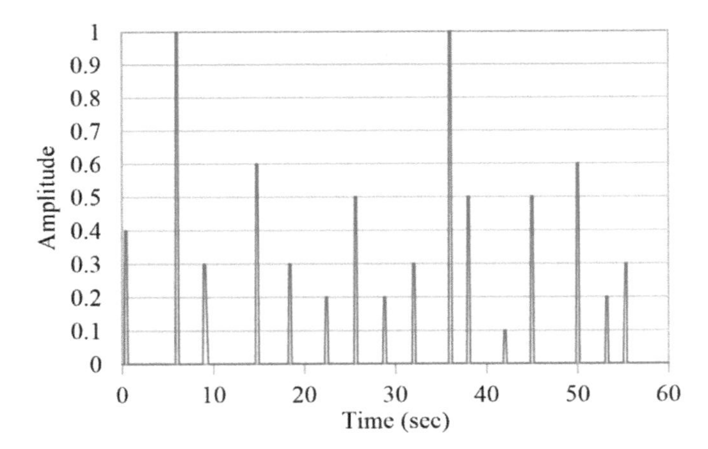

Figure 6. Amplitude curve of the moving traffic load.

earlier. To model the sudden girder fracture, tie constraint was assigned between the elements of the girder web and flange on two sides of the fracture at the first step and was removed for the final analysis. Weigh-in-Motion (WIM) data from a station in the state of Florida collected throughout 4 years (2013–2016) was obtained from the US Federal Highway Administration (FHWA) and was used for simulating the moving traffic loads. Truck weight data format was used as it contains information such as, but not limited to, number of axles, spacing between axles, axle weights and gross vehicle weight (GVW) and exact time of measurement for each recorded vehicle at each location.

Moving traffic load in the model was simulated by defining multiple tire contact area along the bridge only for the lane over the fractured girder (eccentric loading) and assigning amplitude intensity pattern to them based on the WIM data for a period of 60 seconds. Figure 6 shows the amplitude curve assigned to a one tire contact area during this 60-second time period based on the WIM data to simulate the actual traffic consisting of trucks and cars. Amplitude of one (1) on the vertical axis in this figure is equal to the average weight of heavier truck wheel over a month with the average truck weight of 20800 kg (46 kips) and other amplitudes, proportioned with respect to the latter, are a representative of truck loading for a variety of cars and light trucks.

3 RESULTS

Figures 7 and 8 show the time histories and frequency spectrum of girder deflection obtained from the FE model for intact and fractured scenarios at the middle of exterior girder (fractured), respectively, under the moving traffic of 60-sec duration. The simulation necessitated discretization for movement of the loads along the span. Moving a load from one point to the next is modeled by removing the load at one point and adding to the next, as it occurs in actual case with some continuity. Response of the bridge span to this moving load (one of several situated on the bridge) generates oscillations as it has also been measured in similar conditions for actual bridge reported in this paper. It should be pointed out that, in practice, time history curves shown in these figures can be generated by instrumenting the bridge using several types of contacting sensors among which are strain gages, displacement transducers, and accelerometers, as well as non-contacting transducers including vision based equipment and laser vibrometers.

The time history results show that the vibration amplitude of the girder at the middle of the span has almost doubled and at quarter points increased in average about 60% under normal traffic load because of the fracture. Therefore, a threshold can be predefined for the girder bridge deflection amplitude crossing which would be an indication of fracture in the girder. Furthermore, the results

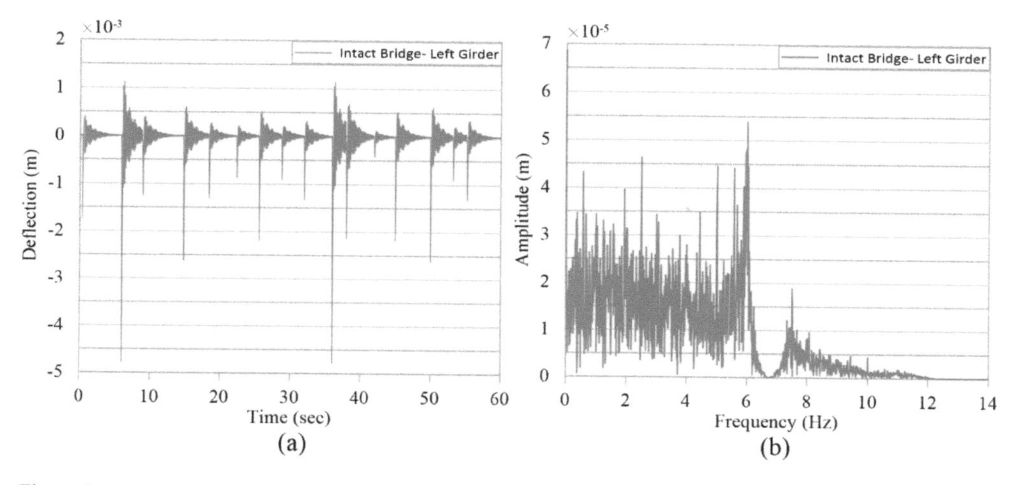

Figure 7. Intact Bridge. (a) Time histories and (b) frequency spectrum of girder deflection.

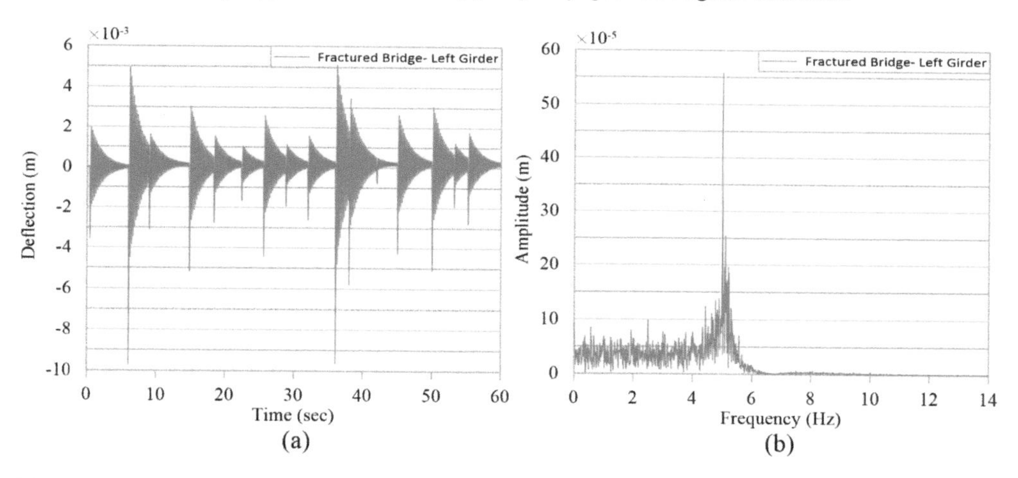

Figure 8. Bridge with Fractured Girder. (a) Time histories and (b) frequency spectrum of girder deflection.

in frequency domain indicate two major changes after the fracture has occurred. One is a major shift in the fundamental/dominant mode vibration frequency of the bridge. The major peak in the frequency spectra can be used to determine this frequency. For the intact condition, the frequency stands at 6 Hz. For the fractured condition, this frequency has shifted lower to 5 Hz. This shift in dominant frequency can be exploited for signaling occurrence of a major event, in this case, fracture of the girder. In a continuous monitoring system, a safe buffer zone can be established and specified for the dominant modal frequency based on expected minor variation in the frequency due to ambient effects. If the monitoring system indicates that the dominant frequency has fallen outside this buffer zone, it will be indicative of a significant event, i.e., girder fracture.

3.1 *Sudden fracture under dead load*

Figure 9 shows the time histories of the girder deflection due to a sudden bottom flange and full web fracture of the girder under dead load. In this case, the fractured girder (exterior girder) has 15.2 mm vertical displacement at the middle of the span under the bridge self-weight. The results obtained for fracture of girder under dead load shows very noticeable amplitudes in all three girders.

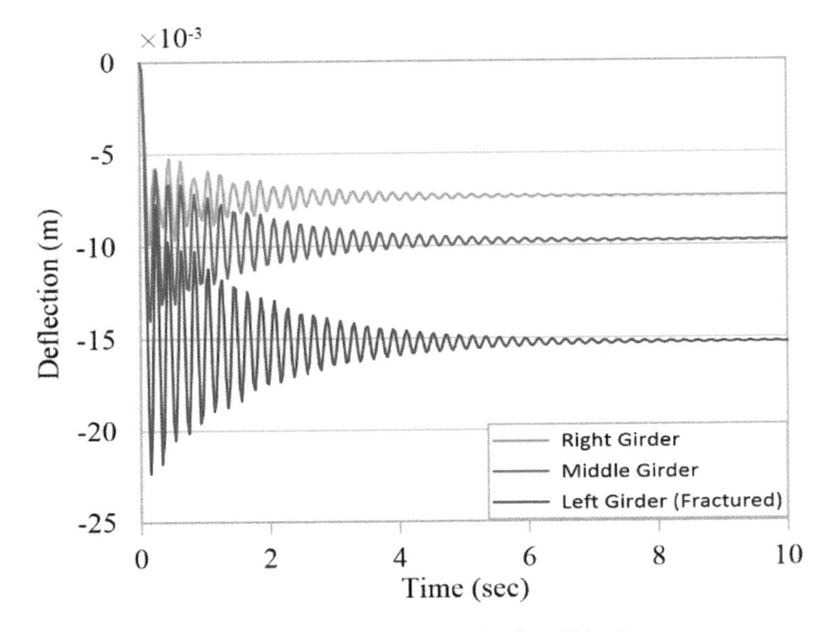

Figure 9. Time-deflection curve for the fractured bridge under the self-load.

Any such vibration signature can be interpreted as major event in the bridge. The signatures can be pre-identified and through a pattern recognition software correlated to fracture in a certain girder at certain location. Differences between amplitudes of the three girder is clearly indicative of which girder has suffered the fracture.

3.2 *Strains under moving traffic*

To investigate the strain variation of the steel plate girders resulted from fracture of a girder, time histories of girders longitudinal strain were calculated for mid- and quarter-length along the bridge for the bottom flange. The strain record corresponding to the middle of span for all three girders are presented in Figure 10.

The strain results clearly show highly noticeable changes in the response of all girders for fracture scenario. The strains in the left girder (fractured) show the highest drop in the fracture scenario. This is indicative of presence of fracture in this girder where the strain near the fracture would relax because of significant reduction in stiffness and load distribution to other girders. As expected, the middle girder shows increase in the strain indicating absorbing some of the load from the fractured girder. The right girder shows the least of variation in strain. This pattern is indicative of firstly the occurrence of the fracture and secondly the girder and location where the fracture is located. Table 1 summarizes the test results for all three girders in different scenarios and in three locations along each girder; middle and quarter points. This table shows average of peaks in strain records shown in Figure 10. A pattern similar to that described above for strain changes can be seen in Table 1. Along the same girder, changes in the strains vary depending on the proximity of the measurement point to the fracture zone. Percentage of strain drop for location closer to the fracture is higher than farther points. Again providing valuable data on the occurrence and location of the fracture.

3.3 *Adequacy of vibration energy for powering sensors*

The results of the finite element analysis were also reviewed to investigate the feasibility of using self-powered wireless sensors for continuous monitoring of steel plate girder bridges for detection

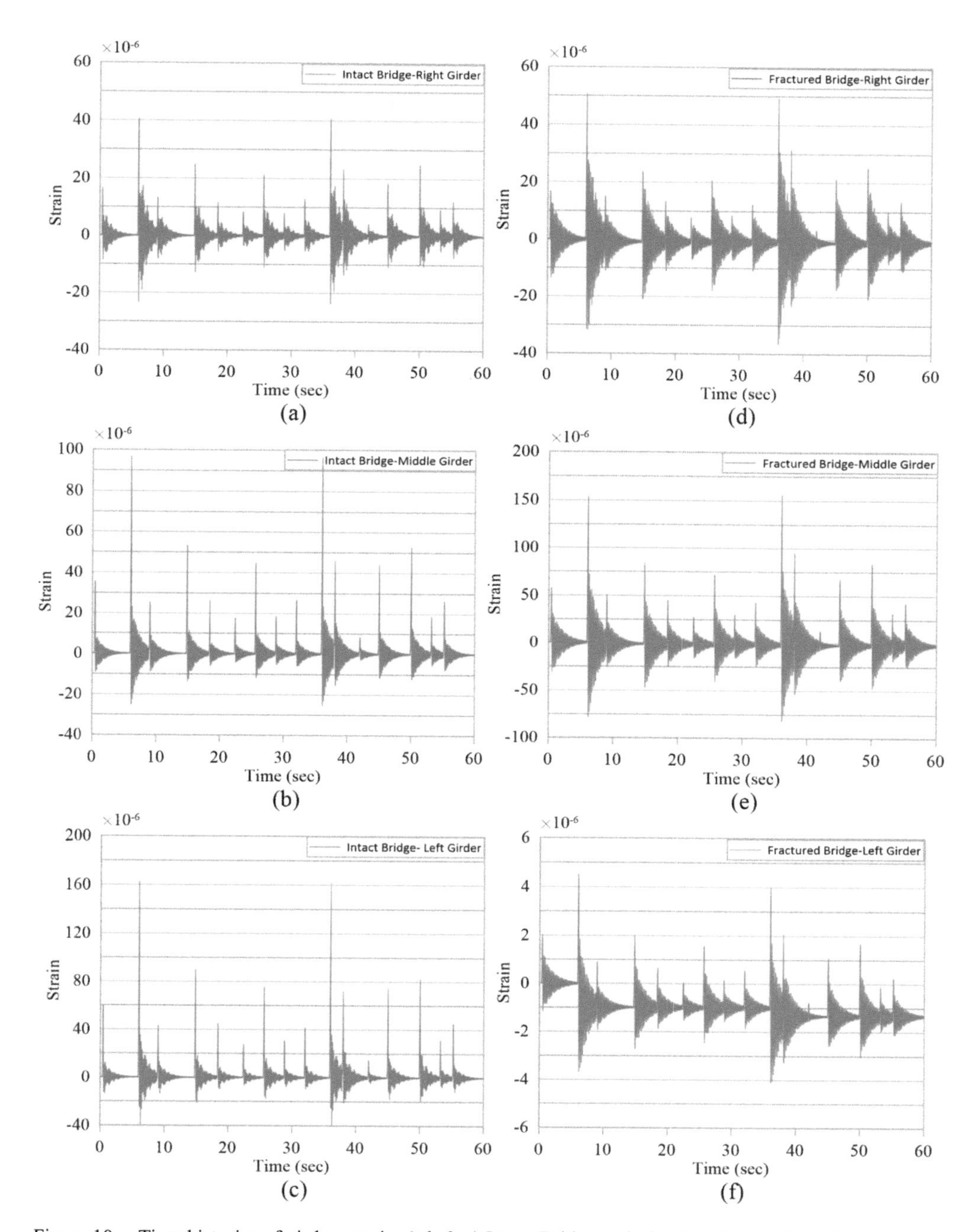

Figure 10. Time histories of girders strain. (a,b & c) Intact Bridge and (d,e & f) Fractured Bridge.

of fracture. Sazonov et al. (2009) in their study demonstrated self-power operation of the bridge sensor in a short-term test. They installed the sensors on one of the girders of a simple three span steel plate girder bridge and a linear electromagnetic generator was used to harvest vibration energy created by passing traffic. The simple electromagnetic energy harvester with a displacement mass

Table 1. Summary of the test results.

	Average Peak Strain mm/mm $\times 10^{-6}$						Average Peak Deflection mm					
	Intact Bridge			Fractured Bridge			Intact Bridge			Fractured Bridge		
Girders	Left	Middle	Right	Left	Middle	Right	Left	Middle	Right	Left	Middle	Right
One-Quarter	29.9	27.1	10.1	35.3	39.2	12.8	−1.5	−1.1	−0.6	−2.4	−1.6	−0.6
Mid-Span	42.6	40.1	16.9	1.4	63.1	20.4	−2.1	−1.5	−0.8	−4	−2.2	−0.9
Three-Quarter	29.1	25.1	9.9	35.1	38.9	12.6	−1.5	−1.1	−0.7	−2.2	−1.6	−0.7

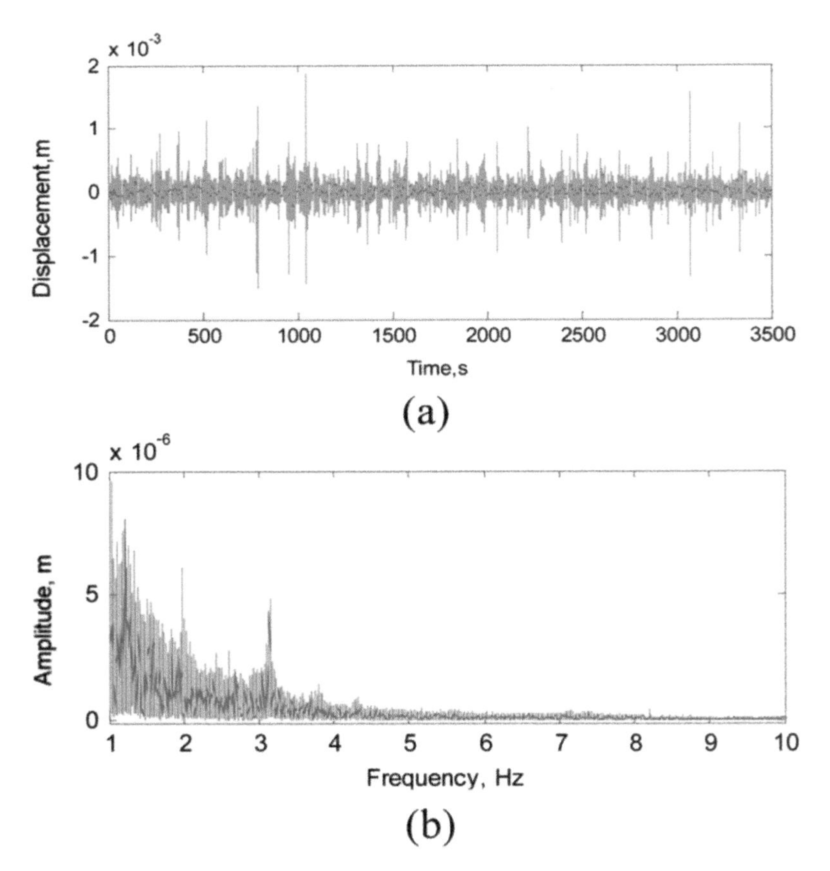

Figure 11. (a) Time history and (b) frequency spectrum of girder displacement obtained from the study by Sazonov et al. (Sazonov et al. 2009).

of 0.09 kg used in that study allowed harvesting energy of up to 12.5 mW at 10 mm displacement amplitude in the resonant mode with the frequency of excitation at 3.1 Hz. Moreover, the generator could deliver up to 1 mW at a low amplitude of 3 mm, which is sufficient for trickle charging of storage. The results of field test indicated the feasibility of self-powered operation for long-term monitoring of a rural highway bridge with low traffic volume. A vibration time history and frequency spectrum obtained from their investigation is shown in Figure 11.

Comparing these results with those obtained from FE analysis conducted in this study (Time histories and frequency spectrum of girder displacement) show that the vibration in the intact and fractured girders in this study is higher or comparable to the results reported by Sazanov et al. (2009), therefore, indicating the feasibility of designing electromagnetic generator (or a similar energy harvesting system) that could feed the sensors for monitoring for detection of fracture in steel girders.

4 CONCLUSIONS

The goal of this study was to demonstrate that dynamic characteristics of the fracture critical steel girder bridges would change significantly once a fracture occurs on one of the girders so that the variation can be used to detect the occurrence and location of the fracture, and that the frequency and displacement amplitude of the fractured bridge under the traffic load would be sufficient for feeding the sensors by using energy harvesting techniques. Moreover, the location of fracture can be captured by comparing the variation of strain/stress before and after fracture.

A multiple steel plate girder bridge tested at the University of Nebraska–Lincoln was selected to investigate the bridge dynamic response for intact and fractured scenarios. Finite element analyses were conducted using a model of this bridge under moving traffic load as well as under dead load. The time history results from analysis of the intact bridge and bridge with one fractured girder showed that the vibration amplitude of the girder at the middle of the span has almost doubled and at quarter points increased in average about 60% under normal traffic load because of the fracture. Therefore, a threshold can be predefined for the girder bridge deflection amplitude crossing which would be an indication of fracture in the girder. The results also showed that the shift in dominant frequency recorded for the bridge and increase in its amplitude can be exploited for signaling occurrence of a major event, in this case, fracture of the girder. In a continuous monitoring system, a safe buffer zone can be established and specified for the dominant modal frequency based on expected minor variation in the frequency due to ambient effects. If the monitoring system indicates that the dominant frequency has fallen outside this buffer zone, it will be indicative of a significant event, i.e., girder fracture.

When strain records for the bridge under moving traffic were compared for intact and one girder fracture scenarios, the results clearly demonstrated a highly noticeable changes in the response of all girders for fracture scenario. The strains in the fractured girder showed the highest drop in the fracture scenario. This is indicative of presence of fracture in this girder where the strain near the fracture would relax because of significant reduction in stiffness and load distribution to other girders. As expected, the middle girder showed increase in the strain indicating absorbing some of the load from the fractured girder. The right girder shows the least of variation in strain. This pattern can be used as indication of the occurrence of the fracture and determining the girder and location where the fracture is located.

By comparing the time histories and frequency spectrum of girder displacement obtained from this study with the results of field experiment conducted by others for investigating the energy required harvesting techniques showed the feasibility of using self-powered wireless sensors for monitoring the steel bridges for their fracture. The test results indicate that the amount of energy available for self-powered sensors depends on the sensor location along the span, and that the displacement amplitude and frequency are sufficient for energy harvesting in the middle of the simple span bridges.

A very important conclusion drawn from the FE tests is that the self-powered sensors that may be dormant for the intact bridge for a long period of time (because of lower vibration amplitudes) can start monitoring once the bridge is fractured (with higher vibration amplitudes), and send a warning to the owner for taking action. In summary, the results show a great potential for the use of self-powered wireless sensors for Structural health monitoring (SHM) of fracture critical steel girder bridges with accuracy and economy as an alternative to detailed and costly inspection.

REFERENCES

AASHTO. 2017. AASHTO LRFD Bridge Design Specifications. *American Association of State Highway and Transportation Officials, 8th Edition, Washington, D.C.*

American Concrete Institute, 2014. *Building Code Requirements for Structural Concrete (ACI 318-14): Commentary on Building Code Requirements for Structural Concrete (ACI 318R-14): an ACI Report.* American Concrete Institute. ACI.

Antunes, P., Lima, H., Varum, H. and André, P., 2012. Optical fiber sensors for static and dynamic health monitoring of civil engineering infrastructures: Abode wall case study. *Measurement*, 45(7), pp.1695–1705.

Connor, R.J., Martín, B., Francisco, J., Varma, A., Lai, Z. and Korkmaz, C., 2018. *Fracture-Critical System Analysis for Steel Bridges* (No. Project 12-87A).

Ding, Y.L., Zhao, H.W. and Li, A.Q., 2017. Temperature effects on strain influence lines and dynamic load factors in a steel-truss arch railway bridge using adaptive FIR filtering. *Journal of Performance of Constructed Facilities, 31*(4), p.04017024.

Elvin, N.G., Lajnef, N. and Elvin, A.A., 2006. Feasibility of structural monitoring with vibration powered sensors. *Smart materials and structures, 15*(4), p.977.

Farhangdoust, S., Mehrabi, A. and Younesian, D., 2019, April. Bistable wind-induced vibration energy harvester for self-powered wireless sensors in smart bridge monitoring systems. *In Nondestructive Characterization and Monitoring of Advanced Materials, Aerospace, Civil Infrastructure, and Transportation XIII* (Vol. 10971, p. 109710C). International Society for Optics and Photonics.

Hebdon, M.H., Singh, J. and Connor, R.J., 2017. Redundancy and Fracture Resilience of Built-Up Steel Girders. In *Structures Congress 2017*.

Huang, H.B., Yi, T.H. and Li, H.N., 2016. Canonical correlation analysis based fault diagnosis method for structural monitoring sensor networks. *Smart Struct. Syst, 17*(6), pp.1031–1053.

Kathol, S., Azizinamini, A. and Luedke, J., 1995. *Strength capacity of steel girder bridges. Final Report* (No. RES1 (0099) P469).

Lubliner, J., Oliver, J., Oller, S. and Oñate, E., 1989. A plastic-damage model for concrete. *International Journal of solids and structures, 25*(3), pp.299–326.

Moschas, F. and Stiros, S., 2013. Noise characteristics of high-frequency, short-duration GPS records from analysis of identical, collocated instruments. *Measurement, 46*(4), pp.1488–1506.

Rahimi, A., Azimi, G., Asgari, H. and Jin, X., 2019. Clustering Approach toward Large Truck Crash Analysis. *Transportation Research Record*, p.0361198119839347.

Samimi, A., Rahimi, E., Amini, H. and Jamshidi, H., 2019. Freight modal policies toward a sustainable society. *Scientia Iranica*.

Sazonov, E., Li, H., Curry, D. and Pillay, P., 2009. Self-powered sensors for monitoring of highway bridges. *IEEE Sensors Journal, 9*(11), pp.1422–1429.

Simulia, D.S., 2013. ABAQUS 6.13 User's manual. *Dassault Systems, Providence, RI.*

Sohn, H., 2014, May. Noncontact laser sensing technology for structural health monitoring and nondestructive testing (presentation video). In *Bioinspiration, Biomimetics, and Bioreplication 2014* (Vol. 9055, p. 90550W). International Society for Optics and Photonics.

Yi, T.H., Li, H.N. and Gu, M., 2011. Optimal sensor placement for structural health monitoring based on multiple optimization strategies. *The Structural Design of Tall and Special Buildings, 20*(7), pp.881–900.

Yuen, K.V. and Kuok, S.C., 2015. Efficient Bayesian sensor placement algorithm for structural identification: a general approach for multi-type sensory systems. *Earthquake Engineering & Structural Dynamics, 44*(5), pp.757–774.

Chapter 19

A convolutional cost-sensitive crack localization algorithm for automated and reliable RC bridge inspection

S.O. Sajedi & X. Liang
Department of Civil, Structural and Environmental Engineering, University at Buffalo, New York, USA

ABSTRACT: Bridges are an essential part of the transportation infrastructure and need to be monitored periodically. Visual inspections by dedicated teams have been one of the primary tools in structural health monitoring (SHM) of bridge structures. However, such conventional methods have certain shortcomings. Manual inspections may be challenging in harsh environments and are commonly biased in nature. In the last decade, camera-equipped unmanned aerial vehicles (UAVs) have been widely used for visual inspections; however, the task of automatically extracting useful information from raw images is still challenging. In this paper, a deep learning semantic segmentation framework is proposed to automatically localize surface cracks. Due to the high imbalance of crack and background classes in images, different strategies are investigated to improve performance and reliability. The trained models are tested on real-world crack images showing impressive robustness in terms of the metrics defined by the concepts of precision and recall. These techniques can be used in SHM of bridges to extract useful information from the unprocessed images taken from UAVs.

1 INTRODUCTION

Reduced time to recovery is one of the fundamental characteristics of resilient systems. Bridge infrastructures, as critical components in a transportation system, play an important role in large communities. Such structures should maintain their functionality in harsh environments especially after extreme events (e.g., earthquakes). Given the number of aging infrastructure across the united states, structural health monitoring (SHM) techniques have been widely used to periodically monitor the condition of bridges. Visual inspections are one of the most common ways of condition assessment where this task is conventionally performed by dedicated teams. There are several drawbacks for human inspections. Having dedicated teams for this purpose requires time and monetary resources that may not be readily available after disasters. Moreover, bridges are commonly built in harsh geographical locations to facilitate transportation. Critical structural components may not be easily accessible for manual investigations of damage. That being said, most visual inspections are inaccurate and biased (Phares 2001) while reliable information about bridge condition is essential to the decision makers. To address these issues, automated SHM has been the topic of interest in many studies (Spencer et al. 2019). Camera-equipped unmanned aerial vehicles can be effectively used in this regard. However, obtaining useful information from raw images is still challenging.

With the rapid progress of the research in the field of artificial intelligence, recent deep learning models have been capable of classifying object within raw images. Proposed algorithms are mainly designed to detect common objects such as pedestrians, roads, etc. With a similar approach, damage detection can be performed in SHM to extract meaningful information from raw images. Recently there has been an interest in implementing deep learning for the given task (e.g., Liang 2018, Yang et al. 2018). Detecting cracks, as one of the dominant damage types in reinforced concrete (RC) bridges is an important task that can be effectively handled with deep learning. However, the performance of data-driven image classifiers is highly dependent on the training data. In pictures taken from the cracked structural components, unlike most object detection problems, there exist a significant imbalance between background and crack pixels. This imbalance may result in models that perform well in classifying background pixel while showing poor performance in identifying cracks. Nonetheless, the correct prediction of classes that correspond to damage (e.g., cracks) is an equally or more important objective in SHM.

In this paper, a pixel-wise crack segmentation algorithm is proposed to label individual pixels in real-world images. The deep learning model is inspired by SegNet (Badrinarayanan et al. 2015), a successful, yet computationally efficient architecture. In the following sections, the effect of different hyperparameters for the task of crack segmentation is studied. Moreover, utilizing Bayesian optimization, three different strategies are investigated to improve the model's robustness against severe class imbalance.

2 CRACK SEGMENTATION MODEL

In this paper, model training and evaluation are performed on a series of real-life unprocessed images. *Crack Forest* (Shi et al. 2016) is a publicly available dataset that includes 118 images with ground truth labeling of cracks and background pixels. This dataset is randomly shuffled with 80%-20% splits, respectively, for training and testing of the model. Moreover, 20% of the training data is held out for the validation which is utilized in evaluating the objective function of Bayesian optimization. The task of crack segmentation is performed using a fully convolutional encoder-decoder neural network. The deep learning architecture is inspired by SegNet which accepts colored 320 × 480 input images (as in *Crack Forest*) and outputs softmax probabilities of crack and background classes for individual pixels. These probabilities will be later modified or directly used in the decision rule.

Considering the significant imbalance in the distribution of existing classes, global mean accuracy may not be a proper metric to monitor the segmentation performance. For example, high global accuracies may result from a model that labels all pixels as background while the main focus of this study (and SHM) is to provide an accurate prediction of crack patterns. To avoid misleading evaluations, the concepts of precision (P) and recall (R) are adopted with the following definitions:

$$P = \frac{t^{\mathrm{p}}}{t^{\mathrm{p}} + f^{\mathrm{p}}} \tag{1}$$

$$R = \frac{t^{\mathrm{p}}}{t^{\mathrm{p}} + f^{\mathrm{n}}} \tag{2}$$

where t^{p} is the true positive as the number of crack pixels identified correctly, f^{p} is the total number of background pixels that are misclassified as crack, and f^{n} is the number of crack pixels misclassified as background. To better distinguish the difference between one from the other, one may consider the information presented in Table 1.

The metric P shows that among all pixels classified as crack, how many actually belong to this label. In contrast, the percentage of crack pixels in the ground truth that is correctly detected compared with the ones that were missed is evaluated with the recall (R). It can be implied that recall is a measure of completeness in crack segmentation. Theoretically speaking, a high precision

Table 1. Different types of pixel segmentation outcomes.

Type of pixel classification	Ground truth label	Predicted label
t^{p}	Crack	Crack
f^{p}	Background	Crack
f^{n}	Crack	Background
t^{n}	Background	Background

value does not necessarily correspond to a high recall ratio while an ideal model will present, i.e., $P = R = 1$. To provide an extreme example, labeling all pixels as crack will yield to $R = 1$ where the precision is poor. In this case, although $f^n = 0$, all the background pixels are misclassified which is an excessive overestimation. Given that the majority of pixels have the ground truth label of background, f^p will be a relatively large number in (1) which yields to a very low accuracy (i.e., precision) for this example. In common datasets, there is usually a trade-off between the two metrics. $F1$ score can also be used as a single scalar that contains combined information about both precision and recall and can have values between 0 and 1 as two extremes. This metric is obtained by calculating the harmonic mean of P and R as follows:

$$F1 = \frac{2PR}{P + R} \tag{3}$$

It should be noted that the output of a neural network is the prediction probability of each label (e.g. crack and background). To label each pixel, one may select the one with the highest probability. However, in binary classification, it is possible to define certain threshold for the cracks. For example, instead of taking the label with max probability, one may consider all the pixels that have at least 20% probability of being crack as such. In this case, P and R can be calculated considering different probability thresholds to segment cracks. Given these values, different combinations of (R, P) can be obtained by considering a series of thresholds. This will result in a precision-recall curve. The area under the curve is known as mean precision accuracy (*MPA*) which can also be used as a single metric to evaluate the crack segmentation performance. P, R, $F1$-score, and *MPA* are later used in this paper to investigate the effects of hyperparameters in crack segmentation.

As mentioned earlier, training the deep learning architecture with uniform weights (UW) for all observations and then taking the maximum a-posteriori probabilities (MAPs) can result in a model with near perfect global accuracy. However, precision and recall values may be low for the classification of crack pixels. In this paper, UW-MAP will be utilized as the baseline strategy while two other approaches are investigated to improve the crack segmentation performance. Training observations can be weighed depending on the true class of each pixel. For example, median frequency weight (MFW) assignment proposed by Eigen & Fergus (2015) has been originally adopted in SegNet. It should be noted that such weights differ from the network learnable parameters and should be selected as hyperparameters prior to the training. With a different approach in a recent publication, Chan et al. (2019) modified the decision rule by using the maximum likelihood (ML). In this case, softmax probabilities are modified based on the prior distribution of damage classes in each pixel. Therefore, maximum likelihood probabilities are used to identify a pixel as crack or background. Depending on the location, the frequency of each class is illustrated in Figure 1. By pixel-specific normalization, prior probabilities used in the ML approach are obtained for crack and background classes. In this paper, we investigate these two strategies and their effect by referring to them as MFW-MAP and UW-ML, respectively. The first term implies the way that the training observations are weighted while the second one refers to the adopted decision rule.

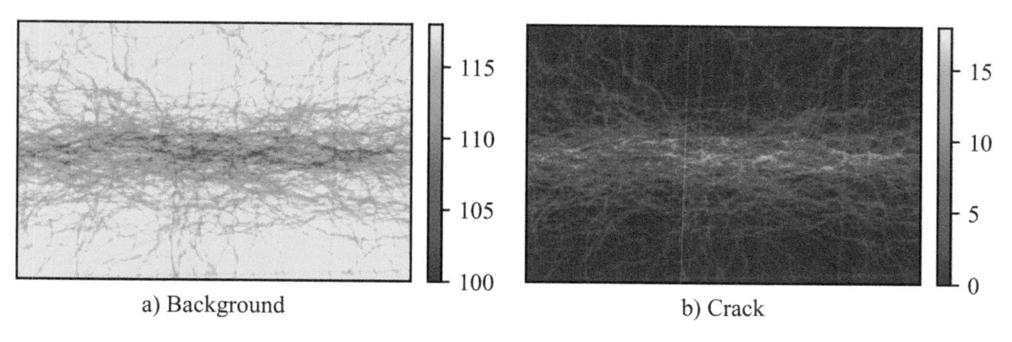

Figure 1. The frequency of two classes in the Crack Forest dataset.

3 TUNING HYPERPARAMTERS

Learnable parameters of the deep learning model are the sliding kernel weights that are set through training in different epochs. Model training is accelerated by running epochs on an NVIDIA GeForce GTX 1080 GPU @ 8 GHz with 2560 CUDA cores, using Keras API (Chollet 2018). 80 epochs are found to be appropriate for training with the mini-batch size of 4 images. This deep learning architecture includes 4 pairs of encoder-decoder computation blocks. Padding, strides, the number of extracted filters, etc. are selected in a similar fashion to the original implementation of SegNet. However, given the input-output shapes of the crack segmentation model, proper adjustments are made. The architecture of the segmentation deep learning model is illustrated in Figure 2.

Considering the default hyperparameters for each optimizer, model is independently trained for UW and MFW strategies. After completing the training process, the best set of learnable parameters were selected for further investigations. The criterion to select such parameters is the minimum validation obtained from the cross-entropy loss function. To minimize the loss, stochastic gradient descent (*SGD*), *RMSprop*, *Adagrad*, *Adadelta*, *Adam*, *Adamax* and *Nadam* optimizers are investigated. Assuming the default hyperparameters in the Keras API, a case study is performed to compare the performance of models trained with different optimizers. Given the three imbalance techniques mentioned earlier, average values of the performance metrics are documented for each weight optimizer. Results are shown in Figures 3 and 4. It should be noted that the *P*, *R*, *F*1-score, and *MPA* are calculated by assuming the crack as the positive class.

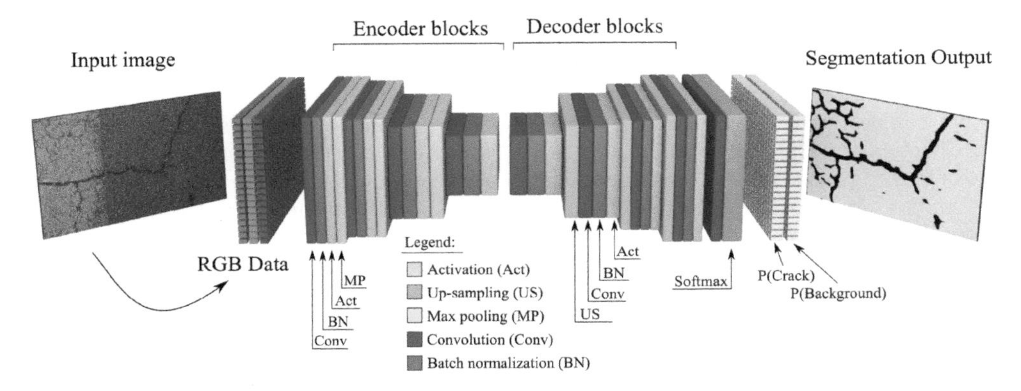

Figure 2. The Deep learning used in the semantic segmentation of cracks.

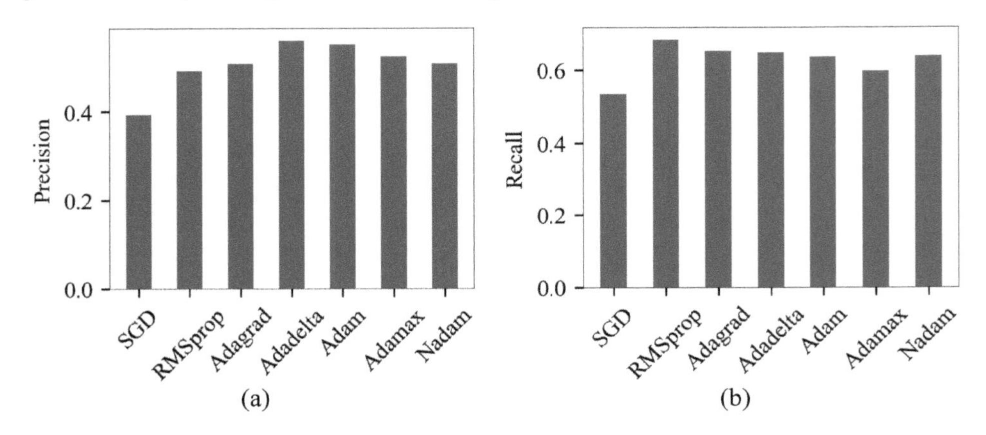

Figure 3. Precision and recall values for different Solvers (average value of UW-MAP, UW-ML, and MFW-MAP).

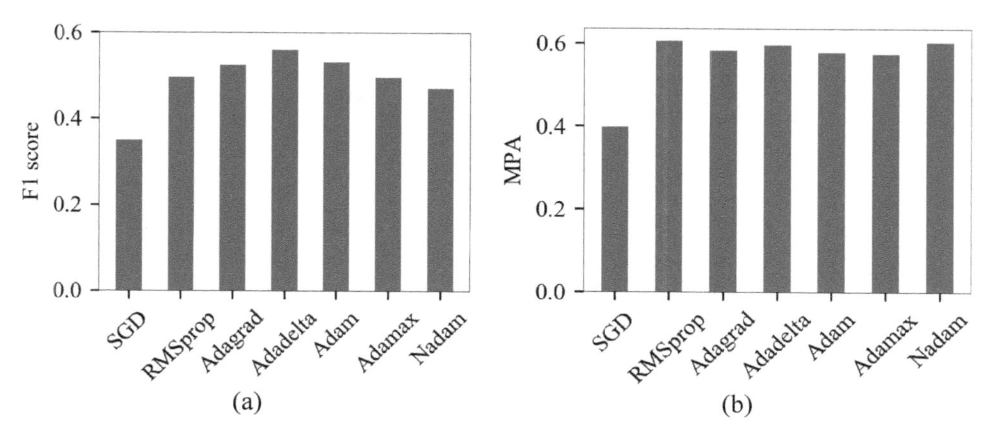

Figure 4. Performance of different optimizers (average value of UW-MAP, UW-ML, and MFW-MAP).

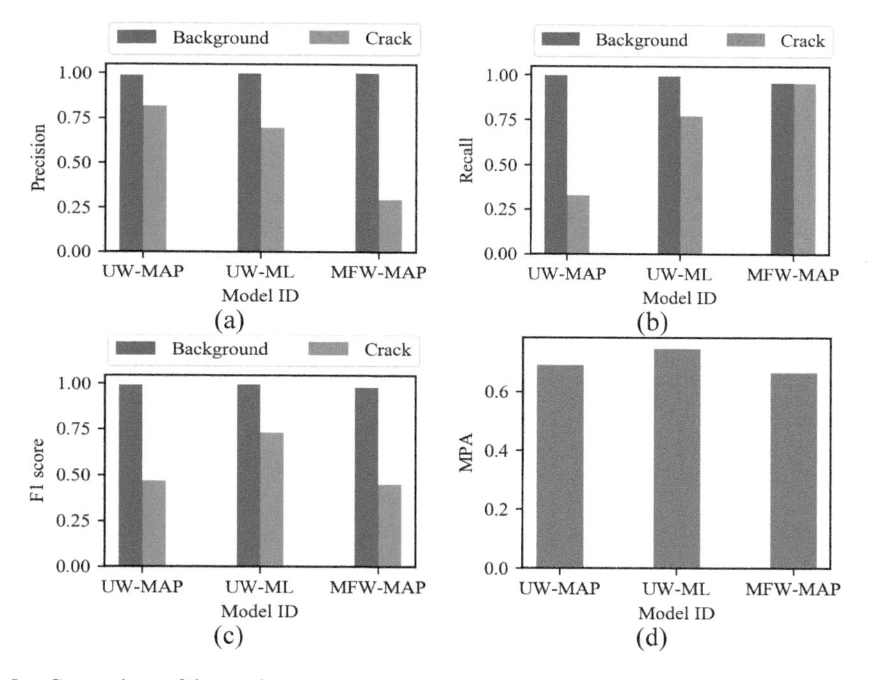

Figure 5. Comparison of the crack segmentation strategies after Bayesian optimization.

It can be observed that *Adadelta* yields relatively better performance compared with the others. However, the default learning hyperparameters of this algorithm may be tuned for enhanced performance. Bayesian optimization (Snoek et al. 2012) is used to adjust *Adadelta*'s hyperparameters to maximize *MPA* as the objective function. In this case, optimization is performed individually for UW-MAP, UW-ML and MFW-MAP strategies. The earlier metrics are documented for both crack and background pixels where the results are shown in Figure 5. Moreover, the precision-recall curves used to calculate *MPA* are shown in Figure 6.

For the background pixels, as expected, the performance of the deep learning models is nearly perfect with respect to both precision and recall. From the baseline strategy, recall is relatively small, indicating that the majority of pixels corresponding to crack are not detected; however, it

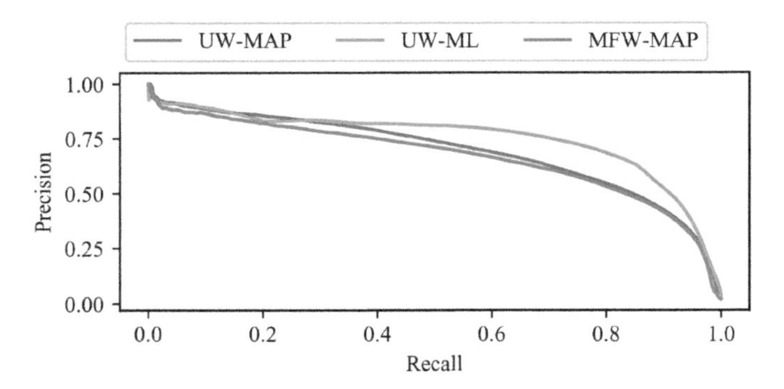

Figure 6. The precision recall curve for three different strategies.

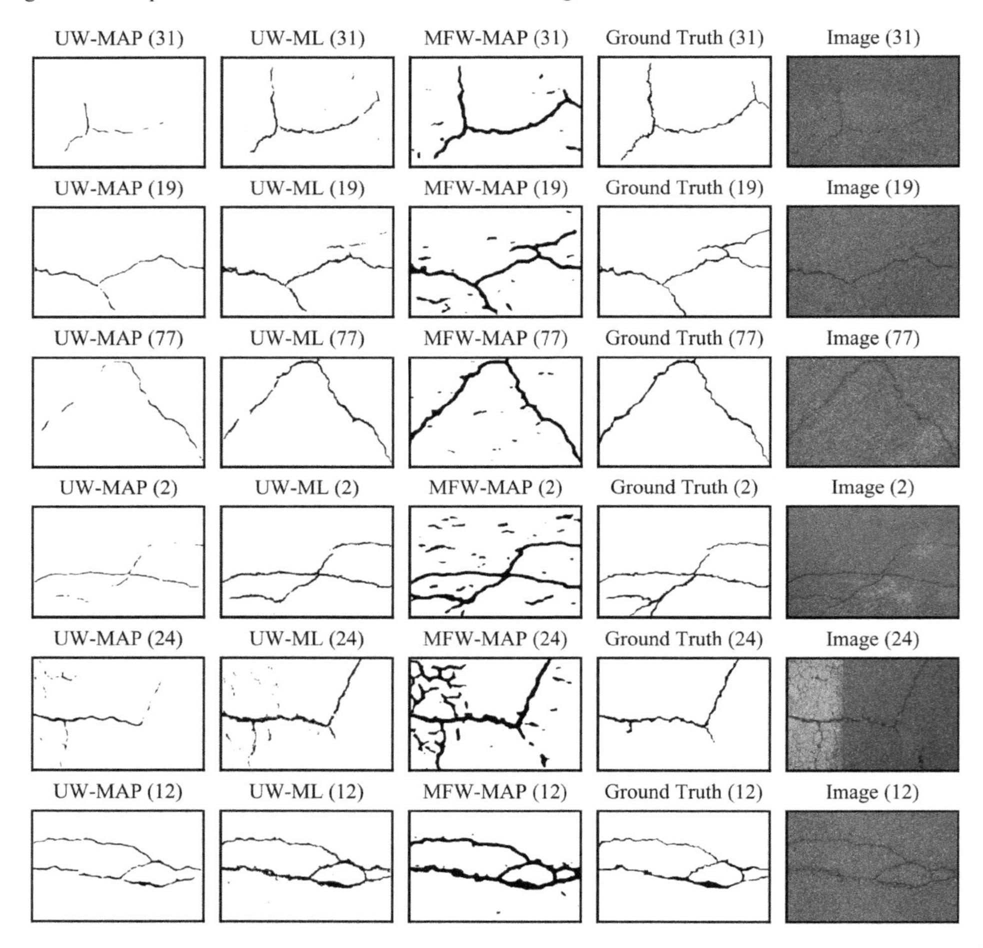

Figure 7. Crack segmentation results for example test observations.

has the highest precision. In contrast, MFW-MAP shows a conservative prediction of cracks where recall values are the highest but with the lowest precision of the three strategies. UW-ML shows the best *F1*-score and *MPA* with moderate precision and recall values. The test set examples of detecting cracks in real-world images are provided in Figure 7.

The first three columns in this figure represent the segmentation results by the three strategies. Ground truth segmentation from the original *Crack Forest* dataset is also provided. The numbers inside the parenthesis indicate to the image ID in *Crack Forest.* An interesting finding of this study is that in some cases (e.g., image 24), there exist cracks that are missed by the human-generated ground truth. However, the deep learning models, more or less, are able to detect such patterns. Some of these cracks are located on a more complex background (white paint). Yet, the performance is reasonable. The predictions from MFW-MAP are more conservative compared to the other methods while the predicted crack widths commonly appear thicker than ground-truth. In addition, misclassified background pixels (e.g., stains) are more likely to occur in this approach. In contrast, the predictions from the UW-MAP appear relatively incomplete as inferred from the recall values shown in Figure 5.b. Overall, the baseline model and MFW-MAP show inferior performance for, respectively, recall and precision metrics. Modifying the decision rule (UW-ML) provides better performance with a reasonable trade-off in precision and recall.

4 CONCLUSIONS

Reliable information is critical in maintaining the structural health of bridges. Automated SHM techniques are developed to obtain such information in an efficient manner. Surface cracks are a useful source of information regarding certain structural defects. However, processing raw real-world images may be challenging and biased when human inspections are performed. This paper proposes a fully automated semantic segmentation model to effectively detect cracks and overcome the challenge of the highly imbalanced dataset. Three strategies of UW-MAP, UW-ML, and MFW-MAP are considered for the task of crack segmentation. Moreover, Bayesian optimization is utilized to tune the hyperparameters for enhanced robustness. It is shown that by changing the training weights and modifying the decision rule, segmentation can be performed more effectively. Such improvements are investigated using P, R, $F1$-score, and MPA. UW-ML strategy shows better results compared with the others while maintaining a reasonable performance considering these metrics. Given the robustness of the deep learning models for the unseen raw images, these techniques can be effectively used for automated SHM inspection of bridge infrastructure.

REFERENCES

Badrinarayanan, V., Kendall A. & Cipolla, R. 2015. Segnet: A deep convolutional encoder-decoder architecture for image segmentation. *arXiv preprint*. arXiv:.00561.

Chan, R., Rottmann, M., Hüger, F., Schlicht P. & Gottschalk, H. 2019. Application of Decision Rules for Handling Class Imbalance in Semantic Segmentation. *arXiv preprint*. arXiv:.08394.

Chollet, F. 2018. Keras: The python deep learning library. *Astrophysics Source Code Library*.

Eigen, D. & Fergus, R. 2015. Predicting depth, surface normals and semantic labels with a common multi-scale convolutional architecture. *Proceedings of the IEEE International Conference on Computer Vision*.

Liang, X. 2018. Image-based post-disaster inspection of reinforced concrete bridge systems using deep learning with Bayesian optimization. *Computer-Aided Civil and Infrastructure Engineering*, 34(5), 415–430.

Phares, B. 2001. Highlights of study of reliability of visual inspection. *Proceedings: Annual Meeting of TRB Subcommittee A2C05* (1): Nondestructive Evaluation of Structures, FHWA Report No FHWARD-01-020 and FHWA-RD-01-0212001.

Shi, Y., Cui, L., Qi, Z., Meng, F. & Chen, Z. 2016. Automatic Road Crack Detection Using Random Structured Forests. *IEEE Transactions on Intelligent Transportation Systems*. 17(12): 3434–3445.

Snoek, J., Larochelle, H. & Adams, R. P. 2012. Practical bayesian optimization of machine learning algorithms. *Advances in neural information processing systems*.

Spencer, B. F., Hoskere V. & Narazaki, Y. 2019. Advances in Computer Vision-Based Civil Infrastructure Inspection and Monitoring. *Journal of Engineering*.

Yang, X., Li, H., Yu, Y., Luo, X., Huang, T. & Yang, X. 2018. Automatic Pixel-Level Crack Detection and Measurement Using Fully Convolutional Network. *Computer-Aided Civil and Infrastructure Engineering*. 33(12): 1090–1109.

Railway & light rail bridges

Chapter 20

Large-scale vulnerability analysis of girder railway bridges

D. Bellotti, A. Famà, A. Di Meo & B. Borzi
European Centre for Training and Research in Earthquake Engineering (EUCENTRE), Pavia, Italy

ABSTRACT: This work concerns the large-scale study of the seismic vulnerability of girder railway bridges in relation to the level of the acquired knowledge (complete or partial). The seismic vulnerability of such infrastructures is estimated using fragility functions. A detailed modeling has been developed for bridges with complete knowledge of the data. A finite element model of the bridge in which non-linear behavior occurs for piers and supports has been developed. Fragility curves are obtained by performing several time history analyses based on the selected return periods, i.e. 9 in agreement with Italian technical standards (D.M. 17/01/2018) using spectrum-compatible accelerograms. In the case of partial knowledge a simplified procedure for seismic evaluation for girder bridges based on displacements is used. Each pier-support subsystem is considered through a SDOF equivalent system, for each return period the probabilities of exceeding the limit states have been calculated and then the fragility curves is calculated.

1 INTRODUCTION

This paper deals with the development of an application for the vulnerability analysis through the determination of fragility curves for large-scale assessment of reinforced concrete girder railway bridges.

Several studies on seismic risk assessment of infrastructures have been undertaken in the past on Italian infrastructures; they mainly concerned the seismic assessment of highway and road bridges (Borzi et al. 2015, Calvi et al. 2010, Modena et al. 2015, Pinto et al. 2009, Pinto & Franchin 2010). Hence, the need to deepen the seismic assessment of railway bridges through a methodology that allows a large-scale study of the seismic vulnerability of such structures through the construction of fragility curves. The fragility curves are defined by cumulative probability distributions which allow to estimate the probability of reaching or exceeding a given level of damage for a given severity of ground shaking. The methods proposed in this work are based on the definition of the bridge's seismic behavior through a more (or less) simplified mechanical modeling in relation to the level of the acquired knowledge. The goal of the work is to create a tool that is able to generate a numerical model, of one or more bridge/s, which is capable to derive bridge fragility curves through either the execution of numerous non-linear time-history analyses or by using a simplified procedure. The investigation of railway bridges' seismic vulnerability analysis methods started by studying the most widespread structural typologies in Italy. Among all typologies, girder bridges were taken into account.

This work will only describe the development of the application and its features as the study is still in progress. The concept was to have two different tools for girder bridges based on the available data: one for full and one for partial bridge knowledge.

A detailed model has been developed for bridges with full data knowledge. The aim is to provide specific fragility curves for individual railway bridges whose complete data are known. This procedure provides an automatic finite element model of the bridge in which non-linear behavior occurs for piers and supports. The application performs the structural calculations using OpenSEES

(McKenna et al. 2000) finite element analysis program and generates a digital model of the analyzed structure, automatically constructing the representative objects of its parts, such as deck, piers and mutual relationships between all elements. This digital model makes it possible to convert the description of the bridge's individual components into a useful and functional finite element model able to carry out simulations for the fragility curves calculation. Fragility curves are obtained by performing several time history analyses based on the selected return periods (i.e. 9 return periods in agreement with Italian technical standards (D.M. 17/01/2018)) and on the spectrum-compatible accelerograms automatically generated by the application.

To evaluate the seismic vulnerability of girders bridges for which limited number of data are available, a simplified procedure based on displacements (Direct Displacement Based Assessment – DDBA) proposed by Priestley et al. (2007) and Priestley (1997) is used. The aim of simplified modeling is to provide fragility curves for railway bridges with partial construction details (partial knowledge), as a result of an expeditious survey, or for classes of bridges for which detailed data are not known. Each pier-support subsystem is considered through a single degree of freedom (SDOF) system that is equivalent to the multi degree of freedom (MDOF) system in terms of fundamental period, displacement and dissipation capacity. For each subsystem's component, the force-displacement curve is determined; then, for each of the 9 return periods (D.M. 17/01/2018) the probabilities of exceeding the limit states are calculated in order to obtain the points that define the fragility curves.

Girder bridges are generally formed by a deck constituted by multiple beams which can be simply supported or continuous, so to have an isostatic or an hyperstatic structural scheme. Due to static and seismic loading, beams are subjected to shear and bending forces. In girder bridges, the deck is supported by unreinforced or reinforced concrete piers and it can be made of reinforced concrete, prestressed reinforced concrete or steel.

Between the deck and the vertical structure of the bridge there are the support devices, the choice of which depends mainly on the deck typology and on the type of constraint required. In the case of a reinforced concrete deck, the adopted construction technologies are mainly dependent on the total length of the bridge and on each span.

2 MODELING

The goal of the work is to create a tool able to generate a numerical model of one or more bridge/s that allows to derive bridge fragility curves through either the execution of numerous non-linear time-history analyses or a simplified procedure. The methods proposed in this work are based on the definition of the bridge's seismic behavior through a more (or less) simplified mechanical modeling in relation to the level of the acquired knowledge.

2.1 *Detailed modeling (full knowledge)*

The application divides the bridge into three areas: super-structure, connections and sub-structures (where connection area is all that is between the super-structure and the sub-structures). These three areas are treated and processed in sequence because the geometric construction of the bridge is carried out from top to bottom and the underlying parts depend on those above them.

The elements constituting the super-structure are organized in a rigid hierarchy: the bridge tracking is composed by decks, which are composed by one or more spans, divided into deck segments. It is assumed that at the beginning and at the end of each span there is a sub-structure (pier or abutment). It is not necessary to specify either the planimetric position or the orientation of these sub-structures, since it can be automatically deduced from the terminal position of spans (the latter derives from the planimetric layout already known at the modeling of the sub-structures). The decks' division into segments is useful for several reasons: to change the structural section within a single span; to follow a plan curve layout; to obtain a discretization of the masses along the span to capture the dynamic response of super-structure modes.

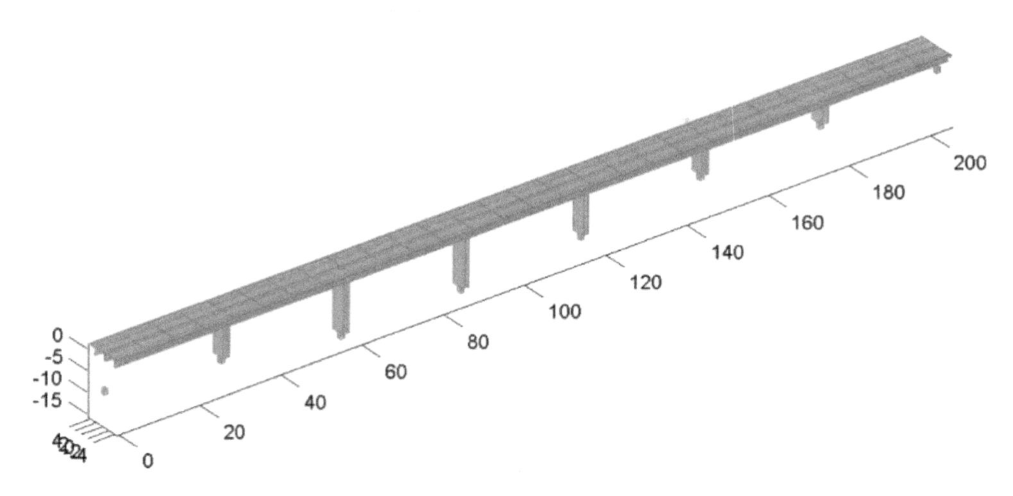

Figure 1. Extruded view of a modeled bridge.

The application allows the maximum flexibility in the definition of the super-structure: it allows, actually, to alternate both supported and continuous decks as well as to alternate straight or curved sections (mixed pattern), even within the same span.

As summarized below, it is possible to model concrete bridges with simply supported, continuous beams or rigid frame static scheme. The application allows to model steel decks, provided that the bridge has concrete piers.

Each railway line of the bridge is modeled with an elastic single-line model consisting of only finished beam elements (Figure 1).

For the geometric construction of the path, the application automatically calculates the planimetric angle that is tangent to the path in each node of the superstructure. This procedure allows to define an intrinsic reference in the same point (i.e. a vertical axis and two horizontal axes, of which one tangent and the other orthogonal to the bridge direction) that is automatically assigned as a natural reference also for the connecting elements and for the substructures. In other words, sub-structures are naturally and automatically rotated in their local reference to follow the trend of a curvilinear path. The abutments and the foundations are modeled as constraints.

The connection between the superstructure and the substructure is made by two objects: connections and supports. The connection has the function of rigidly transferring the load transmitted by the end node of a span to the substructure.

In the case of portal, frame or multiple-frame piers, different assumptions are made for the connection between the nodes of the pier and the lower ones of the supports. In both articulated and simplest cases, it is always assumed that the axis of the substructure is placed in the center of gravity of the master nodes of the respective connections that converge to it.

Figure 2 shows the connection between the nodes of the stack and the lower ones of the supports in the presence of pier top, for the case of single-frame or multiple-frame piers. Each portion of the pier top is divided into two elements. The base node of each support device is automatically connected to the nearest plan top pier node.

The application allows to define a non-linear behavior for piers and supports, while the deck is defined elastic.

The element used for modeling the piers is the *BeamWithHinges* element of the OpenSEES library (McKenna et al. 2000). It is formulated with an optimized integration scheme along its development that employs non-linear fiber sections at the ends and elastic sections in the central part. The model considers a non-linear behavior that occurs in piers and supports. This limit is imposed by the need to automatically manage numerous non-linear finite element time-history analyses.

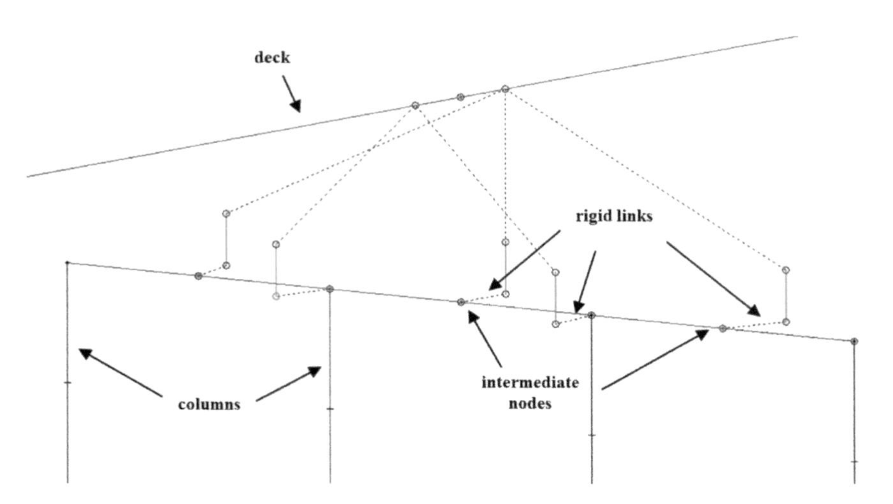

Figure 2. Connection modeling scheme.

To define the concrete material, the application employs a very small set of parameters.

For the definition of elastic properties, it is necessary to specify the Young's modulus, the Poisson's module and the unit weight.

For the definition of the non-linear properties, it is assumed a nonlinear stress-strain law (the application uses OpenSEES's *Concrete01* material) for which, in addition to the weight density, four values need to be specified: the concrete compressive strength at 28 days, the concrete strain at maximum strength, the concrete crushing strength and the concrete strain at crushing strength. It is assumed that the concrete mechanical properties follow a lognormal distribution. For each numerical simulation different sampled values are adopted.

As an example, the case of the concrete compressive strength is illustrated:

$$f_{co\ simulation,i} = f_{co} \times \varepsilon_{f_{co},i} \tag{1}$$

where:

$$\varepsilon_{f_{co}} \sim LN(0, \beta_{f_{co}}) \tag{2}$$

where β_{fco} parameter is the standard deviation of the logarithmic distribution assumed for the concrete's peak resistance.

As for the concrete, the application uses a very small set of parameters to define also the steel material. In this case, as well, for the definition of the elastic properties, it is necessary to specify the Young's modulus, the Poisson's modulus and the unit weight.

For the definition of the non-linear properties, the application uses OpenSEES's *Steel01* material with uniaxial bilinear steel material object with kinematic hardening for which it is necessary to specify the yield stress and the hardening, in addition to the unit weight and the Young's modulus. Again, the mechanical properties follow a lognormal distribution defined in the same way as for the concrete.

The application manages the most common sections used in the engineering practice (i.e. rectangular, box, T, I, U, C, circular, circular hollow, octagonal cross sections) and allows to calculate the mechanical properties (weight, inertias, etc.) for an elastic use of the section.

In the case of complex or different sections from those implemented, the application allows the use of a generic section for which the user defines inertias and mass.

The application also automatically calculates the torsional and shear stiffness of each section using formulas according to the specified type. If the non-linear behavior of the section is required,

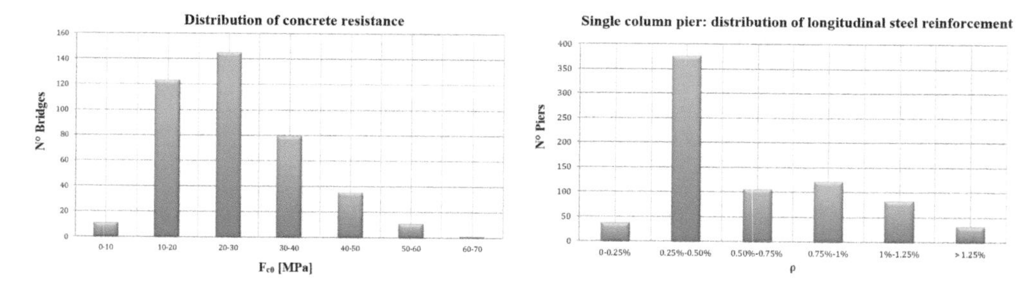

Figure 3. Montecarlo simulations used to consider the uncertainty associated to the capacity of the elements.

such as in the case of the piers, the reinforcement must be specified. The additional variables, to be specified in this case, are: steel type, the longitudinal reinforcement, the stirrups in the two directions and the concrete cover.

The application allows the use of composite sections (to be understood as a composition of the simple sections previously described) whose behavior is only elastic. They have been implemented to facilitate the input phase of the overall inertia of the deck sections. To define a composite section it is necessary to specify: a section for the beams, the number and the spacing of the deck beams, and a slab.

2.2 *Simplified modeling (partial knowledge)*

The aim of simplified modeling is to provide fragility curves for railway bridges with partial construction details (partial knowledge), as a result of an expeditious survey, or for classes of bridges for which detailed data are not available. The procedure implemented in the application is based on Direct Displacement Based Assessment – DDBA proposed by Priestley et al. (2007) and Priestley (1997). A preliminary phase of collection of data related to main properties of bridges (i.e. materials resistance, loads, geometric data, amount of longitudinal and transverse steel reinforcing bars etc.) should be conducted. Subsequently, a significant number of bridges (e.g. 1000) are generated through Montecarlo simulations to consider the uncertainty associated to the capacity of the elements; the unknown variables collected in the first phase are treated as aleatory variables (Figure 3).

It is assumed that the mechanical properties of the concrete and steel follow a lognormal distribution while a normal distribution has been adopted for loads and geometric data. The scope is the definition of the seismic behavior of a bridge, which is representative of one or more bridge/s with partial knowledge. A simplified response has been defined for piers, abutment and supports both in longitudinal and transverse directions, analyzed independently from each other. The interaction between the piers and the girder through the connections is neglected. Each pier-support subsystem is considered through a single degree of freedom (SDOF) equivalent system. The modeling allows the assessment of probabilities of exceeding the limit states for the following subsystems:

- Subsystems #1, that correspond to the squattest pier, the slenderest pier and the pier with average slenderness and related connections;
- Subsystems #2, that correspond to abutments and related connections.

For each subsystem's component, a force-displacement ($F - \Delta$) curve is determined. The pier is defined by an elastic perfectly plastic force-displacement relationship, characterized by a shear capacity V_{rd}, a yield displacement Δ_y and a limit displacement capacity Δ_{LS} that depends on the considered limit state as well as on the failure mechanism (ductile/brittle) of the pier itself. Even the supports are defined by elastic perfectly plastic curve force-displacement as described in paragraph 2.4. Figure 4 (left) shows as example force-displacement curves for a system controlled

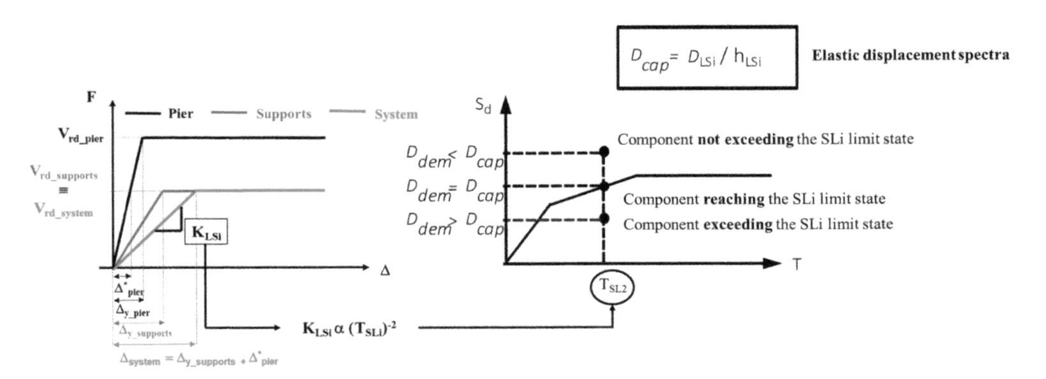

Figure 4. Scheme of simplified procedure to provide fragility curve for railway bridges with partial knowledge. Left: For each subsystem's component, the force-displacementcurve is determined. Right: The properties of the single degree of freedom (SDOF) equivalent system are defined for each limit state.

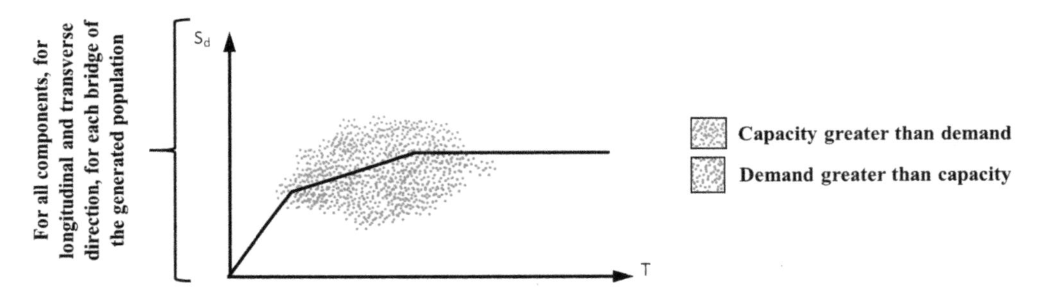

Figure 5. Scheme of simplified procedure to provide fragility curve for railway bridges with partial knowledge: For each return period it is possible to define risk indicators ρ as the ratio between demand and capacity.

by supports resistance. As a result, lower capacity in terms of displacement is obtained for the pier equal to $\Delta^*_{pier} < \Delta_{y\_pier}$.

The properties of the single degree of freedom (SDOF) equivalent system are defined for each limit state: the SDOF system is equal at the starting system (MDOF system) in terms of displacement capacity $D_{cap} = \Delta_{system}$, fundamental period T_{LS} and amount of dissipated energy quantified through an equivalent viscous damping defined starting from ductility demand (Figure 4 right).

Once each subsystem's force-displacement curve is determined, for each return period it is possible to define risk indicators ρ as the ratio between demand and capacity. Then, the probabilities of exceeding the limit states are calculated and the fragility curves are obtained (Figure 5).

2.3 *Definition of loads*

The weights and permanent loads are calculated directly in the hypothesis of standard configurations provided by the application for super-structures and sub-structures. For what concerns super-structures, the application allows two different ways to define the contribution due to the dead load. The first is the direct specification of total weight per meter of deck: this condition allows to generalize the application to all those configurations that are explicitly provided for. The second is the specification of the weight only for the carried permanent loads, assigning to the application the automatic calculation of the weight contribution of the deck's structural section.

The determination of the carried permanent loads referring to the ballast, the railway track and the waterproofing, has been carried out assuming a volume weight equal to $18\,kN/m^3$, applied over the entire average width between the ballast retaining walls for a medium height between the rail level and the upper surface of the deck of 0.80 m, as indicated in §5.2.2.1.1 of Italian technical standards (D.M. 17/01/2018). In particular, the ballast load was defined per unit length of longitudinal development of the bridge (in kN/m), considering an average width of the ballast equal to 4.90 m.

In addition to its self-weights and permanent vertical loads and masses, the application allows to apply a portion of the vertical loads and masses associated with the transit of the train. The loads used in the application are related to real train loads. As indicated in §5.2.2.2.3 of the Italian technical standards (D.M. 17/01/2018), for this type of loads the use of the coefficient φ, which considers the dynamic nature of the trains transit, is denied. The loads associated to the transit of trains are applied to the bridge as loads per unit of longitudinal development length (in kN/m) and assumed equal to $\psi_2 \cdot q_{tr}$, where ψ_2 is the combination coefficient according to Italian technical standards (D.M. 17/01/2018) and q_{tr} is the train load. In the seismic vulnerability calculations, the condition that involves the presence of the overloads and the masses associated with the transit of the train convoys is assumed by adopting ψ_2 equal to 0.2, according to §5.2.2.8 of the Italian technical standards (D.M. 17/01/2018).

The seismic load is based on a user definition of 9 return periods (to choose between 30 and 2475 years) according to Italian technical standards (D.M. 17/01/2018) and, in the case of full knowledge, on the spectrum-compatible accelerograms automatically generated by the application. It is necessary to specify the bridge's site geographical coordinates to determine the parameters of seismic hazard (a_g, F_o, T_c^*).

To perform non-linear analyzes in the time domain, the application uses groups of spectrum-compatible generated accelerograms (Muscolino 2002) independent in the two horizontal directions, applied as synchronous motion on a fixed base, starting from the already mentioned elastic spectra. These generated signals are corrected with respect to the base line, have a predetermined total duration and a similarly prefixed zero intensity tail to allow to exhaust the transient response of the structures.

2.4 *Supports*

The first support devices used for reinforced concrete bridges, with a span less than 18 meters, were those made of two lead sheets of one centimeter each, between which was interposed a steel sheet thick one millimeter (fixed type support) or two steel sheets with graphite contact faces (mobile type support). Nowadays, steel support devices are spherical, hinged or pendulum type and are used in the presence of steel decks or high reactions transmitted by the deck.

The use of bearings vary according to the girder bridges typology (Table 1). They can be classified in relation to: (i) the physical principle used (i.e. rolling, deformation and sliding), (ii) the degrees of freedom (i.e. fixed supports and mobile bearings), (iii) the transmission of the load (i.e. punctual,

Table 1. Correspondence between common types of girders and bearings used for steel and concrete girder bridges.

Macro classes	Static scheme	Types of girders	Types of bearings
Steel girder bridges	1) Simply supported 2) Continuous beams	1) Through truss 2) Deck truss 3) Plate 4) Multiple girder 5) Box	1) Steel fixed 2) Pot PTFE 3) Sliding disc
Concrete girder bridges	1) Simply supported 2) Continuous beams 3) Rigid-frame	1) Slab 2) Box 3) Beams 4) Precast box	1) Elastomeric 2) Confined Elastomeric 3) Pot PTFE 4) Sliding disc

Table 2. Typologies of supports.

Type of support	Description
Fixed	the translational DOF and all the rotational DOF are fixed
Spherical hinge	both the DOF (in plan) are fixed and all the rotational DOF are free
Cylindrical hinge	both the DOF (in plan) are fixed and a single free rotational DOF
Mobile	both DOF (in plan) are free
Unidirectional longitudinal	the longitudinal DOF (in the local reference) is free, the transverse one is fixed
Unidirectional transversal	the transverse DOF (in the local reference) is free, the longitudinal one is fixed
Anisotropic elastoplastic	the two DOF have two elastic perfectly plastic constitutive laws decoupled
Isotropic elastoplastic	the two DOF have isotropic elastic perfectly plastic constitutive law
Friction	the two DOF have a friction isotropic constitutive law

linear and superficial) and (iv) the material of which they are made (i.e. steel, aluminum, natural or synthetic rubbers).

The definition of support devices' properties is similar for detailed modeling (full knowledge) and simplified modeling (partial knowledge). The application foresees that the constitutive law of the two degrees of freedom associated with horizontal translations is specified, since the remaining DOF are either completely rigid (vertical or rotational DOF) or completely flexible (rotational DOF). The types of support that have been implemented in the program have been chosen according to the most common types of support present in Italian railway bridges and are summarized in Table 2.

With reference to the OpenSees calculation engine, the finite elements used for each type of support are constituted by *TwoNodeLink* finite elements: depending on the type of constraint and on the degrees of freedom, materials that represent a linear or fixed DOF are assigned. In the case of anisotropic elastic, an *ElastomericBearing* finite element is used in the two direction, while in isotropic only one typology is defined. In the case of friction support, a *FlatSliderBearing* finite element is used.

3 DEFINITION OF LIMIT STATES

The limit states considered in the calculation of fragility curves vary in relation to structural elements and are:

1) for piers: the ductile and brittle mechanisms. The ductile mechanisms depend on the totalchord rotation amount while the brittle ones depend on the element shear.
2) for decks: the loss of support mechanism towards the supports or the pier cap, controlled by the relative excursion of the support devices.
3) for bearings: the achievement of the maximum capacity in terms of resistance.

For all the limit states considered, two levels of capacity are foreseen during the verification: one related to the necessity to carry out repair interventions (damage limit state) and the other concerning the safety of the bridge (collapse limit state). An exception to this rule is the brittle shear mechanism on the piers, for which only the collapse limit state is defined. The operation limit state is, instead, linked to the deformations limit defined in relation to the train's safe transitability of the bridge and the limits imposed are not connected with the damage to the structures.

For what concerns the bearings, the application considers the achievement of the maximum capacity in terms of resistance for the definition of the damage limit state: this capacity depends on the values declared by the device manufacturer.

The expressions of capacity used for the calculation of the mechanisms inherent to the piers are those reported in Italian technical standards (D.M. 17/01/2018) and described in §3.1.

These capacities are to be separately calculated according to the two directions of the pier's cross section.

When verifying the overall limit state of the section, in detailed modeling it is assumed that the risk indicator ρ, understood as the ratio between demand and capacity, is obtained, for each time step by the SRSS combination of the results of the checks on the two main flexion plans. As an alternative to the SRSS combination rule, the rule for combining the directional maximum is used, for example in the loss of support verification. With reference to the three mechanisms examined (i.e., "θ" ductile mechanism, "V" brittle mechanism, "d" loss support mechanism), is obtained:

$$\begin{aligned}
&\rho_{\theta,x} = \frac{\theta_{demand,x}}{\theta_{capacity,x}} \qquad \rho_{V,x} = \frac{V_{demand,x}}{V_{capacity,x}} \qquad \rho_{d,x,SL} = \frac{d_{demand,x}}{d_{capacity,x,SL}} \\
&\rho_{\theta,y} = \frac{\theta_{demand,y}}{\theta_{capacity,y}} \qquad \rho_{V,y} = \frac{V_{demand,y}}{V_{capacity,y}} \qquad \rho_{d,y,SL} = \frac{d_{demand,y}}{d_{capacity,y,SL}} \\
&\rho_{\theta} = \sqrt{\rho_{\theta,x}^2 + \rho_{\theta,y}^2} \qquad \rho_{V} = \sqrt{\rho_{V,x}^2 + \rho_{V,y}^2} \qquad \rho_{d,SL} = max(\rho_{d,x,SL}, \rho_{d,y,SL})
\end{aligned} \tag{3}$$

The application optionally allows to replace the directional combination rule related to the SRSS shear with the most usual rule that uses the directional maximum:

$$\rho_V = max(\rho_{V,x}, \rho_{V,y}) \tag{4}$$

As described in paragraph 2.2, in the simplified modeling the risk indicator ρ is instead obtained for each subsystem with reference to a limit state. In this case, the application uses the directional maximum value of risk indicator obtained in longitudinal and transverse directions.

The loss-of-support mechanism of decks, although rare, is very dangerous in bridges and occurs when there is an excess of relative excursion between the super-structure and the sub-structure. This eventuality can occur either on alignments of mobile bearings, when they don't guarantee an appropriate restraining retention or they collapse, or on alignments of fixed bearings when they collapse.

The loss of support can have a different degree of severity whether referring to the support or the pier cap. In the first case, the deck falls from its natural seat but does not collapse because supported by the pier cap while, in the second case, the falls of the deck from the pier cap implies the partial collapse of the bridge. This substantial differentiation is maintained during the verification by associating the fall from the support to the damage limit state and the fall from the pier cap to the collapse limit state.

Finally, the assumptions regarding the definition of the limit states concerning the bridge as a whole, and not just its components (piers and supports), are illustrated. As already highlighted, each of the controlled mechanisms on the components is a main mechanism, i.e. the achievement of the corresponding limit state on the component implies the achievement of the same limit state on the bridge. In the theory of systems this hypothesis is known as serial arrangement of components and, although it is the simplest hypothesis that can be formulated, it is quite adherent to the physical reality of bridges, especially in the collapse limit state. It is normal, in fact, to assume a collapsed bridge when one of its piers or girders has collapsed.

3.1 *Ductile and brittle mechanism*

The capacities, in terms of total yield and ultimate rotation, are given by the expressions:

$$\theta_y = \phi_y \frac{L_V}{3v} \tag{5}$$

$$\theta_u = \theta_y + (\phi_u - \phi_y) L_{pl} \left(1 - \frac{L_{pl}}{2L_V}\right) \tag{6}$$

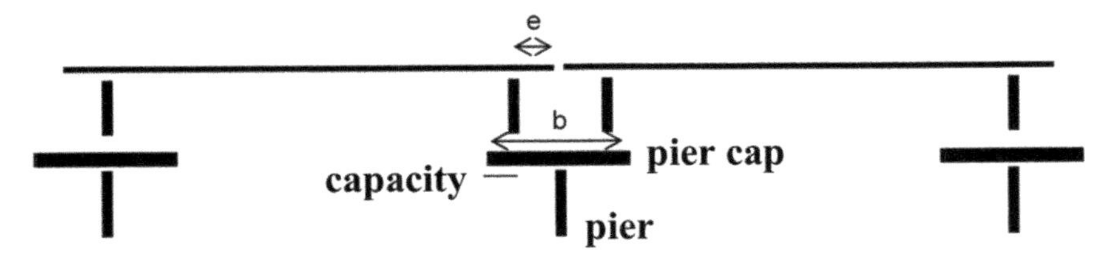

Figure 6. Scheme for the loss in the longitudinal direction.

where the length of plastic hinge was evaluated as:

$$L_{pl} = 0.1L_V \tag{7}$$

where φ_y and φ_u are the yield and the ultimate curvatures; L_V is the shear span at member end and is equal to M/V. L_V can be approximated to L (for single beam and frame piers) and to L/2 (for frame piers) in the longitudinal and transversal direction, respectively.

These capacity expressions are evaluated separately for longitudinal and transverse direction of the base section of each pier.

The application provides for the adoption of different expressions for the shear capacity (resistance): the expressions in the Italian technical standards (D.M. 17/01/2018) and attached to the Circular §C8.8.5.5 (G.U. 2019), the expressions referable to Priestley et al. (2007) and those present in the Eurocode 8 part 3 (CEN 2005).

The shear capacity V_u is always calculated according to the expression:

$$V_u = V_c + V_s + V_N \tag{8}$$

where V_C, V_S and V_N are the contributions of concrete, stirrups and normal stress.

3.2 *Loss of support*

The amount of demand to estimate the risk related to the loss of support mechanism is the relative excursion between super-structure and sub-structures for each alignment of supports, equal to the average deformation of the bearings of each alignment.

This excursion is verified both in the longitudinal direction (parallel to the bridge axis) and in the transversal direction, since the two directions can have different capacity limits. In particular, these capacities must be specified in each alignment, for the two main directions, and for both the verified limit states (SLD and SLC): $d_{SLD,longitudinal}$, $d_{SLC,longitudinal}$, $d_{SLD,transverse}$, $d_{SLC,transverse}$.

Normally, as a capacity to the damage limit state, it is assumed the limit that implies the non-damage of the devices and in any case the fall from any supports. As a capacity of the collapse limit state, instead, the limit of fall from the pier cap is considered.

In the absence of input data, the application estimates the capacities for the loss of support mechanism based on the friction capacity (Magliulo et al. 2011) and on geometrical data of the bridge. In this case, it is recommended to carefully evaluate the engineering consistency of these quantities. As an example, Figure 6 reports a typical case for the loss of support of the span from the pier cap in the longitudinal direction; in this case, the application simply estimates a capacity equal to b/2-e according to Figure 6.

4 GENERATION OF FRAGILITY CURVES

The application generates fragility curves, defined by cumulative probability distributions that allow to estimate the probability of reaching or exceeding a given level of damage for a given severity of ground shaking, for each boundary state analyzed through the following steps.

The seismicity of the national zonation site is acquired, i.e. on the basis of the seismic hazard related to the site of the bridge as well as of the local amplification effects due to the soil. Then, it is possible to select n periods of return of the seismic action and calculate the corresponding elastic response spectra ($n \sim 9$). The maximum number of selectable return periods is 9 in agreement with national technical standards (DM2018).

In detailed modeling, spectrum compatible signals ($m \sim 10$) are generated on the basis of the elastic projections for each return period. In this case, non-linear analyzes in the time domain of bridge model were carried out. Simulations are performed and the application proceeds with the extraction of the time history of the variables and the quantity of demand (N, V, θ for piers and d for supports), where N is the normal stress acting on the elements, V is the shear and d is the excursion of the supports. Subsequently, the capacity vectors are calculated for each verified limit state, generally dependent on the time history of the variables (θ_{SO}, V_{SLC}, θ_{SLD}, θ_{SLC}, d_{SLD}, d_{SLC}). At this point, risk indicators (ρ) are calculated as the ratio between demand and capacity for each limit state verified for each component of the bridge. Each risk indicator depends on five indices: the period of return of the seismic action (i1); the analyzed signal (i2); the time step of the signal (i3); the component of the bridge (pier or support) (i4) in hypothesis of damage state limit (SLD) and collapse (SLC), while monitoring only the displacement at the top of the piers in hypothesis of operation limit state (SLO); the verified direction (i5), indicating more simply X or Y. At this point, a directional combination of the risk indicators is made according to whether they are piers or supports. The Y risk indicators of the system are calculated from the component indicators (envelope of the mechanisms). Subsequently, the normal fit log of the Y risk indicators is performed with respect to the analyzed signals and the calculation of the probability of exceeding the verified limit state, conditioned to the return period (point of the fragility curve).

In simplified modeling, a significant number of bridges (e.g. 1000) are generated through Montecarlo simulations to consider the uncertainty associated at capacity of the elements (i.e. piers, abutments, supports). Once simulations are performed, the application proceeds with the extraction of the quantity of capacity. For each subsystem's component, force-displacement curve is determined. The properties of single degree of freedom (SDOF) equivalent system are defined. Based on the seismic hazard related to the site of the bridge as well as on the local amplification effects due to the soil for 9 return periods of the seismic action, the corresponding elastic response spectra are calculated. The displacement demand for each return period is defined considering the fundamental period and viscous damping of equivalent single degree of freedom (SDOF) system. Once each subsystem's force-displacement curve is determined, for each return period and each limit state, it is possible to calculate risk indicators ρ or, rather, the ratio between demand and capacity for longitudinal and transverse direction. Repeating the phases listed above for all the generated bridges and the nine return periods in a probabilistic framework, it is possible to obtained point-wise the probabilities of exceeding the limit states, or, rather, the probabilities that capacity exceeding the demand and finally calculate the fragility curves (Figure 7).

On the left side of Figure 7, the results in terms of risk indicators (top) and fragility curve (bottom) were obtained for a specific bridge through detailed modeling while on the right side shows the result referred to a taxonomy of bridge modeled by the simplified procedure

Both in detailed modeling (full knowledge) case (Figure 7, top) and in simplified modeling (partial knowledge) each point represents the risk indicators for each analysis for each return period for the analyzed bridge.

Figure 7 (left) shows the risk indicators (top) and the relative fragility curves (bottom) for a specific bridge analyzed for the damage limit state according to the detailed modeling. Furthermore, Figure 7 (right) shows the risk indicators (top) and the relative fragility curves (bottom) for the damage limit state calculated for a taxonomy of bridge modeled by the simplified procedure: in this case a population of bridges (i.e. 1000) are analyzed.

For the calculation of the fragility curves shown in Figure 5, nine return periods of the reference input motion were adopted; in detailed modeling for each of them 10 simulations using different accelerograms were carried out. In the simplified model, the uncertainty of input motion was

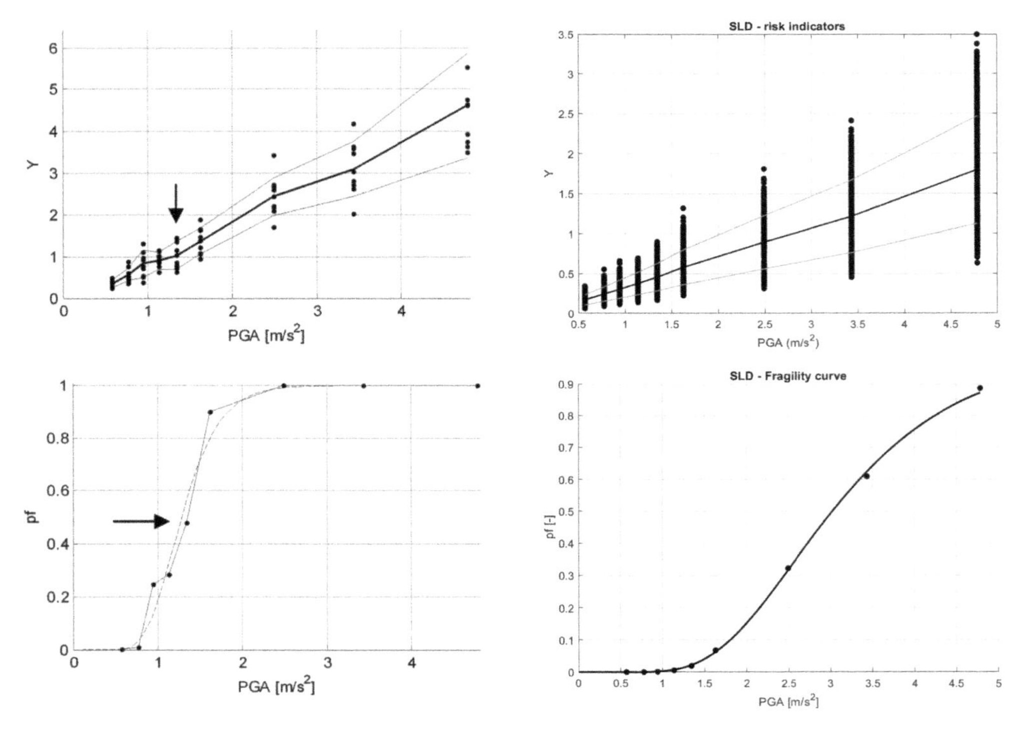

Figure 7. Risk indicator and fragility curve example with reference to damage limit state.

considered adopting different spectral shapes and then associating a variability to parameters of response spectra, as defined in national technical standards D.M. 17/01/2018 (i.e. F0, Tc*).

5 CONCLUSIONS

The paper deals with the development of an application for the automatic calculation of fragility curves of railway bridges in both the hypothesis of complete and partial knowledge of the bridge characteristics. The main hypotheses adopted in two case of structural modeling of bridges are illustrated.

In case of full knowledge, a detailed modeling of bridge has been developed and FEM analyzes are carried out in the time domain with partially degrading hysteretic constitutive laws, using the OpenSEES code (McKenna et al. 2000): a description is given of the main analytical options used in the calculation models. The application automatically manages numerous non-linear finite element time-history analyses after carrying out the static analysis and calculating the curvature moment relationship of each pier. For each simulation, the input motion is obtained from a pair of artificial spectral-compatible accelerograms generated by the application: the seismic action is then applied to the structure, that is fixed at the base, in the directions parallel and perpendicular to the axis of the bridge. The adopted reference spectra are those contained in the current national code (D.M. 17/01/2018).

In the case of limited data, to evaluate the seismic behaviour of girder bridges, a simplified procedure is used. The adopted procedure is based on displacements (Direct Displacement Based Assessment – DDBA), as proposed by Priestley et al. (2007) and Priestley (1997). A significant number of bridges (e.g. 1000) are generated through Montecarlo simulations to consider the uncertainty associated to the elements capacity (i.e. piers, abutments, supports). Each pier-support subsystem is considered through a single degree of freedom (SDOF) equivalent system. For each simulation, the input motion is based on the seismic hazard related to the site of the bridge as well as to the local amplification effects due to the soil for 9 return periods of the seismic action in terms of elastic response spectra, as indicated in Italian technical standards (D.M. 17/01/2018).

The limit states considered in the calculation of the fragility curves essentially concern the deformation and the resistance capacity of the piers (ductile and brittle mechanisms), the capacity of the bearings, as well as the loss of support of the deck towards the supports (bearing device or pier cap), due to the excessive horizontal excursion induced by seismic motion. These limit states are defined and checked for the individual components of the bridge (piers and supports); since these components are all main elements of the bridge, the achievement of a limit state on a component automatically determines the achievement of the same limit state on the bridge (system in series).

The procedure for calculating fragility curves is described with particular reference to the articulated use of risk indicators. These indicators are instantaneous quantities defined as the relationship between the seismic demand and the capacity of the components for each verified limit state. Their elaboration allows the production of fragility curves.

The developed application could allow to define sections of the railway network in which the circulation needs to be slowed down or suspended and the inspection priorities of bridges in the phases immediately following an earthquake.

REFERENCES

Borzi B., Ceresa P., Franchin P., Noto F., Calvi G. M. & Pinto P. E. 2015. Seismic Vulnerability of the Italian Roadway Bridge Stock. *Earthquake Spectra*: November 2015, Vol. 31, No. 4, pp. 2137-2161

Calvi G. M., Pinto P. E., Franchin P. & Marnetto R. 2010. The highway network in the area struck by the event. *Progettazione Sismica 1*, IUSSPress, Pavia, Italy.

CEN 2005. Eurocode 8: Earthquake resistance design of structures Part 3: Assessment and retrofitting of buildings, EN 1998-3:2005.

D.M. 17/01/2018. Norme Tecniche per le Costruzioni. Gazzetta Ufficiale, n. 42 del 20/02/2018, Supplemento ordinario n.8, 2018.

G.U. 11/02/2019. Circolare 21 gennaio 2019, n. 7 C.S.LL.PP. Istruzioni per l'applicazione dell' ≪Aggiornamento delle "Norme tecniche per le costruzioni"≫ di cui al decreto ministeriale 17 gennaio 2018, Supplemento ordinario n. 5 della Gazzetta ufficiale n. 35 del 11 febbraio 2019.

Magliulo G., Capozzi V., Fabbrocino G. & Manfredi G. 2011. Neoprene-concrete friction relationships for seismic assessment of existing precast buildings. *Eng Struct*, 33(2), pp.532–538.

Modena C., Tecchio G., Pellegrino C., Da Porto F., Donà M., P. Zampieri & M. A. Zanini, 2015. Reinforced concrete and masonry arch bridges in seismic areas: typical deficiencies and retrofitting strategies. Structure and Infrastructure Engineering, 11:4, 415–442, 2015.

Muscolino G., 2002. Dinamica delle Strutture, *McGraw-Hill*.

OpenSees – F. McKenna, G. L. Fenves, M. H. Scott. 2000. Open System for Earthquake Engineering Simulation Pacific Earthquake Engineering Research Center – University of Berkeley, CA.

Pinto P.E. & Franchin P. 2010. Issues in the upgrade of Italian highway structures. *Journal of Earthquake Engineering* 14, pp. 1221–1252.

Pinto P.E., Franchin P. & Lupoi A. 2009. Valutazione e consolidamento dei ponti esistenti in zona sismica. *IUSSpress*, Pavia.

Pinto P. E., Giannini R. & Franchin P. 2004. Seismic reliability analysis of structures. *IUSSpress*, Pavia.

Priestley M.J.N., Calvi G.M., Kowalsky M. 2007. Displacement Based Seismic Design of Structures. *IUSS Press*, Pavia.

Priestley M. J. N. 1997. Displacement-Based Seismic Assessment Of Reinforced Concrete Buildings, *Journal of Earthquake Engineering*, 1:1, pp. 157–192, DOI: 10.1080/13632469708962365

Chapter 21

Nonlinear rail-structure interaction effects for multi-frame, multi-span curved Light Rail Bridges

E. Honarvar & M.K. Senhaji
Jacobs Engineering Group

ABSTRACT: Although use of non-ballasted continuous welded rail (CWR) has become inseparable portion of light-rail transit (LRT) projects due to maintainability, safety, and passenger comfort, the CWR could adversely affect the bridges supporting the LRT vehicles through rail-structure interaction (RSI). Due to temperature variations, significant axial rail stresses/deformation may develop affecting the track serviceability by increasing the probability of rail fracture. Additionally, the substructure forces caused by the RSI effects are not addressed by the current design codes, resulting in inaccurate estimation of these. These effects are aggravated by using long multi-frame multi-span bridges with curved alignments. In this paper, a nonlinear RSI analysis was undertaken for five long curved bridges with a combined length of more than 2.5 mi (4.02 km) to accurately compute rail stresses, rail gap, and substructure forces. As a result, modifications to the existing simplified approaches are proposed to improve the estimate of the RSI effects.

1 INTRODUCTION

With aging infrastructure across the nation, the demand for light-rail transit (LRT) system and associated infrastructure will continue to grow as a means to deliver an improved, cost effective, and environmentally sensitive public transportation solution. Due to maintainability, safety, and passenger comfort, non-ballasted continuous welded rail (CWR) has become a widespread alternative to support railways, including LRT systems. Non-ballasted tracks are categorized into: direct fixation (DF) tracks; embedded rail tracks; and embedded block tracks. In DF tracks, a rail is held in place on a support using a plate assembly, as demonstrated in Figure 1. The support is generally concrete, steel, or other materials.

The CWR typically extends several miles depending on the alignment, covering various terrains with different constraints. When the LRT crosses over existing features on the ground, including roadway and railways, bridges are utilized to support the LRT. The structural performance of these bridges is adversely affected by the non-ballasted CWR through rail-structure interaction (RSI). As a result of temperature variations, significant axial rail stresses/deformation may develop affecting the track serviceability by increasing the probability of rail fracture. In addition, the substructure forces caused by the RSI effects are not addressed by the current design codes, resulting in inaccurate estimation of these forces when simplified methods are employed. These effects are aggravated by using long multi-frame multi-span bridges with curved alignments.

In this paper, the RSI effects were investigated for bridges designed for one of the most eminent ongoing LRT projects in the nation, known as Durham-Orange LRT (D-O LRT) project in North Carolina. A nonlinear RSI analysis was undertaken for five long curved bridges with a combined length of more than 2.5 mi (4 km) to accurately compute rail stresses, rail gap, and substructure forces. As a result, modifications to the existing simplified approaches are proposed to improve the estimate of the RSI effects and avoid unnecessary conservatism.

Figure 1. DF tracks and fasteners.

1.1 *Rail structure interaction*

For railway bridges, the interaction between the bridge structure and rails occurs due to thermal effects, live load effects, and time-dependent properties, including concrete creep and shrinkage. The thermal effects are typically caused by variation in ambient temperature and rail lay temperature. This paper examines the thermal effects on RSI by computing axial rail stresses and quantifying rail gaps to ensure the allowable limits are satisfied. Furthermore, the finding of RSI analysis is used to better estimate the substructure forces caused by the restraint of superstructure to the rail deformations.

1.2 *Project description*

To meet the increasing transportation demand in the Triangle area in North Carolina, GoTriangle Authority has proposed D-O LRT project. The D-O LRT Project is a 17.7-mi (28.49 km) LRT system with 18 stations that will provide over 26000 trips per day to the residents and commuters in Durham and Chapel Hill in North Carolina, as shown in Figure 2. This LRT will connect three major universities, including University of North Carolina at Chapel Hill (UNC), North Carolina Central University (NCCU), and Duke University. In addition, three major medical facilities are connected through the proposed LRT system.

The track system of D-O LRT bridge structures is DF tracks with CWR. This track system minimizes the number of joints, reduce maintenance costs, and improve the ride quality.

1.3 *Description of selected bridges*

To study the RSI effects, five different bridges from the D-O LRT project with various configuration were selected, as given in Table 1. The bridge lengths vary from 801 ft (0.24 km) to 4259 ft (1.30 km), with multiple spans and frames. The superstructure of these bridges includes both prestressed concrete girders and steel plate girders. Concrete single column bents, two columns bents, and hammerhead bents are used to support the superstructure. The foundation consists of a combination of footings and drilled shafts.

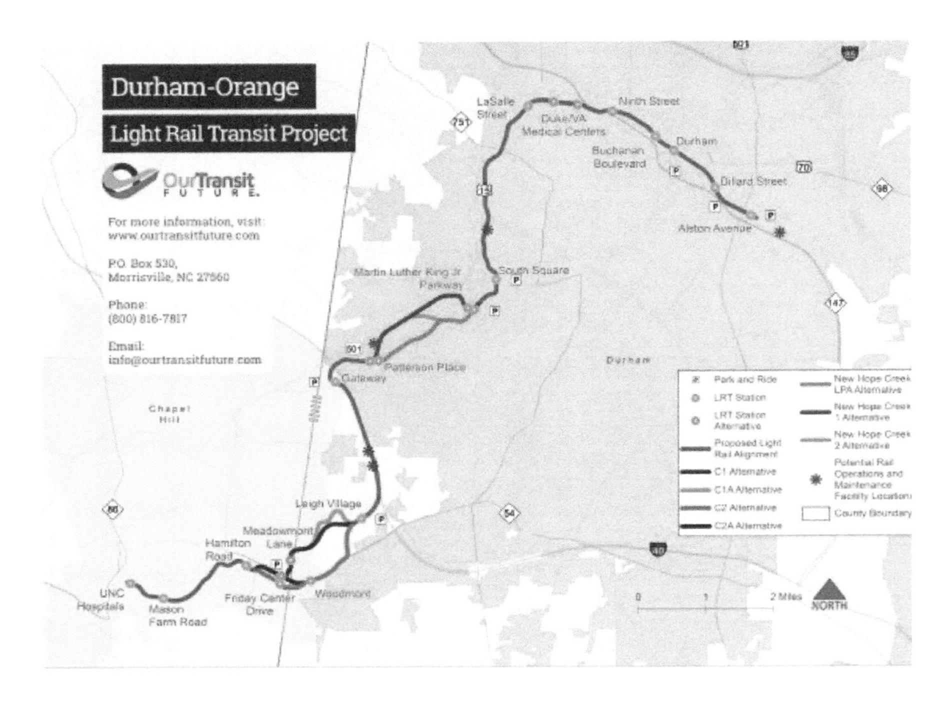

Figure 2. DOLRT Project Footprint (DOLRT 2017) (Courtesy of GoTriangle website).

Table 1. Description of bridges used for RSI analysis

Bridge	Total Length (ft)	Number of Frames	Number of Spans	Superstructure Type	Substructure Type
B-1	801 (0.24 km)	4	8	Prestressed concrete beams	Two-column bents supported on drilled shafts
B-2	3788 (1.15 km)	12	27	Prestressed concrete beams; Steel plate girders	Concrete two-column bents with and without footing supported on drilled shafts; Concrete Hammerhead bents supported on footings and drilled shafts
B-3	1915 (0.58 km)	7	14	Prestressed concrete beams; Steel plate girders	Concrete two-column bents with and without footing supported on drilled shafts; Concrete Hammerhead bents supported on footings and drilled shafts
B-4	2467 (0.75 km)	10	21	Prestressed concrete beams; Steel plate girders	Concrete two-column bents with and without footing supported on drilled shafts; Concrete Hammerhead bents supported on footings and drilled shafts
B-5	4259 (1.30 km)	16	39	Prestressed concrete beams	Concrete single column bents supported on footings and drilled shafts

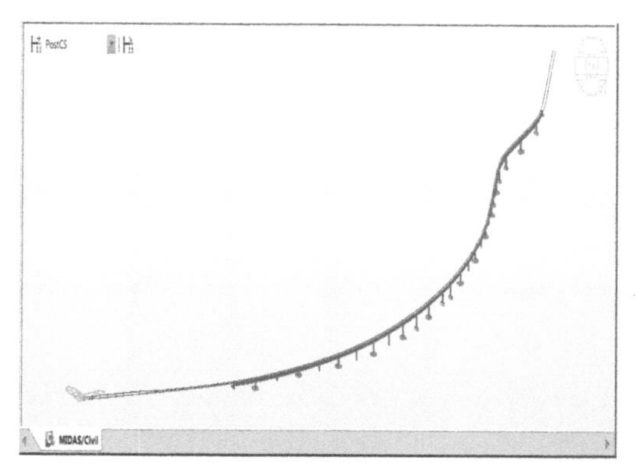

Figure 3. The 3D developed FEM of B-2 (left) reflecting the bridge curved alignment – Rendering, Courtesy of Jacobs (right).

Table 2. Material properties used in the FEM.

Deck concrete compressive strength	4 ksi (27,579 kN/m^2)
Prestressed concrete compressive strength	8.5 ksi (58,605.4 kN/m^2)
Structural steel ultimate tensile strength	AASHTO M270 Grade 50, 50 ksi (344,738 kN/m^2)*
Steel thermal coefficient	6.5e-06 1/°F (117e-07 1/°C)
concrete thermal coefficient	6.0e-06 1/°F (108e-07 1/°C)
Steel modulus of elasticity	29000 ksi (2.0 e+08 kN/m^2)

* AASHTO LRFD 2016.

2 ANALYTICAL MODEL OF BRIDGES

A detailed nonlinear 3D finite element model (FEM) was developed for each bridge, including the rails on the bridge with adequate rail extensions on the embankments using Midas Civil software (Midas 2016). In general, plates and beam elements were developed to represent different bridge structural members. These rails were connected to the bridge/embankment through use of fasteners, represented by springs in the FEM. In the longitudinal direction, bilinear springs were used to characterize the fasteners longitudinal stiffness while elastic springs characterized the vertical and transverse stiffness.

2.1 *FEM Overview*

The FEM incorporates the horizontal alignment of the bridges including skews and curves. Figure 3 shows an example of 3D model developed for B-2. Moreover, 300 ft (91.44 m) plus 10% of the bridge length is extended on each end of the bridge to model the embankments. In addition, the material properties used in the FEM are presented in Table 2. Beam elements were used to model the bridges structures.

The bridges analyzed are constituted of two expansion end bents and intermediate bents with alternate expansion and fixed support connections. These connections were modeled through elastic links with specified stiffness properties. The elastic links connect the bridge superstructure frames to the bents. The behavior of these elastic links is assumed to be linear. The point of fixity was assumed at the base of footing and at a distance of 3D for drilled shafts, where D is the drilled shaft diameter, as shown in Figure 4.

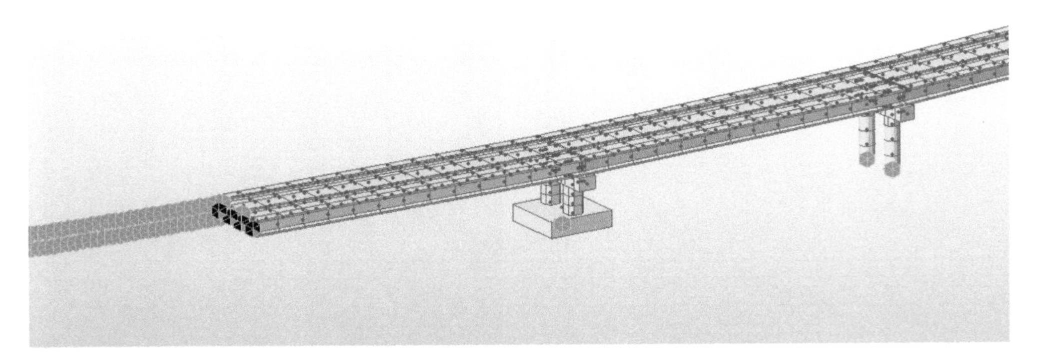

Figure 4. Bents boundary conditions.

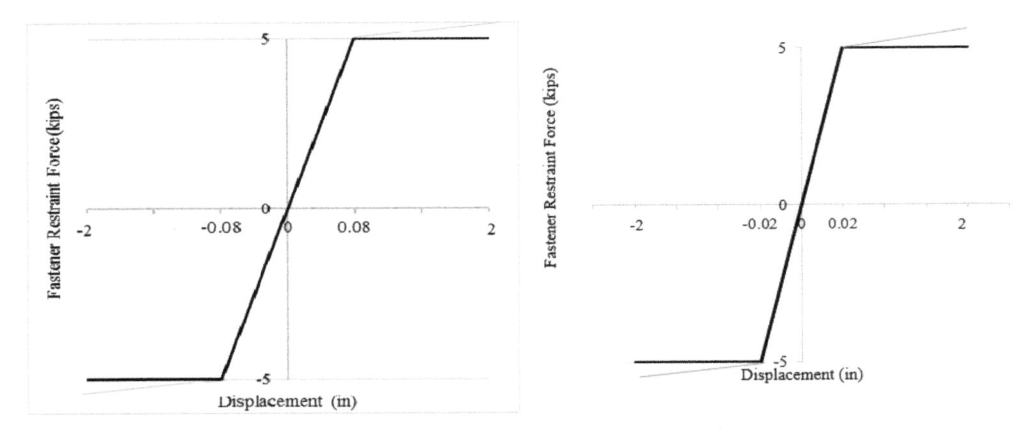

Figure 5. Rail fastener bilinear springs for ballasted tracks (left) and non-ballasted (right).

2.2 *Stiffness of rail fasteners*

Rails are connected to the deck or track slab by fasteners that retain the rail in place through friction (TCRP Report 155 2012) and (TRB 2015). The fasteners were generally distributed every 30 inches (76.2 cm). Bilinear springs were used to characterize the fasteners longitudinal stiffness, as shown in Figure 5. The fastener longitudinal slip was considered to be 1 kip (4.45 kN) per rail per spacing of 1 ft (30.48 cm) (i.e. 2.5 kip (11.12 kN) per rail for a spacing of 30 inches (76.2 cm)). The slip displacement for non-ballasted track along the bridge and ballasted tracks outside the bridge was assumed 0.02 inches (0.05 cm) and 0.08 inches (0.2 cm), respectively. In addition, elastic springs characterized the vertical and transverse stiffness. Figure 6 shows the use of multilinear elastic links to represent the bilinear behavior and linear behavior of fasteners in the FEM.

2.3 *Temperature loads*

Temperature effects were considered in the FEM during all stages of rail construction from the rail lay to service condition of the bridge system. The maximum and minimum rail lay temperatures were considered as the rail is heated before the rail clips are installed. This is conducted to to lock the rail in position and to induce a tension force in the rail as to the rail temperature falls to cool to

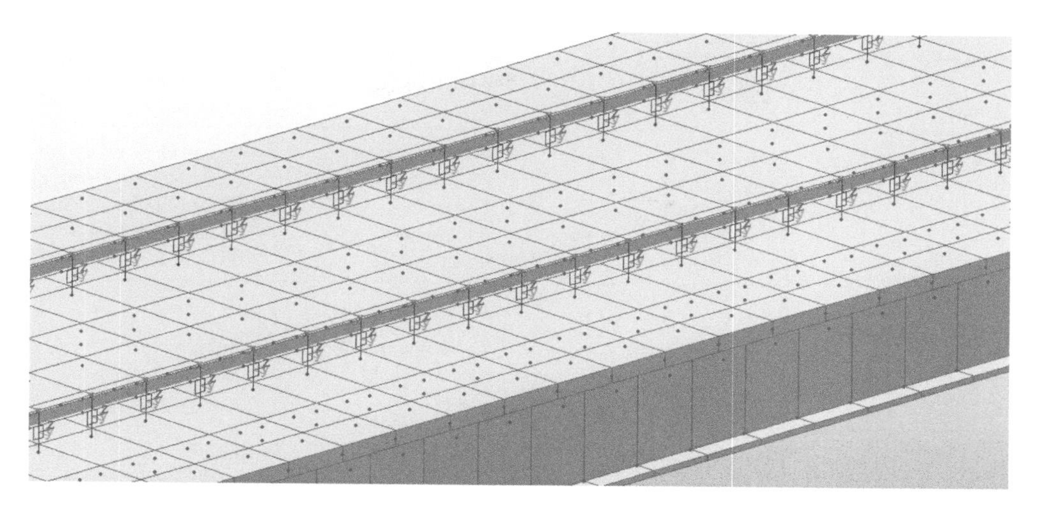

Figure 6. Rail Fastener bilinear springs connected to bridge superstructure in the FEM.

Table 3. Temperature Static Load Cases

Min rail lay temperature	95°F (35°C)
Max rail lay temperature	110°F (43°C)
Temperature rise for concrete superstructures	45°F (7°C)
Temperature rise for steel superstructures	50°F (10°C)
Temperature rise for rails	50°F (10°C)

ambient temperature. Also, the condition related to seasonal temperature variation is accounted by using maximum summer and minimum ambient temperatures. The temperature load cases analyzed in the model are given in Table 3.

2.4 *Construction stage analysis*

Construction stage analysis was carried out using the FEM to model applicable loads and boundary condition from rail lay through the service condition of rail. Rail gap was modeled by physically removing a rail element of 1 to 2 inches (2.54 to 5.08 cm) from the FEM through deactivation of the element in construction stage analysis.

3 RESULTS

The rail break and axial stress results were extracted at the rail break stage for the summation load case which include the maximum and minimum of rail lay temperature, temperature rise, and temperature fall of the structures.

It was found that the rail gaps and axial stresses are the within the allowable limits and are proportional to the bridge length as shown in Table 4. The results were used to develop a linear regression model as presented in Figure.

Table 4. Summary of RSI analysis results.

Bridge	Length ft	Rail Gap (in.)	Axial Stress (ksi)
B-1	801 (0.24 km)	1.45 (3.68 cm)	16.9 (116,521.4 kN/m^2)
B-2	3788 (1.15 km)	1.87 (4.75 cm)	32.8 (226,148 kN/m^2)
B-3	1915 (0.58 km)	1.65 (4.19 cm)	21.7 (149,616.2 kN/m^2)
B-4	2467 (0.75 km)	1.55 (3.94 cm)	22.7 (156,511.0 kN/m^2)
B-5	4259 (1.30 km)	1.94 (4.93 cm)	34.1 (235,111.2 kN/m^2)

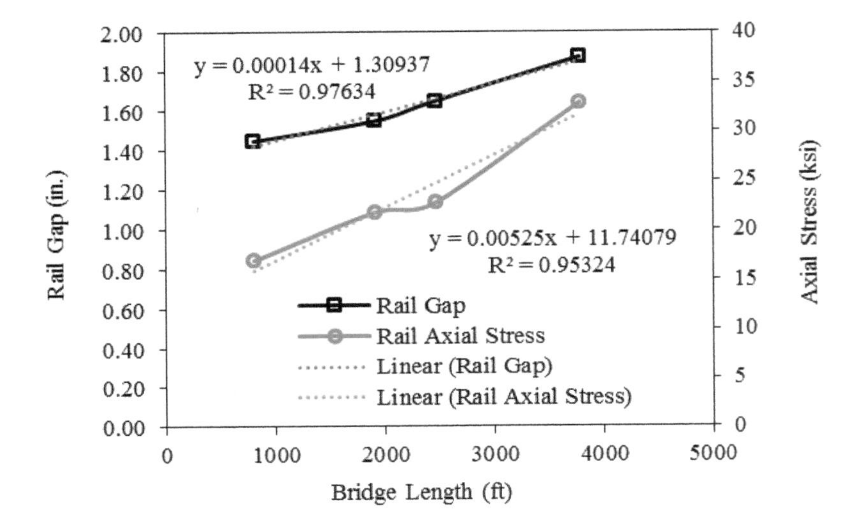

Figure 7. Developed linear regression model for the rail response of D-O LRT bridge structures.

4 CONCLUSIONS

A study on the response of a direct fixation track on multi-frame, multi-span curved bridges was conducted at a rail break event under temperature change, using 3D FEMs of five D-O LRT bridge in MIDAS Civil. The effects of a direct fixation track on curved bridges were captured through a non-linear analysis including the transverse and longitudinal stiffness of the superstructure and bilinear behavior of the fasteners. The study indicated a proportional relationship between the bridge length, the rail gap and axial stresses of DF tracks along the length of the bridge and approaches at the abutment locations.

REFERENCES

American Association of State Highway and Transportation Officials (AASHTO) LRFD Bridge Design Specifications, 7th Edition with 2016 Interims hereinafter referred to as AASHTO LRFD.

Direct-Fixation Track Design Specifications, Research, and Related Material, Transportation Research Board (TRB), 2015.

DOLRT Structures Design Guide dated October 2017.

Midas Civil 2016 User Guide.

Transit Cooperative Research Program (TCRP) Report 155 Track Design Handbook for Light Rail Transit, Second Edition, 2012

Chapter 22

Nonlinear analysis of the first concrete network tied arch bridge for California High-Speed Rail

E. Honarvar, M. Kendall & H. Al-Khateeb
Jacobs Engineering Group

ABSTRACT: While California high-speed rail (CAHSR) project encompasses various types of infrastructure to accommodate high-speed trains (HST), a single-span concrete network tied arch bridge was proposed for the first time to support HST as a part of CAHSR Construction Packages 2–3. Such a system was utilized to accommodate a long span and to satisfy the minimum horizontal and vertical clearance requirements to the existing features on the ground. Due to compound nature of a network tied-arch bridge in combination with CAHSR seismic and track-structure interaction (TSI) analysis requirements, a comprehensive nonlinear analysis is crucial to ensure satisfactory performance of the bridge. This paper presents an overview of the nonlinear analysis required to design the bridge, including nonlinear seismic, TSI, time-dependent, and geometry analyses using a detailed 3D finite-element model of the bridge. Recommendations are provided to effectively perform different analyses based on the finding of this study.

1 INTRODUCTION

California High-Speed Rail (CAHSR) project has been planned to connect San Francisco to Los Angeles in California in less than three hours at speeds over 200 miles per hour. The CAHSR project includes various types of infrastructure to support and accommodate high-speed trains (HST), including aerial guideways, culverts, roadway bridges, and embankments. Among all the infrastructure, a single-span concrete network tied arch bridge was proposed for the first time to support HST as a part of CAHSR Construction Packages 2–3. Such a system was utilized to accommodate a long span and to satisfy the minimum horizontal and vertical clearance requirements to the existing features on the ground.

Due to complex nature of a network tied-arch bridge in combination with CAHSR seismic and track-structure interaction (TSI) analysis requirements, a comprehensive nonlinear analysis is crucial to ensure satisfactory performance of the bridge. This paper presents an overview of the nonlinear analysis required to design bridge, including nonlinear seismic, TSI, time-dependent, and geometry analyses. A detailed 3D finite-element model of the bridge was developed using a variety of solid, plate, beam, and truss/cable elements to satisfy the objectives of different types of analyses and associated design elements. Through time-dependent construction stage analysis, the effects of creep and shrinkage on the design of post-tensioned concrete tie beams and reinforced concrete arch ribs and knuckles were investigated. The elastic behavior of the superstructure, including the arch ribs and braces were validated by performing nonlinear time-history analysis. As a part of the TSI analysis, the floor-beam sizes, deck thickness, and longitudinal diaphragms were determined to satisfy live load deflection and vertical deck acceleration limits. Possible sag (catenary) effects due to inclined hangers were examined using nonlinear geometry analysis. Additionally, recommendations are provided to effectively perform different analyses and designs based on the finding of this study.

1.1 *Bridge description*

Bridge 1H-03, State Route-43 Underpass (hereafter, SR-43 UP), carries two ballasted tracks of the proposed High-Speed Rail (HSR) over SR-43. This bridge consists of a single span of 236 ft (71.93 m) between bearings, approached from constructed embankments to both sides. The single-span structure accommodates the existing two-lane state route (SR-43) but the span will also accommodate a six-lane widened expressway in the future. The superstructure is a single-span concrete network tied-arch. The floor system of the tied-arch consists of a 10 in. (254 mm) thick concrete deck supported by wide flange prestressed precast concrete beams spaced at 8′-0″ (2.44 m) to 6′-0″ (1.83 m) spacing. The cast-in-place (CIP) reinforced concrete arch ribs are connected to the CIP posttensioned tie beams through inclined hangers. These hangers are 2.5″ (63.5 mm) partially threaded (PT) bars located inside a 5″ (127 mm) diameter stainless steel seamless pipes. Arch rib struts are also used to provide lateral bracing for the arch ribs. Adjacent to the two abutments, CIP post-tensioned concrete end diaphragms are used. The hanger attachments to the arch ribs and tie beams are located within the depth of concrete members. Adjustment to the loads in the hangers will be made from jacking locations on the top surface of arch ribs.

The clear bridge width between tie beams of 42′-0″ (12.8 m) will be constructed with an allowance for additional horizontal clearance between the edges of the tie beams to account for the curved horizontal alignment of the HST track across the bridge.

The abutments are full height conventional seat type abutments. Abutment 1 is pinned and Abutment 2 provides expansion supports at the top of the abutment seat with high load expansion bearings. The foundation for each abutment is composed of 15, 4 ft (1.22 m) diameter Cast in Drilled Hole (CIDH) piles with a CIP pile cap. Shock Transmission Units (STU) or Lock Up De4vices (LUD) will be utilized at Abutment 2. This will allow for gradual movements due to temperature variation, creep and shrinkage at Abutment 2, but will still be effective in resisting longitudinal forces from seismic, wind, traction, and braking. Figure 1 shows a rendering and a plan view of the proposed bridge.

1.2 *Shock Transmission Units (STU) and Lock Up Devices (LUD)*

Shock Transmission Units (STU) or Lock Up Devices (LUD) are utilized to engage both Abutments longitudinally, without impeding gradual movements at Abutment 2 caused by temperature variation or creep/shrinkage. The STU/LUD is a type of fluid viscous damper, with a tight orifice to provide full restraint under transient movements. The STU/LUD will provide a longitudinally fixed connection for all transient loads (such as braking and wind) as well as seismic events.

The STU/LUD is sized to behave elastically for all strength and service conditions. For MCE events the devices are considered "fused" and the bridge is longitudinally free to engage the backwalls. The abutments are therefore designed for the larger of the capacity of the devices (a 1.75 overstrength factor was utilized) or the shear capacity of the backwalls, per Caltrans Seismic Design Criteria (CSDC 2013) and (MTD 2014). Connections to both the abutments and the arch are also designed for the capacity of the device.

2 GEOTECHNICAL INFORMATION

Since the Geotechnical Engineering Design Report (GEDR) for the SR-43 UP has not been yet completed, the geotechnical parameters used for structural design are based on the Preliminary Foundation Report (PFR). A site-specific analysis was not been completed at the SR-43 UP site at this time, due to access issues. Current design is based on Cone Penetration Tests from two boreholes, which are located 0.20 miles (0.32 km) north and 0.25 miles (0.40 km) south of the proposed structure. These tests were part of the Geotechnical Baseline Report. When geotechnical parameters could not be found in the PFR, geotechnical data developed for an adjacent bridge were assumed to be applicable to SR-43 UP to perform preliminary analysis due to proximity of

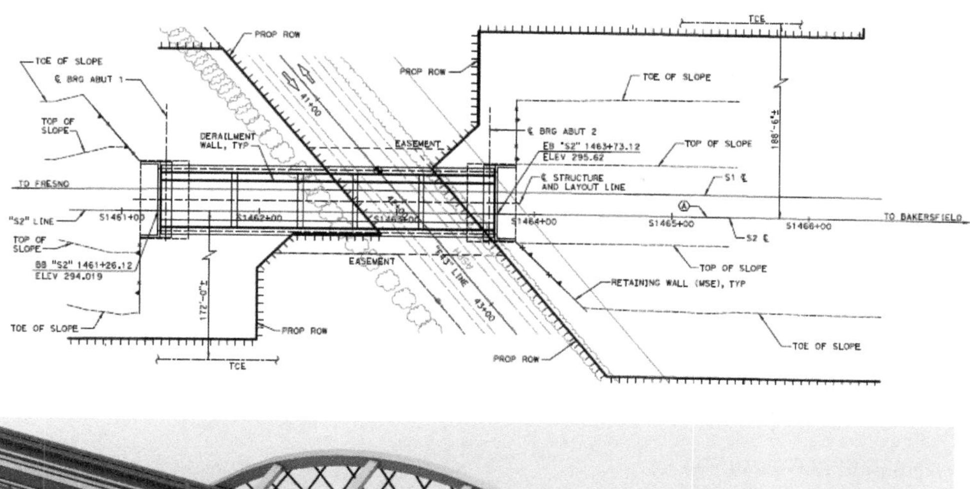

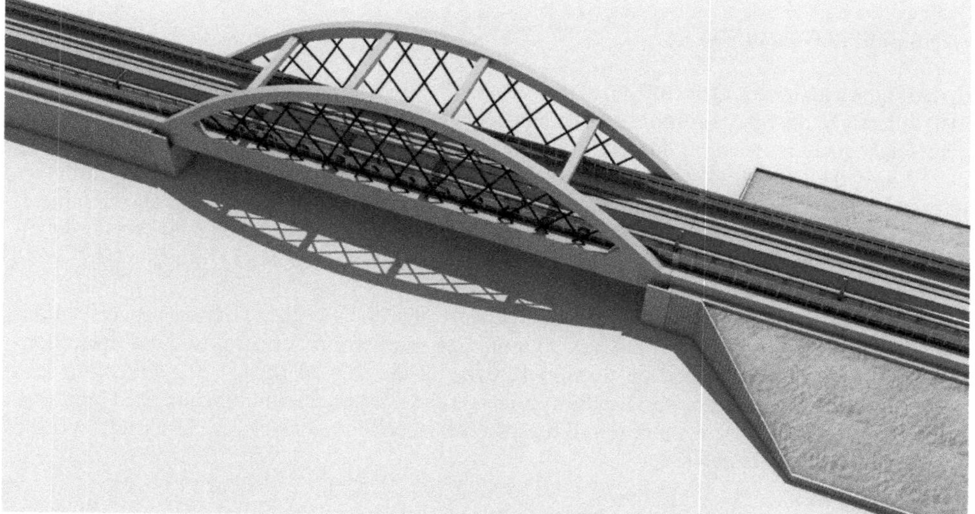

Figure 1. An overhead view and rendering of the SR-43 UP.

the two bridge sites. The following sections present the current geotechnical information used to accomplish the structural design of the bridge.

2.1 *Liquefaction potential*

Since the current groundwater level is at a depth of more than 75 feet, the cohesionless soils at the site are not considered susceptible to liquefaction based on both OBE and MCE ground motions.

2.2 *CIDH pile lengths*

Based on the preliminary foundation loads, Group Delta provided the proposed CIDH pile lengths of 91.0 ft for the structure.

2.3 *Soil profile*

Using the adjacent bridge GEDR, the soil profile was used to carry out preliminary substructure analysis and design for SR-43 UP.

2.4 *Soil structure interaction*

Using LPILE files developed for the adjacent bridge, p-y curves were generated along the shaft length to satisfactorily account for the soil structure interaction. The p-multiplier was adjusted to account for the number of rows in each direction, thereby properly capturing the group action for the drilled shafts. Since the number of output depths for p-y curves are limited in LPILE, the output locations along the pile were chosen such that the pile lateral behavior could be adequately captured. Therefore, a finer mesh (i.e., number of LPILE p-y curves output) was used at the top compared to the bottom of pile since the lateral pile behavior is mainly dominated by the top portion of the pile.

3 ANALYSIS OVERVIEW

Detailed finite-element models (FEMs) of the SR-43 UP were developed using Midas Civil software (2016) to analyze the bridge response when subjected to a wide range of loading conditions.

The loads include but not limited to: static, thermal, live, seismic, and dynamic loads. Table 12-4 of the CAHSR Project Design Criteria Manual (PDCM 2015) presents the different load cases needed to produce different load combinations and limit states, as given in Table 1. These load combinations are used to satisfactorily design the bridge components for service conditions and seismic events. The seismic design includes the Operating Basis Earthquake (OBE) and Maximum Considered Earthquake (MCE).

In addition to load combination described in Table 1, the PDCM (2015) outlines specifications to appropriately complete the TSI analysis to minimize excessive deformations and accelerations due to high-speed trains. Based on Section 12.6.4.1 of the PDCM (2015), the following load combinations are used for track serviceability analysis and rail-structure interaction (RSI) analysis. Load Groups 1 (a and b) to 3 are required for track serviceability analysis, while Groups 4 and 5 are required for the RSI analysis.

- Group 1a: $(LLRM + I)_1 + CF_1 + WA$
- Group 1b: $(LLRM + I)_2 + CF_2 + WA$
- Group 2: $(LLRM + I)_1 + CF_1 + WA + WS + WL_1$
- Group 3: $(LLRM + I)_1 + CF_1 + OBE$
- Group 4: $(LLRM + I)_2 + LF_2 \pm T_D$
- Group 5: $(LLRM + I)_1 + LF_1 \pm 0.5T_D + OBE$

In order to systematically address the PDCM (2015) requirements, various FEMs of the bridge with specific objectives were developed in Midas Civil software. For each FEM, Table 2 summarizes the analysis type, objective, and applicable limit states.

3.1 *Nominal properties*

Elastic behavior is defined for concrete, mild steel reinforcement, and pre-stressing steel using modulus of elasticity and Poisson's ratio. Concrete and mild steel are also expected to behave elastically due to seismic loads, as discussed in seismic design. The concrete compressive strength and the corresponding modulus of elasticity were defined in the FEM for different structural components following Section 12.8.1.5 of the PDCM (2015). In the calculation of modulus of elasticity, plain concrete unit weight was estimated following AASHTO LRFD Table 3.5.1-1. The unit weight of reinforced concrete was taken as 0.005 kcf greater than that of plain concrete, per AASHTO LRFD Section C.3.5.1.

Table 1. Load combinations for design of structures.

Load Combinations and Load Factors, γ_i Load Combination/ Limit State	DC DW DD EV EH ES EL PS CR SH	LLP LLV + I LLRR + I LLH + I LLS LF NE CF SS	WA FR	WS	WL	TU	TG	SE	DR	CL	OBE WA D ED	MCE WA D ED
Strength 1	γ_P	1.75	1.00	--	--	0.50/ 1.20	--	γ_{SE}	--	--	--	--
Strength 2	γ_P	--	1.00	1.40	--	0.50/ 1.20	--	γ_{SE}	--	--	--	--
Strength 3	γ_P	--	1.00	--	--	0.50/ 1.20	--	--	--	--	--	--
Strength 4	γ_P	1.35	1.00	0.65	1.00	0.50/ 1.20	--	γ_{SE}	--	--	--	--
Strength 5	γ_P	γ_{EQ}	1.00	--	--	--	--	--	--	--	1.0	--
Extreme 1	1.00	1.00	1.00	--	--	--	--	--	1.40	--	--	--
Extreme 2	1.00	0.50	1.00	--	--	--	--	--	--	1.00	--	--
Extreme 3	1.00	γ_{EQ}	1.00	--	--	--	--	--	--	--	--	1.00
Service 1	1.00	1.00	1.00	0.45	1.00	1.00/ 1.20	γ_{TG}	γ_{SE}	--	--	--	--
Service 2	1.00	1.30	1.00	--	--	1.00/ 1.20	--	--	--	--	--	--
Service 3	1.00	1.00	1.00	--	--	1.00/ 1.20	γ_{TG}	γ_{SE}	--	--	--	--
Buoyancy @ Dewatering Shutoff	0.80	--	1.10	0.45	--	--	--	--	--	--	--	--
Fatigue	--	1.00	--	--		--	--	--	--	--	--	--

Yield strength of 60 ksi (414 MPa) was used for the mild steel reinforcement conforming to ASTM 706, Grade 60 steel. The yield and ultimate tensile strength of pre-stressing steel was specified to be 243 ksi (1675 MPa) and 270 ksi (1862 MPa), respectively. When post-tensioned bars are used, a minimum ultimate tensile strength of 150 ksi (1034 MPa) is required.

For thermal analysis, the coefficient of thermal expansion of 6.0×10^{-6}/°F and 6.5×10^{-6}/°F were used for concrete and steel, respectively.

Three-inch (76 mm) diameter PT bars were used for hangers.

3.2 *Expected properties*

For seismic and TSI analysis, the expected concrete material properties were calculated per Section 3.2.6 of CSDC (2015), Version 1.7. For reinforcing steel, the expected properties were defined following PDCM (2015) Section 11.7.4.5 and CSDC (2015) 3.2.3.

3.3 *Time-dependent properties*

Time-dependent material properties, including concrete creep and shrinkage, and relaxation of prestressing steel were defined using the CEB-FIP (1990) code. For creep and shrinkage, a relative humidity of 65% was assumed to calculate the creep coefficient and shrinkage strains per CEB-FIP

Table 2. Summary of the FEMs developed to complete the different analyses per the PDCM.

FEM	Analysis type	Objective	Applicable limit state(s)/ load group(s)
Non-seismic (general)	Linear static and live load analysis	Superstructure and substructure force demands	Strength 1 to 4, Service 1 to 3, and Extreme 1 to 2
Non-seismic – construction stages	Nonlinear time-step analysis	Time-dependent forces/ deformation	Strength 1 to 4, Service 1 to 3, and Extreme 1 to 2
Non-seismic – geometry analysis	Nonlinear analysis	Sag effects and hanger pre-tensioning	Strength 1 to 4, Service 1 to 3, and Extreme 1 to 2
Non-seismic – Hanger loss analysis	Nonlinear analysis	System redundancy for loss of a hanger	Extreme 1
Seismic OBE	Nonlinear time-history analysis	Force demands due to OBE	Strength 5
Seismic MCE	Nonlinear time-history analysis	Displacement demands due to MCE	Extreme 3
TSI- Frequency	Eigenvalue	Vertical, transverse, and torsional frequency	–
TSI-Track serviceability	Live load and nonlinear time-history analysis	Superstructure and track deformations	Group 1 to 3
TSI-RSI	Nonlinear time-history analysis	Rail deformations and stresses	Group 4 and 5
TSI- Dynamic	Nonlinear time-history analysis	Vertical accelerations, dynamic amplification factor	–

(1990). These properties are used to perform construction stage analysis and camber calculations for prestressed members.

4 ANALYTICAL MODELS

Detailed 3D FEMs of the bridge were developed using a variety of solid, plate, beam, and truss/cable elements to satisfy the objectives of different types of analyses and associated design elements. Figure 2 shows the FEM developed to perform general analysis. Through time-dependent construction stage analysis, the effects of creep and shrinkage on the design of post-tensioned concrete tie beams and reinforced concrete arch ribs and knuckles were investigated. The elastic behavior of the superstructure, including the arch ribs and braces were validated by performing nonlinear time-history analysis. As a part of the TSI analysis, the floor-beam sizes, deck thickness, and longitudinal diaphragms were determined to satisfy live load deflection and vertical deck acceleration limits. Possible sag (catenary) effects due to inclined hangers were examined using nonlinear geometry analysis.

Four types of elements were used to model the superstructure and substructure members, as follows:

- Beam elements: CIDH piles, tie beams, floor beams, longitudinal diaphragms, arch ribs, arch rib struts
- Plate/shell elements: Footings, abutments, deck
- Solid elements: Transverse diaphragms
- Tension only truss/cable elements: Hangers

While four arch rib struts are currently shown in the FEM, the number and location of these struts will be refined as a part of the final seismic analysis and design. Besides, the knuckle analysis and design were further investigated through a local model using solid elements.

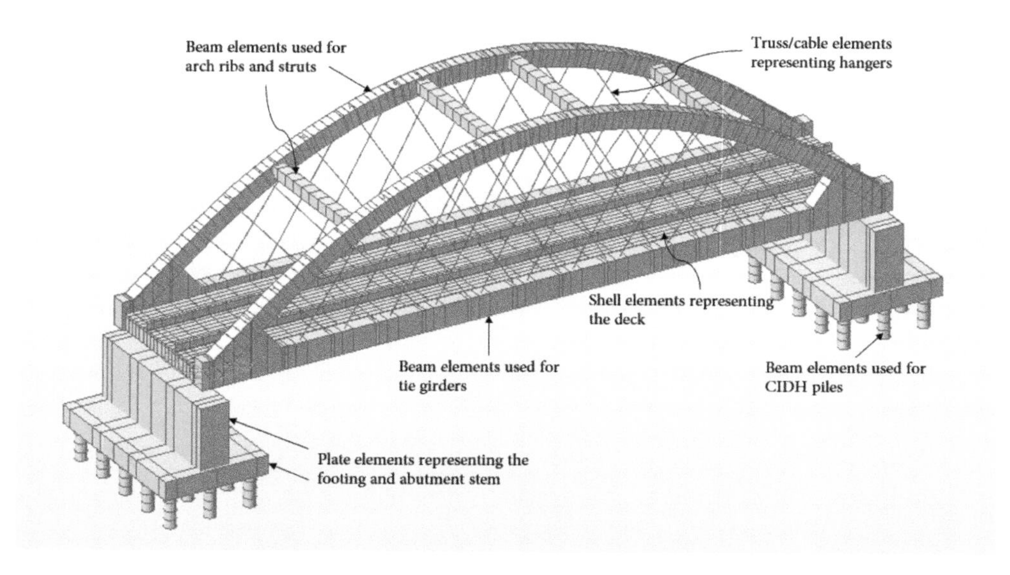

Figure 2. The FEM of the SR-43 UP developed in the Midas Civil software.

4.1 *Geometry*

The bridge geometry was assembled in the FEM using the latest design plans. A parabolic curve was followed to create the arch rib, with a maximum height of 56ft (17.07 m) at the centerline of the arch rib, measured from the working line of the tie beams to the working line of the rib.

4.2 *Model discretization*

Joints and nodes were used to discretize elements such that the bridge response could satisfactorily be captured at the locations of interest. In the model discretization, the minimum number of elements for the superstructure and substructure required by the CSDC were taken into consideration.

4.3 *Boundary conditions*

The boundary conditions for the structural elements follow the bridge superstructure and substructure data. The composite action between the floor-beams and the deck is captured by connecting the center of gravity of the deck and floor-beam elements using rigid links. Similarly, rigid links are used to create fixed connections between the floor-beams and tie beams. The arch ribs are monolithically connected to the struts, hangers, and tie beams using shared nodes among these members.

At Abutment 1, elastic links represent the pinned connection between the tie beams and the abutment by restraining all of the translational degrees of freedom, while allowing rotations. Conversely, expansion bearing allowing for gradual longitudinal movements due to temperature effects, creep and shrinkage are modeled by elastic links. While accommodating gradual movements, the expansion bearing at Abutment 2 will be replaced by elastic links restraining the longitudinal displacement due to seismic, traction, and braking forces in order to replicate the LUD behavior in the FEM. At each abutment, two bearings which transmit primarily the vertical force between the diaphragm and abutment to reduce the vertical deck acceleration, are modeled using elastic links. The bottom of the CIDH piles is assumed to be fixed at a length equal to three times the diameter of the shaft (i.e., the fixity length). After deriving p-y curves using LPILE analysis, the fixed support condition at the bottom of piles will be substituted by multi-linear elastic links along the entire length of the piles.

4.4 *Knuckle*

Different techniques were explored to model the knuckles in the global model, including use of beam elements and solid elements. It was found that the knuckle behavior could be adequately captured by extending the beam elements representing the tie beams and arch ribs into the knuckle region, while accounting for the tapered sections. In addition, diagonal beam elements (struts) were used in the knuckle region to replicate the connection between the arch ribs and tie beams through the knuckles.

4.5 *Hanger pre-tensioning*

The hangers were pre-tensioned to achieve ±0.25 in. (±6.35 mm) deflection in the tie beam when subjected to permanent loads. Any sag effects and geometric nonlinearity were found to be negligible.

4.6 *Construction stage analysis*

In pre-stressed concrete structural elements, deformations and stresses vary as a function of time due to time-dependent properties of pre-stressing steel and concrete and their associated pre-stress losses. Therefore, the construction stage analysis is performed using the FEM to accurately capture the time-dependent effects on the SR-43 UP. This analysis will account for the different loading and boundary conditions at different construction stages in combination with the time-dependent properties calculated by the CEB-FIP (1990) Code.

4.7 *Time-history model*

Using p-y curves, the SSI was included in the time-history model. These curves were approximated by multi-linear curves through use of general links in the FEM. At each depth, two nodes were defined: one node placed at the center of gravity of the pile; and the other node placed on the perimeter (two ft from the center). These two adjacent nodes were connected to each other through a general link, while the circumferential nodes were modeled as fixed supports. Additionally, inelastic plastic hinge properties (calculated based on the expected material properties) were used along the columns and CIDH piles to evaluate their response to ground motions.

5 LOADS

The loads were calculated according to Table 12-4 of the PDCM. The resulting loads were input into the FEMs to compute the force responses for the SR-43 UP structure. The following sections summarize the load calculations and their applications in the FEM of the network tied arch developed using the Shell Model.

5.1 *Dead load (DC, DW)*

The dead load was computed for all structural components and non-structural attachments. The total self-weight (DC1) of the frame (e.g., abutments, tie beams, pre-tensioned floor-beams) was calculated using the FEM's. The permanent loads (DC2) resulted in a dead load equal to 3.06 klf (44.66 kN/m) on one side of guideway, and 0.30 klf (4.38 kN/m) on one track centerline.

5.2 *Non-structural attachment loads (DW)*

The dead load was found to be equal to 5.40 klf (78.81 kN/m) per track centerline.

5.3 *Creep and Shrinkage (CR, SH)*

The displacement-induced forces due to creep and shrinkage were included in the FEM of the bridge using the time-step method adopted in Midas Civil, given consideration to the time-dependent material properties and construction sequences.

5.4 *Live Loads (LL)*

Live loads due to Cooper E-50 (LLRR) in addition to the high-speed train set (LLV) were incorporated into the live load analysis in Midas Civil based on the PDCM (2015). Two lanes were defined in the FEM to replicate the S1 and S2 lines, with a wheel spacing of five ft along the bridge. Multiple presence factor of one was used for both one-track and two-track loaded lanes.

5.5 *Impact Effect (I)*

For non-ballasted track with span length longer than 40 ft (12.19 m), the impact factor is calculated using the AREMA (2014) equation for ballasted track.

5.6 *Centrifugal Force (CF)*

The centrifugal force was calculated for LLV and LLRR, and applied horizontally towards the outside of the curve at six ft above the top of rail (TOR). Since the loads were applied to the center of gravity of the track slab elements, the vertical eccentricity was adjusted in the FEM accordingly.

5.7 *Traction and Braking Forces (LF)*

Due to LLV were applied at the TOR elevation as uniform loads. The larger value causing a more critical response was subsequently used. The braking force due to LLRR was applied at eight ft above TOR elevation, while a traction force per track due to LLRR was calculated and applied at three ft above the TOR elevation as a uniform load. The most critical (larger) response was subsequently used.

5.8 *Nosing and Hunting Effects (NE)*

The lateral load due to NE was, applied as a nodal load per track at the TOR elevation at the midspan in the FEM.

5.9 *Wind Load on Structure (WS)*

The load was applied to the exposed areas of the arch rib and the tie beams, parapets, sound-walls, and OCS poles. In doing so, the shielding of the downwind elements from the upwind elements was disregarded for the parapets and sound-wall. Wind loads were applied in different angles to the structure to maximize the structural demands of all components. Wind loads were considered for 0, 15, 30, 45, and 60 degrees.

5.10 *Wind Load on Live Load (WL)*

A uniform load per track was applied at 8 ft (2.44 m) above the top of rail (TOR) elevation throughout the guideway.

5.11 *Slipstream Effect (SS)*

The aerodynamic forces due to slipstream effects caused by LLV and LLRR were approximated by equivalent distributed loads in the FEM.

5.12 *Uniform Temperature (TU)*

The ranges of temperature were determined using procedure A, outlined in AASHTO 3.12.2.1 with California Amendments. To approximate the increased force demands due to the RSI with a uniform temperature (TU), an additional temperature differential of ± 40 (°F) between rails and deck was applied to the superstructure. This was applied to all the load combinations comprising of TU load case.

5.13 *Temperature Gradient (TG)*

Positive and negative TG were estimated across the tie beam height in addition to the pre-tensioned floor-beams including deck and haunch. These profiles were assigned to the beam elements representing the tie beams and PC/PS floor-beams using "Beam Section Temp." moduli in Midas Civil software.

5.14 *Imbalance of Live Load (ILL)*

Due to rocking of the train was estimated using a $\pm 20\%$ imbalance of live load. These loads were applied as a uniform torsional moment along the centroid of the track slabs in the direction that produced the most unfavorable effect in the member under consideration for LLRR.

5.15 *Construction Loads*

A construction live load of 20 psf (957.6 N/m^2) was used in the FEM.

5.16 *Derailment Loads (DR)*

Three cases were considered to calculate the derailment loads.

6 CONCLUSIONS

This paper presented an overview of the various nonlinear analysis required to design the bridge, including nonlinear seismic, TSI, time-dependent, and geometry analyses using a detailed 3D finite-element model of the bridge. Recommendations were also provided to effectively perform different analyses based on the finding of this study.

REFERENCES

AREMA Manual for Railway Engineering, 2014.

California High Speed Train Project HSR13-57 Design Build Contract Design Criteria and Scope of Work-Project Design Criteria Manual (PDCM), 2015.

Caltrans Memo to Designers (MTD), 2014.

Caltrans Seismic Design Criteria (CSDC), Version 1.7 Dated April 2013.

California High Speed Train Project HSR13-57 Design Build Contract, Book IV, Part C.4- Preliminary Ground Motions Guidelines, 2014.

CEB-FIP Model Code for Concrete Structures, 1990.

Midas Civil Reference Manual, 2016.

Bridge analysis & design

Chapter 23

Experimental and FEM studies on the innovative steel-concrete hybrid girder bridge

M. Hamid Elmy
Nangarhar University, Jalalabad, Nangarhar, Afghanistan

S. Nakamura
Tokai University, Hiratsuka-shi, Kanagawa, Japan

ABSTRACT: The girder bridge using steel rolled H-beams is competitive and economical for short span road bridges due to low material and fabrication costs. However, the applicable span length is only 20 m to 25 m because the maximum web height is about 900 mm. To extend the span length a new steel/concrete composite bridge was developed using the steel rolled H-beam. The new bridge form has continuous-span steel H-girders which are composite with the RC slabs to resist positive bending moments at span-center. Experiments were conducted with the partial bridge model, showing that a new SRC structural form has high bending strength and good ductile property. FEM model was developed to simulate the experiments, showing that displacements and strains obtained by FEM agreed well with test results. A design was conducted with a highway bridge model with a maximum span length of 50 m, showing that the proposed bridge satisfied the required safety and serviceability. This study showed that the proposed girder bridge was structurally rational, feasible and economical.

1 INTRODUCTION

The girder bridge using steel rolled H-sections is competitive and economical in short spans due to low material and fabrication costs. The rolled steel H-beam is a compact section and no stiffener is necessary for girders. However, the maximum available web height of the rolled H-beam is about 900 mm, the applicable span length for a road bridge is about 20 m for simple spans and 25 m for continuous spans. In a continuous structural form, the intermediate support area is subjected to negative bending moment and is usually more critical than the span center, which limits the span length.

For a simply supported girder where the bending moment is positive along the span length, the composite girder consisting of the steel H-girder and the reinforced concrete (RC) slab is rational. The steel girder is in tension and the RC slab is in compression. However, for a continuous composite girder, the RC slab is tension and does not contribute much at support joints, where large negative bending moment exists.

A new form of steel and reinforced concrete (SRC) composite girder bridge was proposed using a rolled steel H-section, (Nakamura et al. 2002). The superstructure is a continuous girder with the rolled steel H-section which is composite with the RC slab. Steel/concrete composite bridges are commonly used all over the world because of the attractive appearance and efficient structural rationality. The steel girders and the RC piers are rigidly connected by reinforced concrete at the pier top. The rolled H-beams are strengthened around the joints by being covered with reinforced concrete, which forms the SRC section and increases the resisting capacity of the section at support joints. The proposed SRC bridge using rolled steel H-section is basically a multi-span rigid frame bridge structure and is expected to be competitive and economical compared with the welded plate girder bridges.

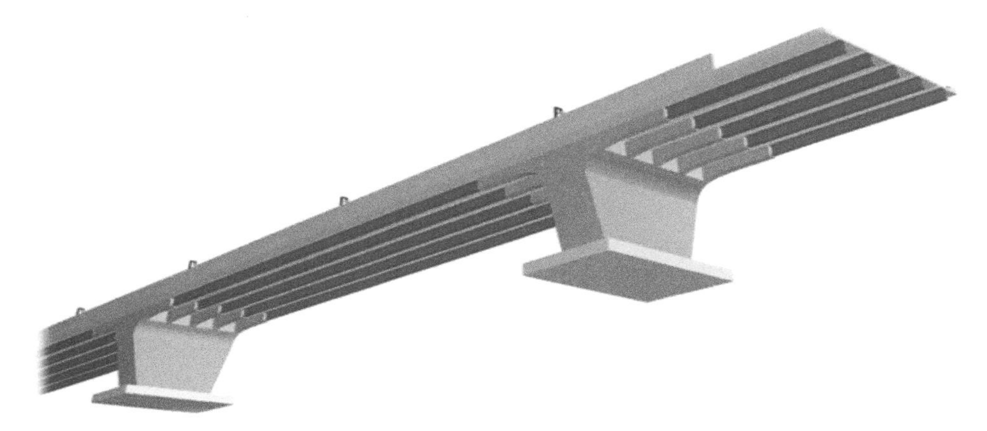

Figure 1. The SRC girder bridge with steel rolled H-section.

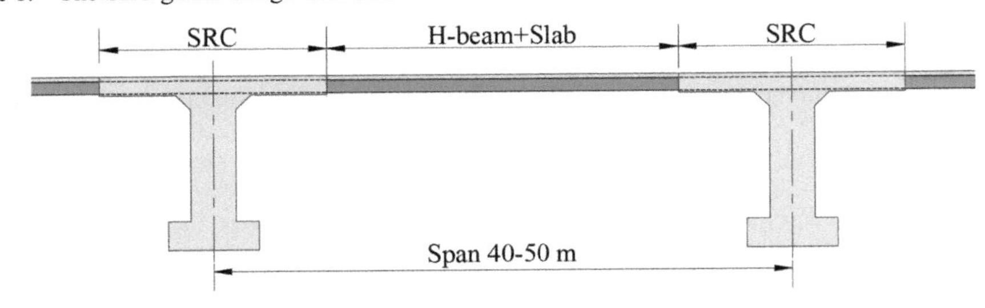

Figure 2. Side view of the SRC bridge.

Experiments were conducted with the partial SRC bridge model, and the bending strength and the load transfer mechanism at the steel-concrete rigid joint was investigated, (Takagi et al. 2003). Finite element model was then developed, considering material non-linearity and shear connections between steel and concrete. The FE model was applied to the experiments and compared with the test results to clarify the behaviours of the rigid joints and to verify the FE model. A trail design was conducted with a road bridge with the maximum span length of 50 m, and safety verification of the sections was performed by the limit states design method. Furthermore, the structural performance of the proposed bridge form against ultra-strong earthquakes was studied by dynamic analysis. These studies were intended to confirm that the new SRC bridge using rolled H-girders is feasible for a much longer span bridge than the existing one.

In this SRC bridge with rolled H-girders, the web height is only about 900 mm and the steel girder is composite with relatively thick RC slab, which can provide sufficient stiffness against transverse forces without cross beams. Therefore, the cross beams could be unnecessary in the new SRC bridge, which was also verified by FEM analysis with the full bridge model.

2 STRUCTURAL FORM

The structural form of the proposed steel and reinforced concrete composite girder bridge with rolled H-beam section is illustrated in Figure 1 to Figure 3. The super-structure is a continuous girder with the rolled steel H-section and the RC slab, and the substructure is the RC piers. The girder and the pier are rigidly connected by concrete and reinforcements. The concept of this new structural form is that the steel/concrete composite girder resists the positive bending moment at the span center and the steel girder covered by reinforced concrete section (SRC) resists the negative bending moment at the rigid joints. As shown in Figure 2, this new SRC bridge is basically a

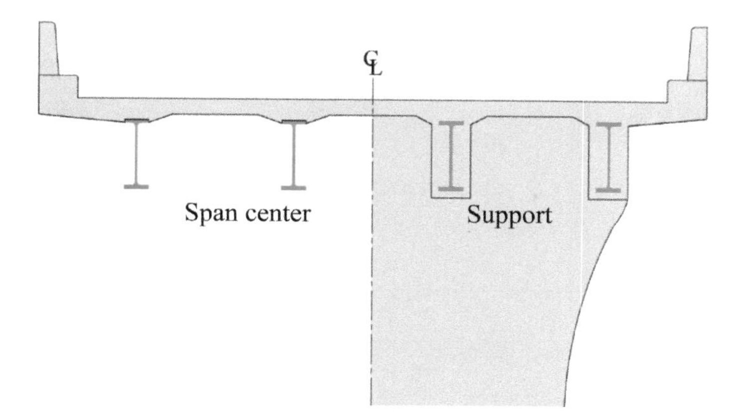

Figure 3. Cross-section of the SRC bridge.

Figure 4. Test set up.

multi-span rigid frame structure and the applicable span length is expected to reach almost double the existing ones.

In order to achieve a reliable composite rigid frame bridge, shear forces must be transferred between the steel girder and the RC part. Perfo-bond rib (PBL) shear connectors were welded to the upper and lower flanges of the steel girders in the rigid joint area. The PBL consists of the steel plate with holes, into which reinforcements go through. The steel girders were covered by reinforced concrete at around 15% of the span length near the support joints. Sufficient amount of reinforcements in the cross-section were installed longitudinally and stirrups were also arranged at proper intervals.

3 EXPERIMENTAL INVESTIGATION

Bending tests were conducted with the partial SRC specimen, which has a reverse L-shape shown in Figure 4. The objective of the test was to investigate the bending strength and ductility of the steel-concrete rigid joint and to clarify the load transfer mechanism from the girder to the pier at the rigid joint. The test specimen was scaled down to half of the actual bridge model, using rolled H-beam section. The embedded length of the steel girder at the rigid joint was approximately twice the girder height to provide adequate embedded length. A half of the length of the steel girder was covered by RC to form SRC.

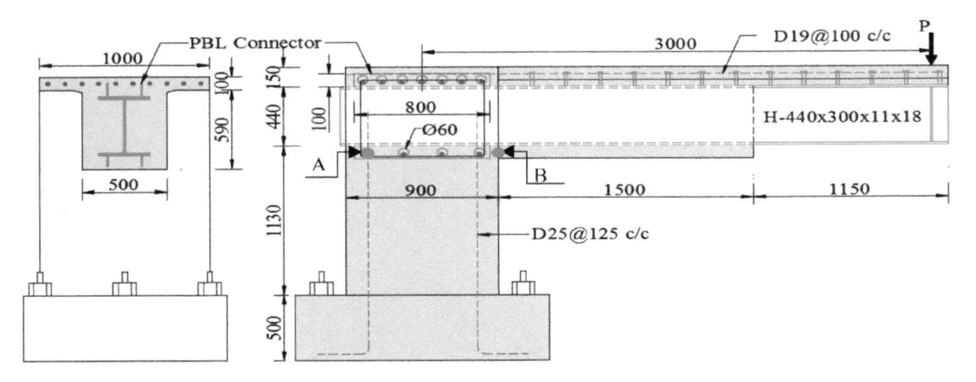

Figure 5. Dimensions and set up of test model (mm).

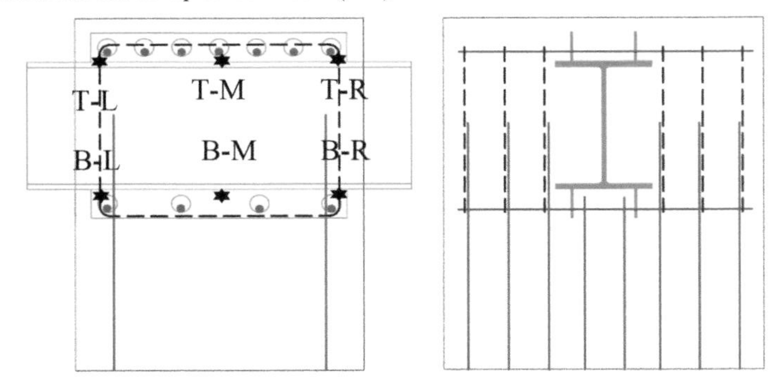

Figure 6. Details and arrangement of PBLs.

3.1 *Experimental setup*

The structural steel used in the test was mild steel SS400 with a yield stress of 338 MPa. The specimens had a length of 3,500 mm and a height of 1,720 mm. The RC slab had a width of 1,000 mm and a depth of 100 mm, with an additional 50 mm depth of the haunch. The concrete slab had reinforcement steel bars of SD295 with a yield stress of 388 MPa. The reinforcement bars were 19 mm in diameter and spaced at a pitch of 100 mm. It was arranged in one layer at the slab center. The rolled H-beam consisted of a web with 440 mm in height and 11 mm in thickness, and upper and lower flanges with 300 mm in width and 18 mm in thickness. About half length of the steel girder was covered by RC with a section of 500 × 590 mm, forming the SRC section. The measured cylinder strength of concrete was 30.0 MPa. The RC pier had a cross section of 1,000 mm × 900 mm. The reinforcement was SD345 with a yield stress of 374 MPa and with a diameter of 25 mm at a pitch of 125 mm (D25@125). The general layout and dimensions of the designed specimen is shown in Figure 5.

The Perfo-bonded rib (PBL) shear connectors were used at the beam-column rigid corner joint to prevent shear slippage between the steel and concrete interface. The PBL consisted of steel plates with a height of 100 mm, a thickness of 12 mm and a length of 800 mm. There were 60 mm diameter holes at the center. The PBLs were welded to the upper and lower flanges of the rolled H-girder. Reinforcements (D22) were installed in the holes. Stirrups were also used to bond the steel girder and the RC pier. The PBL arrangement and details are shown in Figure 6.

3.2 *Test procedure*

The specimen was incrementally loaded at the end of a cantilever beam and the vertical and horizontal displacements were monitored. The capacity of the loading machine was 625 kN and the

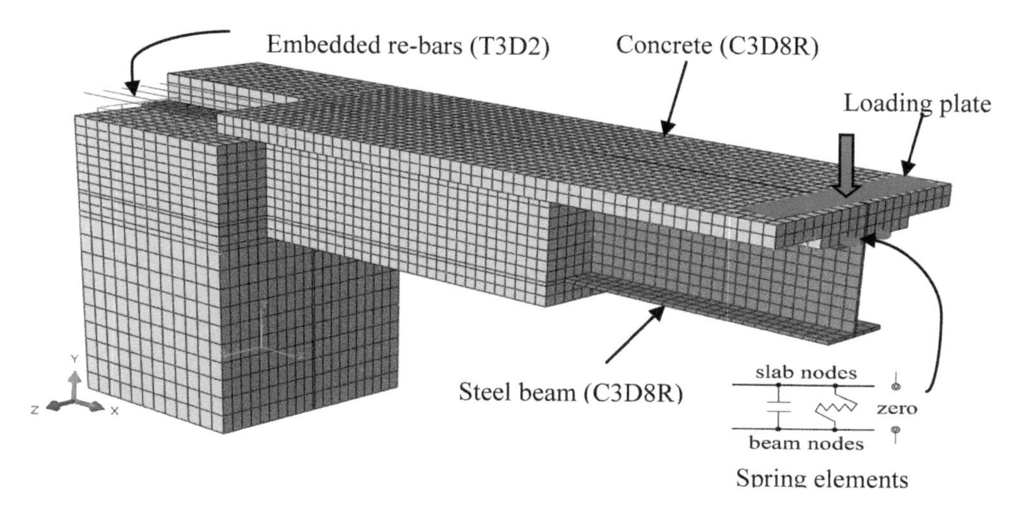

Figure 7. Finite element model of SRC girder.

test was terminated when the jack reached the maximum capacity. Concrete cracks were carefully checked at each loading step at the slab and around the beam-column rigid joint. Vertical and horizontal displacements were measured at the loading position. Strains of the steel girder, the RC slab, the RC rigid joint, the RC column and the reinforcements were measured by strain gauges. The results obtained by experiments are presented with the FEM analysis in Section 5.

4 FINITE ELEMENT MODEL

4.1 *FE model*

A finite element model was developed using the FEM program ABAQUS (2014) and applied to the experimental specimen to clarify the behaviors of the rigid joints. The steel girder and the concrete slab, column and beam were modelled as a solid element (C3D8R), and the steel reinforcements as a truss element (T3D2) with two nodes and three translational degrees of freedom were adopted, (Liu et al. 2016). The shear connectors (PBLs and headed studs) were modelled as spring elements, (Ellobody 2014). Figure 7 shows the developed finite element model.

A parametric study was conducted about the finite element mesh size to obtain accurate results with the minimum computational time. It was found that a mesh size of approximately 50 × 50 mm (length by width of the element) provides adequate accuracy in modelling the concrete slab and steel beam, while a mesh size of approximately 100 × 75 mm for the concrete pier. The smallest mesh size was equal to the steel plate thickness (50 × 18 mm). The total number of elements used in the model shown in figure 7 was 28,168. The vertical load was applied incrementally as a concentrated load at the top surface of the concrete slab along the full width of the slab, which is identical to the experimental procedure. The total load was divided into several steps in the same way as the experiment and analysis was conducted at each step. Geometrical non-linearity was considered in the FE analysis to account for large deformation.

4.2 *Contact and interaction models*

A mechanical interaction, the surface to surface contact in ABAQUS, was used between the steel girder and the concrete. In the tangential direction, the friction coefficient of 0.45 was specified in the interaction model. In the normal direction, the steel girder and the covered concrete were jointed together. The rigid constraint was used between the steel girder and the vertical stiffeners

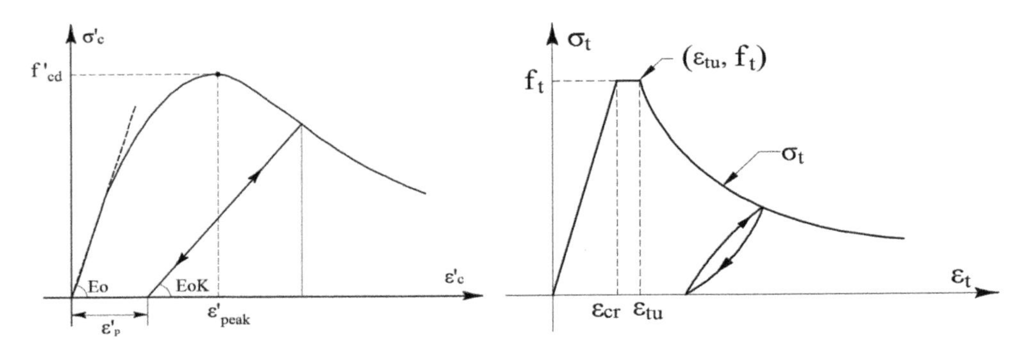

Figure 8. Stress-strain relationship in compression. Figure 9. Stress-strain relationship in tension.

at the welded parts. The reinforcements were embedded into the concrete, which enabled a perfect bond between the reinforcing bars and the surrounding concrete.

4.3 *Material modeling*

As for the concrete in compression, the non-linear elastic-plastic behaviour was expressed by an equivalent uniaxial stress-strain curve including the softening effect after the post-peak point based on the JSCE specification. The modulus of elasticity and compressive behaviour of concrete in compression is expressed by the following Eqs. (1), (2), (3) and (4) and the curve is shown in Figure 8.

$$\sigma' = E_0 K\left(\varepsilon'_c - \varepsilon'_p\right) \geq 0 \tag{1}$$

$$E_0 = \frac{2f'_{cd}}{\varepsilon'_{peak}} \tag{2}$$

$$K = \exp\left\{-0.73\frac{\varepsilon'_{max}}{\varepsilon'_{peak}}\left(1-\exp\left(-1.25\frac{\varepsilon'_{max}}{\varepsilon'_{peak}}\right)\right)\right\} \tag{3}$$

$$\varepsilon'_p = \varepsilon'_{max} - 2.86\varepsilon'_{peak}\left\{1-\exp\left(-0.35\frac{\varepsilon'_{max}}{\varepsilon'_{peak}}\right)\right\} \tag{4}$$

Where, E_0: initial modulus of elasticity, ε'_{peak}: strain corresponding to compressive strength (in general, a value of 0.002 is used), σ'_c: compressive concrete stress, K: residual rate of elastic stiffness, ε'_p: plastic strain, ε'_{max}: maximum strain in the past and f'_{cd}: design compressive strength of concrete (N/mm^2), f'_{ck}: characteristic compressive strength (N/mm^2). Fig. 8 illustrates the compressive behaviour of concrete. The density and Poisson's ratio for the concrete was assumed to be 2400 kg/m^3 and 0.2, respectively.

The stress-strain relation of concrete in tension is assumed that the tensile stress linearly increases with the tensile strain up to the concrete cracking stress and decreases non-linearly because of the tension stiffening effect, as expressed by Eqs. (5), (6), (7) and (8). Figure 9 illustrates the behaviour of concrete in tension.

$$\sigma_t = E_0\,\varepsilon_{cr} \qquad 0 \leq \varepsilon_t \leq \varepsilon_{cr} \tag{5}$$

$$\sigma_t = f_t \qquad \varepsilon_{cr} < \varepsilon_t \leq \varepsilon_{tu} \tag{6}$$

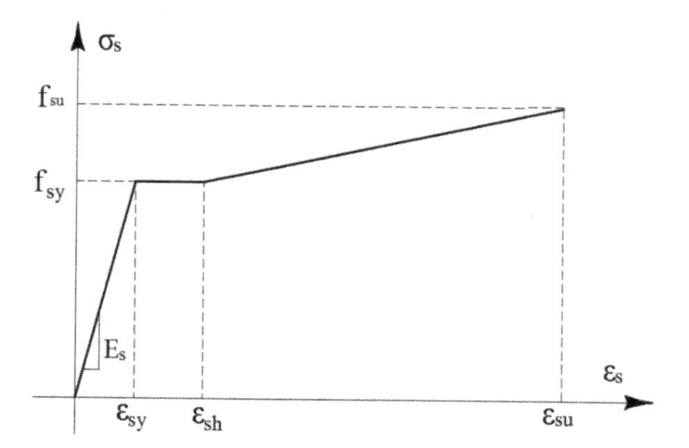

Figure 10. Stress-strain relationship for steel.

$$\sigma_t = f_t \left(\frac{\varepsilon_{tu}}{\varepsilon_t} \right)^{0.4} \qquad \varepsilon_{tu} < \varepsilon_t \tag{7}$$

$$f_t = 0.23 \left(f'_{cd} \right)^{\frac{2}{3}} \tag{8}$$

Where, f'_{cd}: design compressive strength of concrete (N/mm^2), f_t: design tensile strength of concrete, ε_t: average of concrete tensile strain, ε_{cr}: initial cracking strain, ε_{tu}: crack developing strain (in general, a value of 0.0002 is used). The complex non-linear material behavior of the concrete material both in compression and tension was modelled using damage plasticity model available in the ABAQUS material library, (Lubliner 1989).

A tri-linear relation was used as constitutive law for the steel beam, shown in figure 10. The yield strength f_{sy}, yield strain ε_{sy} and the ultimate strength f_{su} need to be input to define the stress-strain curve, while the strain at the onset of strain hardening ε_{sh} was assumed to be 0.015 and the slope of non-linear hardening state was assumed to be $E_s/100$ (E_s: modulus of elasticity of steel) . The density of the steel in all the components was assumed to be 7800 kg/m^3. The modulus of elasticity and Poison's ratio of steel was taken as 200 GPa and 0.3, respectively. For the steel reinforcement materials, a simple elastic-perfectly plastic model without strain hardening behavior is employed as the constitutive law.

To validate the developed finite element model and clarify the load transfer mechanism, the results by FEM analysis were compared with the experimental results. The load-deflection curve, load-strain relation and strain distribution patterns obtained experiments and FEM using the finite element model were compared.

4.4 *Load vs. displacement relationship*

Vertical displacements at the loading position are shown in Figure 11, indicating that FEM and experimental results are in good agreement with test results. The bending stiffness curve changed when cracks occurred at the RC pier and the steel rebar yielded at the pier/girder joint, and the loading was terminated when the loading jack reached the limit. Numerical and experimental load-deflection curve agreed in the initial elastic region up to about 130 kN, which indicates that no degradation occurs in stiffness of the SRC girder. After then, although there observed a small difference in the curve between the test results and FEM calculations, reasonably good agreement was obtained. The maximum deflection obtained by FEM was nearly the same as the experimental value at the ultimate load.

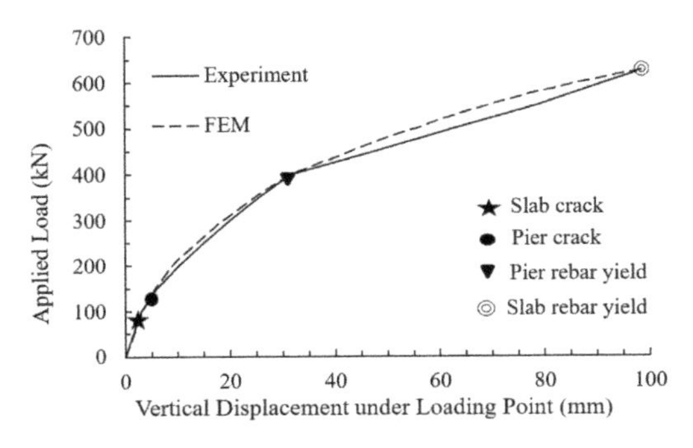

Figure 11. Vertical displacements at the loading position.

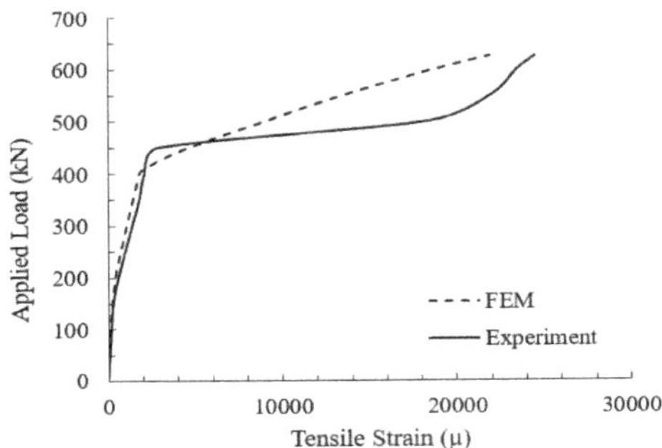

Figure 12. Strains of re-bar at point B in Fig.5.

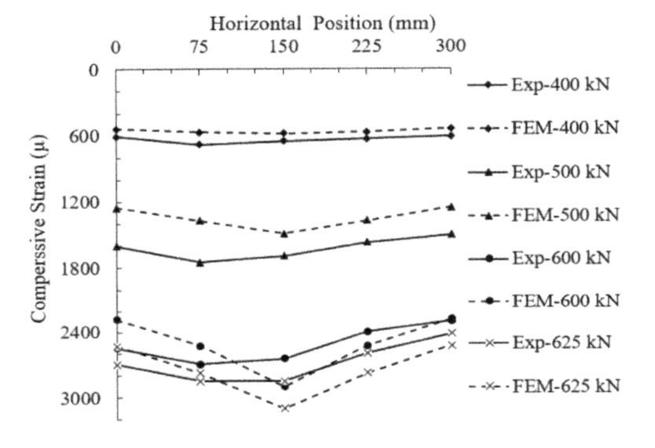

Figure 13. Strains distribution at point C in Fig.5.

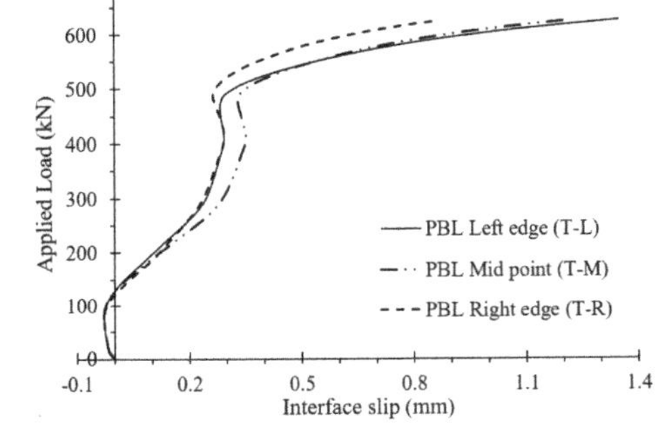

Figure 14. Load vs interface slip at position of upper flange PBLs.

4.5 *Load vs. strain relationship*

The longitudinal strains of reinforcements in the pier (at point A in Figure 5) were measured in the experiment and compared with FEM results (Figure 12). The pier rebar yielded at a load of 400kN in the FE analysis, which almost agreed with the experimental result. Both experiment and FEM curves followed the same tendency until yielding point, after that the experiment showed larger tensile strain compared to FEM results.

Figure 13 shows the compressive strain distribution on the concrete surface of the pier face at 50 mm below the lower flange (point B in Figure 5). It was understood that the strain distribution was almost uniform in the transverse direction at each load stage and the FE results agreed well with those obtained by the experiments.

4.6 *Load vs. slip behavior of interface*

The upper and lower flanges were connected to the concrete by PBLs, which were modelled by a spring in FEM model. The interface slip was computed by relative displacements in the longitudinal direction between the adjacent nodes of the concrete and the steel flanges. The slip between the upper flange and concrete is shown in Figure 14. It is understood that the slip increases with the load and the interface slip is only 1.4 mm even at the ultimate load. Also, the slip along the three points at PBL behaved similarly, which indicated that PBLs worked uniformly. These slip results proved that PBL shear connectors had a significant effect on reducing the interface slip and provided adequate bonding properties for load transfer at the rigid joint.

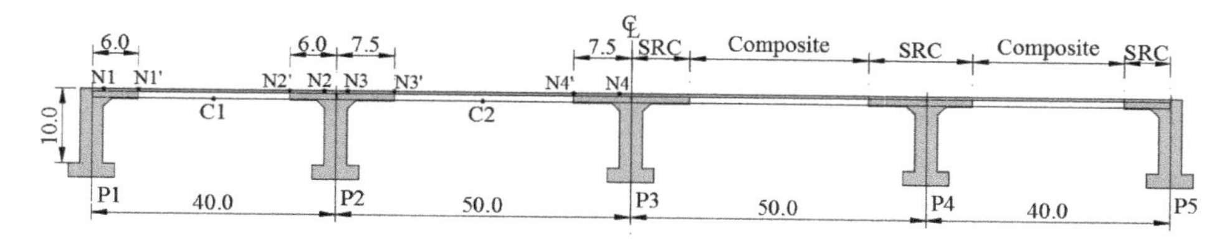

Figure 15. Side view of the bridge model (unit: m).

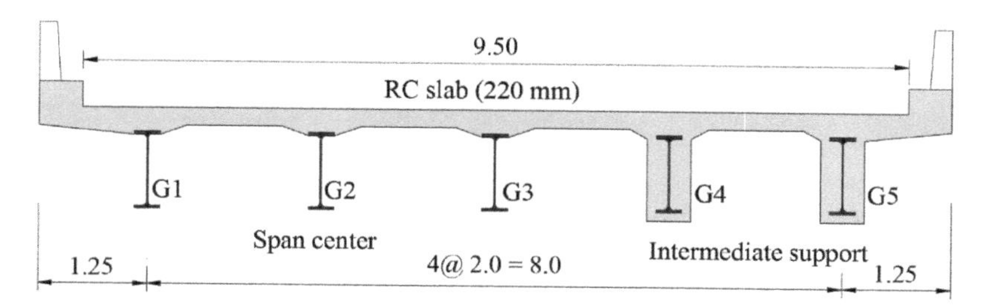

Figure 16. Cross section of the bridge model (unit: m).

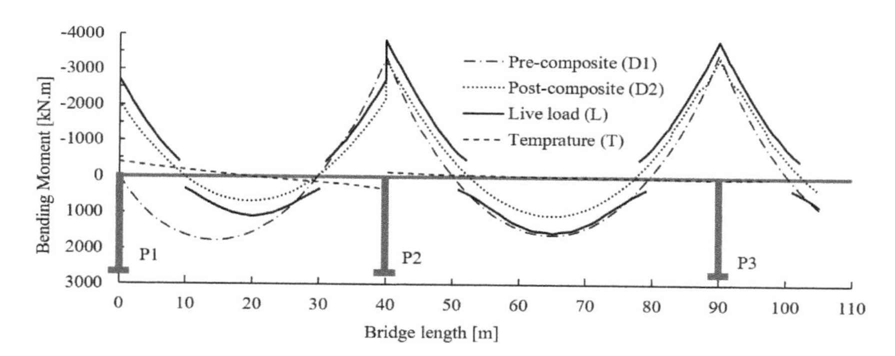

Figure 17. Bending moment diagram (G1).

5 DESIGN EXAMPLE

To confirm the feasibility and the applicability of the proposed structural form, a trial design was carried out with a full SRC bridge model. The bridge was a four-span continuous girder road bridge with the span length of 40 + 50 + 50 + 40 m, accommodating two lanes, and consisting of five steel rolled H-girders. The concrete piers were 10 m high, 3.0 m wide, and rigidly connected to the girders. The layout and dimensions of the model bridge is shown in Figure 15 and Figure 16.

The steel/concrete composite girder was assumed for the positive bending moment part at the span center and the steel girder covered by the reinforced concrete section was assumed for the negative bending moment at the support joints. The concrete slab was assumed 220 mm thick. The steel girders were covered by reinforced concrete at about 15% of the span length near the support joints. The SRC section had reinforcements with about 2% of the total cross-sectional area. The steel with tensile strength of 490 MPa was assumed for H-girders, and the reinforcement with yield strength of 390 and 490 MPa were assumed for the slab and the RC section, respectively. For the RC slab and piers, concrete with a compressive strength of 40 MPa was assumed.

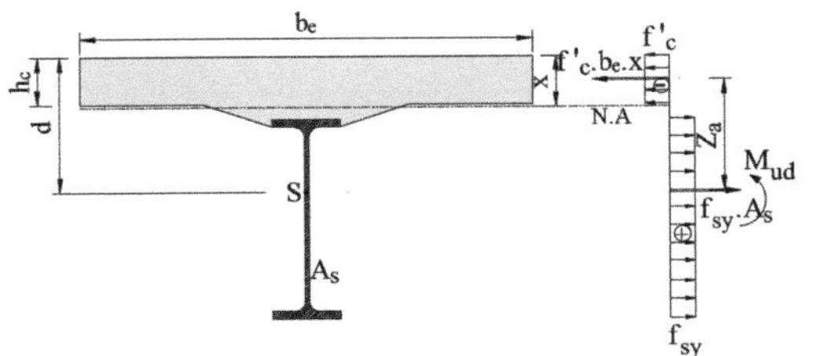

Figure 18. Stress distribution at composite section.

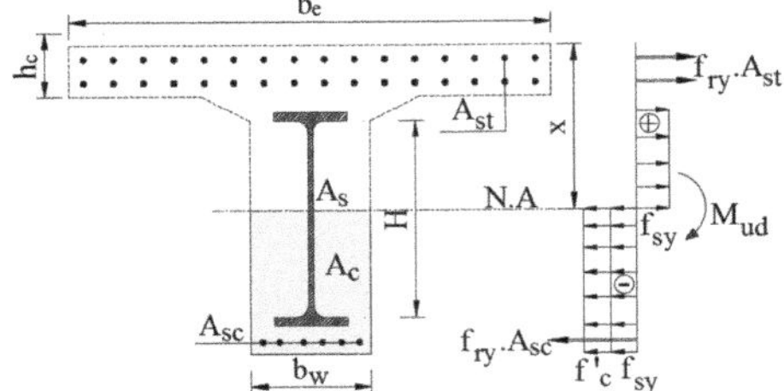

Figure 19. Stress distribution at SRC section.

5.1 *Design loads and sectional forces*

The two kinds of design dead loads were considered: pre-composite dead load (D1) due to the self-weight of girders, fresh concrete of the slab and form work, and the post-composite dead load (D2) due to the pavements, railings and traffic barriers. The design live load (L) and the impact effect (I) was adopted from the Japanese road bridge specification. The design live load consists of the equivalent concentrated load ($p1 = 10\,kN/m^2$) due to a train of large vehicle acting over a length ($D = 10\,m$), and the uniformly distributed load ($p2 = 3.5\,kN/m^2$) which represents all kinds of small vehicles and imposed over the entire length. The p1 and p2 are loaded with a width of 5.5 m and half of them loaded on other parts. The live load impact factor was assumed as 20% based on the span length.

Linear static analysis with a 3D bridge model with beam elements was performed to obtain sectional forces and deformations. The continuous beam model was used for pre-composite case, and the 3D rigid frame model was used for the post-composite load case. Influence line analysis was performed for the live loads, using (SAP 2000 2016). The distributed p1 and p2 line loads was applied as moving loads so that they produced the maximum and minimum sectional forces and deformations.

The sectional forces of the edge girder (G1) were largest among all the girders. The bending moment diagram of G1 is illustrated in Figure 17, which showed the bending moment diagram due to pre-composite dead load (D1), post-composite dead load (D2) and the maximum and minimum bending moments obtained by the influence line analysis due to live load (L). The maximum deflections due to the design live load obtained was 68.6 mm in the P2–P3 span of G1, which was within the allowable value of 100 mm (1/500 of the span length).

5.2 *Safety verification*

Safety verification was carried out by the limit states design method to ensure load carrying capacity of the composite girders throughout the service life of the structure against possible actions. The verification method and safety factors conform to the Standard specifications for hybrid structures. Figure 18 shows the stress distribution of composite section at the span center subjected to positive bending moment to find the resistant design bending moment. Figure 19 shows the stress distribution of the SRC section at the intermediate support subjected to negative bending moment to find the resistant design bending moment.

The fiber model was used to obtain the bending moment-curvature (M-ϕ) relation and the ultimate bending capacity of SRC sections. The cross section was divided into small fibres and each fiber conformed to the constitutive law of steel and concrete explained in Section 4. The safety of member at each section was verified before and after formation of composite action in accordance to the Standard specifications for hybrid structures. The basic verification equation is expressed by:

$$\gamma_i\,(M_d / M_{ud}) \leq 1.0 \tag{9}$$

where γ_i: structure factor which is usually set at 1.1, M_d: design bending moment and M_{ud}: design resistant bending moment. In this equation, the load factor is 1.1 for pre-composite dead load, 1.2

Table 1. Verification of girder due to pre-composite dead loads.

Position	M_{d1} (kN.m)	M_{sud} (kN.m)	Safety check $\gamma_i (M_{d1}/M_{sud}) \leq 1$	Remarks
N1	0	3200	0.00	OK
C1	1950	3200	0.67	OK
N2	−2860	−3200	0.98	OK
N3	−2950	−4100	0.79	OK
C2	1780	4100	0.48	OK
N4	−3080	−4100	0.83	OK

Table 2. Verification of girder due to post-composite loads.

Position	M_d (kN.m)	M_{ud} (kN.m)	N_d (kN)	Safety check $\gamma_i (M_d/M_{ud}) \leq 1$	Remarks
N1	−6435	−14300	0	0.50	OK
N1'	−2775	−5900	0	0.52	OK
C1	4700	6400	0	0.81	OK
N2	−9775	−14300	0	0.75	OK
N2'	−4230	−5900	0	0.79	OK
N3	−13100	−15300	0	0.94	OK
N3'	−4940	−6900	0	0.79	OK
C2	6160	7600	0	0.89	OK
N4	−13270	−15300	0	0.95	OK
N4'	−5070	−6900	0	0.81	OK

for post-composite dead, and 1.98 for live load. The material factor is 1.1 for steel, 1.3 for concrete and 1.0 for steel bars, in addition to the member factor of 1.1 for section overstretching.

The design bending moment due to the pre-composite load (M_{d1}) is taken only by the plastic bending capacity of the steel H-girder (M_{sud}), while the design bending moment due to the post-composite load (M_d) is resisted by the ultimate bending resisting capacities (M_{ud}) of the composite and the SRC sections at the span center and the support joints, respectively. Table 1 and Table 2 are the verification results of G1, showing that all the sections satisfy Eq. 9.

6 SEISMIC ANALYSIS

According to Japanese Seismic Code for Road Bridges, two levels of design earthquake ground motions shall be considered in the seismic design of bridges (JSCE 2002). The strong earthquake level corresponds to earthquakes with high probability of occurrence during the bridge service life (Level-1 earthquake, L1-EQ), and the ultra-strong earthquake level corresponds to earthquakes with less probability of occurrence during the bridge service life but very strong enough to cause critical damage (Level-2 earthquake, L2-EQ).

The structure should satisfy the required performance against the design earthquake forces. Important bridges in highways and national roads must satisfy the seismic performance-1 (SP-1), which specifies that members must be kept elastic and no damage is allowed, for L1-EQ. For L2-EQ, they must satisfy the seismic performance-2 (SP-2), which specifies that members can be elastic and plastic and minor damage is allowed but no major damage, or the seismic performance-3 (SP-3), which specifies that the bridge does not collapse. The push-over analysis and the time history analysis were conducted for L1-EQ and L2-EQ.

6.1 *Structural model for dynamic analysis*

The 3D frame model with beam elements was used for the seismic analysis. The girder and the piers were divided into fiber elements of steel and concrete. The non-linear material constitutive

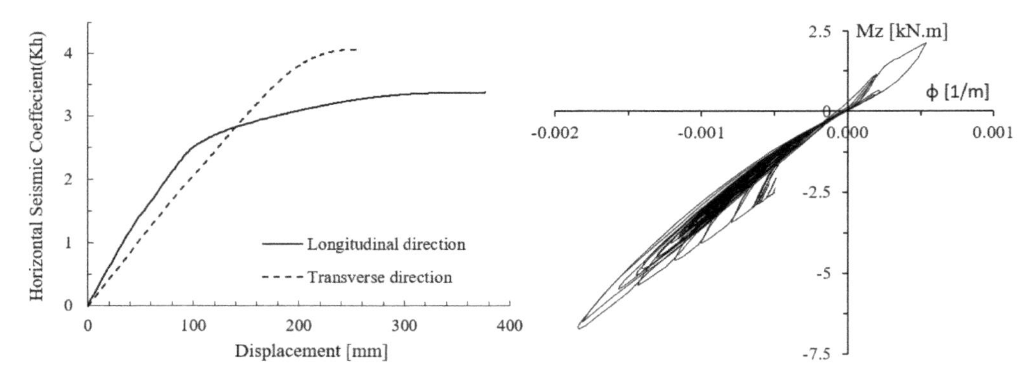

Figure 20. Pushover results of proposed bridge model.

Figure 21. The earthquake input for L-2 wave acceleration.

relation of steel and concrete of each fibre was the same as those used for the FEM analysis in Section 4. The RC slab was replaced by the discrete beam elements. Good and hard grounds such as the diluvium and rock ground was assumed in this design example. The interaction between the sub-structure and the ground was considered by linear springs in the three directions and three rotations. The equivalent stiffness of these springs was obtained by the Design Specification for Railway Structures.

6.2 *Push-over analysis*

Eigen value analysis was conducted, finding the dominant natural period, which was 0.435 sec. in the transverse direction and 0.384 sec. in the longitudinal direction. For L1-EQ, the corresponding design seismic coefficient was 0.2 in two directions. For L2-EQ, the corresponding design seismic coefficient was 2.0 in two directions. The push-over analysis was conducted in the longitudinal and transversal directions. The horizontal force H was incrementally applied to the structure until the structure collapsed.

Material non-linearity was considered in this process. Figure 20 shows the obtained relation of the seismic coefficient with the displacements in two directions. The seismic coefficients at the yield points are 2.34 and 3.05 in the longitudinal and transverse directions, respectively. They are much higher than the design seismic coefficient of L1-EQ. The seismic coefficients at the ultimate points are 3.37 and 4.04 in the longitudinal and transverse directions, respectively. They are also much higher than the design seismic coefficient of L2-EQ. These results show that the proposed structure has sufficient seismic strength and ductility capacity in the longitudinal and transverse directions.

6.3 *Time history analysis*

The proposed bridge is a multi-degrees of freedom structure and needs to be verified by dynamic analysis. The time history analysis was then carried out for the proposed bridge model by the design seismic wave for L1-EQ and L2-EQ (JSCE 2002). Figure 21 shows a typical design acceleration wave for L2-EQ.

The response at the pier top calculated by the time history analysis were observed with the maximum values 93.1 mm and 167.7 mm in the longitudinal and transverse directions, respectively. Figure 22 shows the hysteretic response of bending moment vs. curvature of the SRC girder section near the support joint in the longitudinal direction with the maximum bending moment of 6730 kN.m and the curvature of 0.00184 m^{-1}. The hysteretic response exhibits almost linear. The hysteretic responses of bending moment vs. curvature at the RC pier base (P3) in the longitudinal direction is shown in Figure 23. it shows elastic and plastic behaviors and the area inside the oval shapes absorbed the seismic energy.

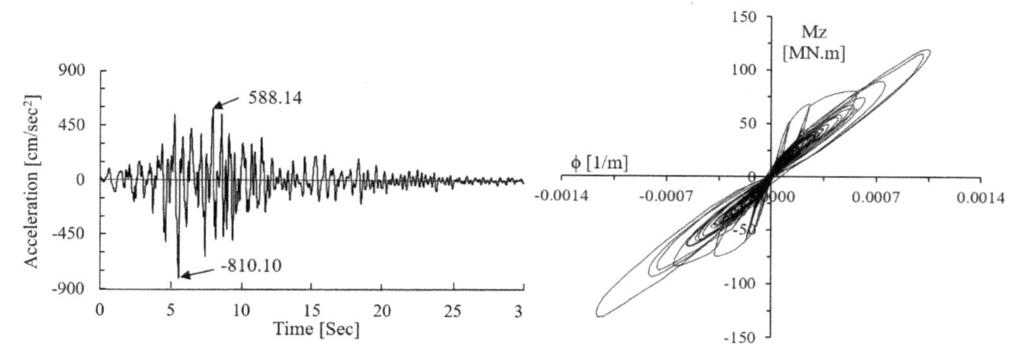

Figure 22. Hysteretic response of SRC section.

Figure 23. Hysteretic response at P3 base.

Table 3. Verification of shear strength by nonlinear dynamic history method.

Direction	Designation	V_{vd} (MN)	V_d (MN)	$\gamma_i \frac{V_d}{V_{yd}} \leq 1.0$	Safety check
Long.	Pier (P2)	30.40	20.8	0.68	OK
Trans.	Pier (P3)	36.75	33.5	0.91	OK

Table 4. Restorability verification of the bridge pier.

		SP-1			SP-2		
Level	Direction	θ_{sd}	θ_{rd}	Verification	θ_{sd}	θ_{rd}	Verification
L1-EQ	Longitudinal	0.0018	0.0094	0.19 < 1			
	Transverse	0.0036	0.0167	0.22 < 1			
L2-EQ	Longitudinal				0.014	0.0579	0.24 < 1
	Transverse				0.031	0.044	0.71 < 1

6.4 *Seismic performance verifications*

The response of the RC pier obtained by time history analysis due to L2-EQ should satisfy the shear force verification and the restorability criteria, according to the JSCE Specification for Concrete Structures. The shear force should satisfy the following equation.

$$\gamma_i \, (V_d / V_{yd}) \leq 1.0 \tag{10}$$

where γ_i: structural factor ($\gamma_i = 1.0$), V_d: design shear force, V_{ud}: design shear resistance capacity. The results are presented in Table 3, which show that the RC piers satisfy EQ.(10).

It is necessary to control the damage level according to the seismic performance grade. Table 4 shows the restorability verification of the bridge piers. Bending moment vs. rotation angle (M-θ) relationship of the bridge pier is used to verify their restorability. The verification equation proposed by limit states design method can be written as in Eq 11.

$$\gamma i \, \theta_{sd} / \theta_{rd} \leq 1.0 \tag{11}$$

where, θ_{sd}: design response angle of rotation at pier base, and θ_{rd}: design limit angle of rotation at pier base. The design limit rotation angles correspond to the yield point and the maximum point for SP-1 and SP-2, respectively. Table 4 shows the restorability verification of the RC piers of the proposed bridge for L1-EQ and L2-EQ, showing good seismic performance and satisfy the design requirements.

7 CONCLUSIONS

A new form of steel reinforced concrete (SRC) girder bridge using the steel rolled H-section was proposed. The super-structure consisted of continuous rolled steel H-girders which was composite with the RC slab, and the sub-structure was RC piers. The girder and the pier were rigidly connected by concrete and reinforcements to resist large negative bending moment at the intermediate support joints. This new bridge was studied by experiments and FEM analysis. A design example and seismic analysis were also conducted. The main conclusions obtained by this study are as follows.

1) Experiments were conducted with the partial SRC bridge model, showing that the new SRC structural form had high bending strength and ductile property. The applied load was smoothly transferred from the girder to the pier through the rigid joint, and the cracks observed on the concrete surfaces were within the acceptable limit.
2) Finite element model was developed and the analyzed results were compared with the experimental results. The displacement and strains obtained by FEM agreed well with the test results, which verified the established FE model. The local behavior of steel/concrete interface at the rigid joint was also simulated with non-linear springs, showing that the shear forces were smoothly and sufficiently transferred.
3) A design example with a four-span continuous road bridge with the span-length of $40 + 50 + 50 + 40$ m was conducted, finding that the proposed bridge satisfied the safety and serviceability requirements.
4) Seismic analysis was carried out by two methods, the push-over analysis and the time history analysis, considering material non-linearity. The proposed SRC bridge was proved to have sufficient ultimate strength and ductility against ultra-strong earthquakes.
5) A new bridge system without cross beams was proposed for the new SRC bridge because the girder height was low and composite with the RC slab. The structural behaviors of this new bridge system due to horizontal forces were studied by FEM analysis, showing that the new system without cross beams were stable and feasible against transverse seismic forces.
6) This study shows that the proposed steel reinforced concrete girder bridge using rolled H-beam was structurally rational, feasible and economical. The applicable span length could be extended to almost double the existing H-girder bridge.

REFERENCES

ABAQUS Business User's Manual. Version 6.14, 2014.

Ellobody E. 2014. Finite Element Analysis and Design of Steel and Steel-Concrete Composite Bridges, first *ed., Butterworth-Heinemann*, USA.

Japanese Society of Civil Engineers. 2012. Standard specification for concrete structures.

Lubliner, J. J. Oliver, S. & Oller, E.1989. A plastic damage model for concrete, *int. J. solid struct.*

Japanese Society of Civil Engineers (JSCE). 2002. Standard specification for concrete Structures, Seismic performance verification.

Liu, X., Bradford, A.M., Chen, Q-J., & Huiyong, B. 2016. Finite Element Modelling of Steel-Concrete Composite Beams with High-Strength Friction-Grip Bolt Shear Connectors, *Elsevier, Journal of Construction Steel Research*, 108(2016), pp. 54–65.

Takagi, M., Nakamura, S., & Muroi, S. 2003. An experimental investigation on rigid frame steel-concrete composite girder bridge, JSCE, Journal of Structural Engineering. 49 (2003) No.32.

Nakamura, S., Momiyama, T., Hosaka T. & Homma K. 2002. New technologies of steel/concrete composite bridges, *Journal of Constructional Steel Research*. 58 (2002) 99–130.

SAP2000 Structural Analysis Program, *Computer and Structures, Inc*. Advanced, 2016.

Chapter 24

Quality specifications for roadway bridges, standardization at a European level

J.R. Casas
UPC-BarcelonaTech, Barcelona, Spain

J.C. Matos
University of Minho, Guimaraes, Portugal

ABSTRACT: Across Europe, the need to manage roadway bridges efficiently led to the development of multiple management systems in different countries. Despite presenting similar architectural frameworks, there are relevant differences among them regarding condition assessment procedures, performance goals and others. Therefore, although existing a complete freedom of traffic between countries, this dissimilarity constitutes a divergent mechanism that has direct interference in the decision making process leading to considerable variations in the quality of roadway bridges from country to country. The need for harmonization is evident. Action TU1406, funded by COST (European Cooperation in Science and Technology), aims to institute a standardized roadway bridges condition assessment procedure as well as common quality specifications (performance goals). Such purpose requires the establishment of recommendations for the quantification of performance indicators, the definition of performance goals and a guideline for the standardization of quality control plans for bridges. By developing new approaches to quantify and assess bridge performance, as well as quality specifications to assure expected performance levels, bridge management strategies will be significantly improved and harmonized, enhancing asset management of ageing structures in Europe. The work developed and the final results and conclusions achieved by COST Action TU1406 will be presented.

1 INTRODUCTION AND BACKGROUND

Infrastructure managers are facing now conflicting requirements to improve the availability and serviceability of aging infrastructure, while the maintenance planning is constrained by budget restrictions. Many research efforts are ongoing, for the last few decades, ranging from development of bridge management systems, optimization models, life cycle cost analysis, to big data analysis and implementation of artificial intelligence models into decision support tools. Since transport infrastructure is presenting crucial factor for the economy and societal development, it is not only subject to technical requirements, but it is also required to keep up with societal and economic requirements.

In the past years, significant worldwide research has been done regarding condition assessment of roadway bridges. Obtained values from visual inspections and/or non-destructive testing, which provide information regarding the assessed bridge condition state, are compared with previously established goals. As a result, there are currently several methodologies to assess the bridge condition. More recently, the concept of performance indicators (PI) was introduced, simplifying the communication between consultants, operators and owners. However, large deviations continue to exist on how these indicators are defined and obtained in different countries and, therefore, specifications are required for a standardization of this procedure. For a standardized procedure, for example, quality control plans (QCP) are further important instruments, which are based on the previously mentioned PIs and pre-specified Performance Goals (PG). However, these PG are

- **Denmark**
  - DANBRO (DANish Bridges and Roads)
- **Finland**
  - FinnRABMS (Finnish National Roads Administration Bridge Management System)
- **France**
  - Advitam
- **Italy**
  - SAMOA (Surveillance, Auscultation and Maintenance of Structures)
- **Netherlands**
  - DISC
- **Norway**
  - BRUTUS
- **Sweden**
  - BMS
- **Switzerland**
  - KUBA
- **United Kingdom**
  - STEG (Structures REGister);
  - HiSMIS (Highway Structures Management Information System)
  - SMIS (Structures Management Information System)
  - BRIDGEMAN (BRIDGE MANagement system)
  - COSMOS (Computerized System for the Management Of Structures)

Figure 1. Bridge Management Systems in some European countries.

difficult to define, as they are highly subjective as they reflect the user and societal requirements. QCP and performance goals are normally stored in the respective Bridge Management Systems (BMS). In Europe, most countries have developed their own BMS in recent years as presented in Figure 1. Looking to Figure 1, it becomes evident an important contradiction. In fact, the different BMS reflect different performance indicators and performance goals among countries. However, do not exist highway traffic restrictions of people or goods between them. The user, travelling across countries may find substantial different quality levels in the bridges he crosses. Therefore, the need for some kind of standardization becomes evident at an European level. As a result, recently, COST (European Cooperation in Science and Technology) Action TU1406 emerged with the main ambition of developing a guideline for the establishment of common QCPs in roadway bridges in Europe, by integrating the most recent knowledge on performance assessment procedures with the adoption of specific goals (Casas, 2016 a&b, Matos et al., 2016). By developing new approaches to quantify and assess bridge performance, as well as quality specifications to assure expected performance levels, it is expected that bridge management strategies will be significantly improved and homogeneized all across Europe, enhancing asset management of aging structures in Europe. The aim of COST Action TU1406 was to agree between all countries in Europe about common performance goals and quality levels that could be implemented all across Europe.

2 OBJECTIVES OF THE ACTION

As mentioned before, there is an increasing need of standardization of the quality control plans for highway bridges in Europe. As a response, the overall intention of the Action was to develop a guideline for the establishment of Quality Control (QC) plans in roadway bridges reachable by pursuing the following 5 objectives:

1. Systematize knowledge on QC plans for bridges in European countries, which will help to achieve a state-of-art report that includes performance indicators and respective goals
2. Collect and contribute to up-to-date knowledge on performance indicators, including technical, environmental, economic and social indicators
3. Establish a wide set of quality specifications through the definition of performance goals, aiming to assure an expected performance level

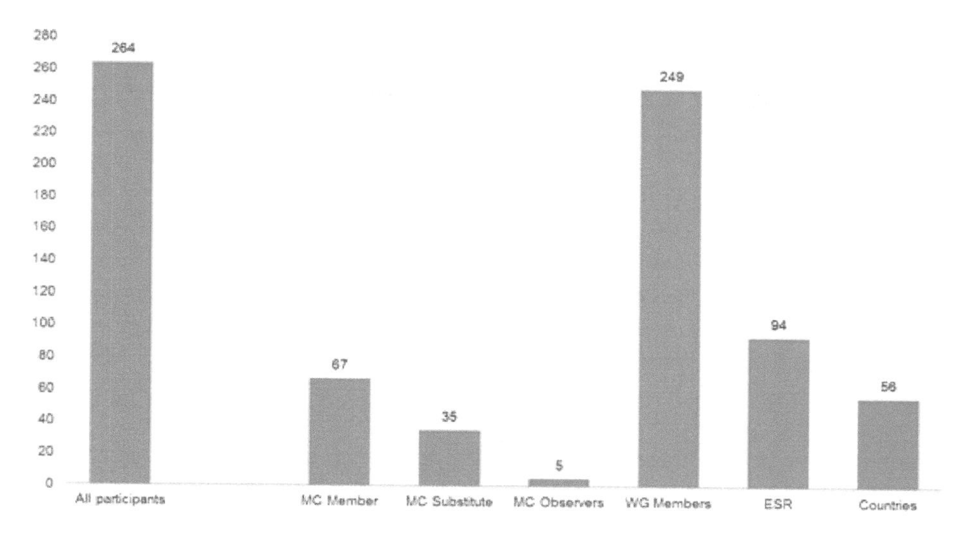

Figure 2. People involved in the Action.

Figure 3. Participating countries. Blue:COST Countries, Red: International Partner Countries.

4. Develop detailed examples for practicing engineers on the assessment of performance indicators as well as in the establishment of performance goals, to be integrated in the developed guideline
5. Create a database from COST countries with performance indicator values and respective goals, that can be useful for future purposes.

The number of participants in the action can be seen in Figure 2. A total of 264 people has participated, with a high number of Early Stage Researches (ESR). A total of 56 countries worldwide (not only European) participated in the project with a geographical distribution as presented in Figure 3. Between the participants a very representative mix of academics, consultants and bridge owners were present. An Industrial Advisory Board (IAB) with 3 bridge owners and 2 consultants was created in order to give the maximum decision level to those that have to latter apply the recommendations. Also international observers (1 per continent) were endorsed in order to compare the results of the Action with worldwide actual state of the art on the subject.

3 DESCRIPTION OF INITIAL WORKS

3.1 *State-of-the art on QC plans in Europe*

As a starting point it was decided to start looking to the available guidelines and documents related to inspection and maintenance today in use by the bridge/highway owners and operators in Europe. The reason for such decision looks quite evident:

1. In most countries the performance of bridges is good. Therefore, the agencies, at least in these countries are doing a good job.
2. We need to know exactly what they are doing in order to improve and enhance (if re-quired) their procedures and rules.
3. The implementation of a common methodology across Europe with flexibility to accommodate country-specific requirements needs to know what is being done now. If too many changes are proposed, reluctance of bridge owners and operators to perform those changes in the daily operation will appear. In addition, the new harmonized methodology can not disregard all the knowledge accumulated by the owners/operators along many years of bridge inspection and maintenance.

Through the WG1 activities, the development of a PIs database had been defined as an essential component of COST TU1406. The preliminary works developed into the working group indicated that an extended systematic screening on practical national inspection and evaluation documents was necessary in order to obtain a consistent and conclusive information associated with PIs, PGs and performance thresholds (PTs) from each COST partner country. The core of the survey process for obtaining PIs and corresponding key performance indicators (KPI) is given in Figure 4.

Taking into account the high number of languages spoken across Europe, it was obvious that language problems will arise as the required documents from the different owners and operators around Europe are written in many different languages. In many cases, a full code or guideline only presents some pages devoted to the subject of interest (performance indicator, performance goal, quality control, maintenance scenarios,. . .). Therefore, it was seen as unnecessary to translate the

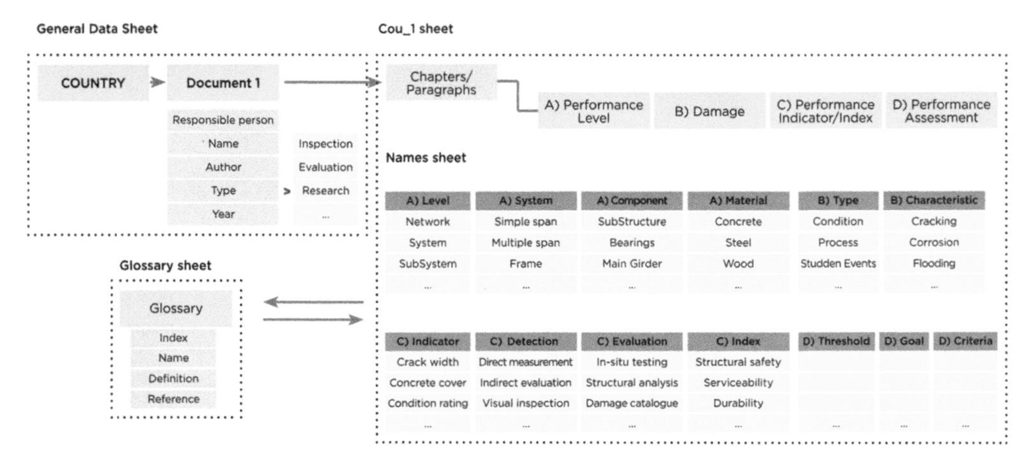

Figure 4. Structure of the Performance indicators database for the survey process.

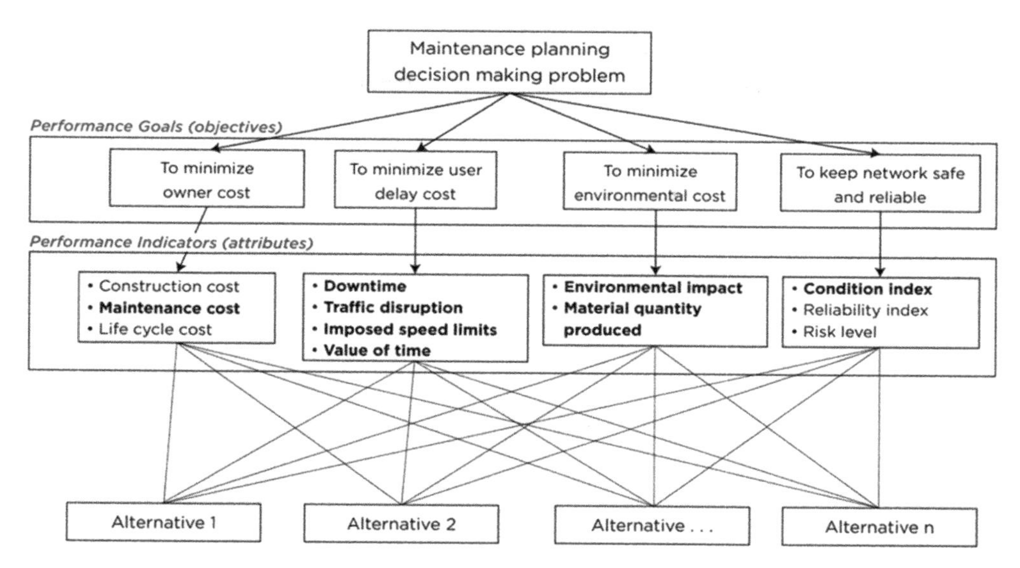

Figure 5. Performance Goals (KPI) and corresponding performance indicators.

whole document. On the contrary, the adopted strategy was to identify a participant in the Action as responsible to collect the relevant parts for the questionnaire of the existing guideline in his country and translate them to English. Of course, the responsible person should be somebody with good knowledge and expertise on inspection/assessment of existing bridges in order to identify the relevant parts. Apart from the language problems related to the translation of the documents, another language problem relates to the definition of several terms, as many times the same operation or concept has different English translations or wording (see, as an example, the concept of condition state where different terms and words are used in different countries to refer to the same idea). For this reason, a glossary of terms was also developed and agreed among the COST participating countries.

The two COST nominated persons per country, together with the infrastructure operators and owners chose beforehand the relevant documents (e.g. inspection, evaluation, research etc.) from which the PIs, and related information were extracted. Additional information on the survey process, the harmonization of the obtained information, as well as the final data-base of performance indicators and glossary of terms elaborated is fully available in WG1 report of the Action, reachable at www.tu1406.eu.

3.2 *Proposed performance goals and KPI*

The second step was to provide an overview of existing performance goals for the indicators previously identified in WG1 and to develop technical recommendations which will specify the performance goals. These goals will vary according to technical, environmental, economic and social factors. As presented in Figure 5, the final adopted performance goals were based on the requirements of safety to the user and minimization of cost (including owner, user and environmental costs). As a result, the following KPI (Key Performance Indicators) were decided: Reliability (safety in the structural sense), Safety, Availability and Cost. The first two are obtained as an static snapshot of the actual bridge condition, whereas the other two depend on the alternative maintenance scenarios adopted (see Figure 5) and therefore become dynamic along time.

As sometimes PGs could be contradictory (increasing of safety also derives in increase of cost, for instance), bridge performance goals should be set as a multi-objective system, taking into account different aspects of bridge and network performance. Also, the described performance goals are defined at a system (bridge) level. In Figure 6, the process of the multiple performance

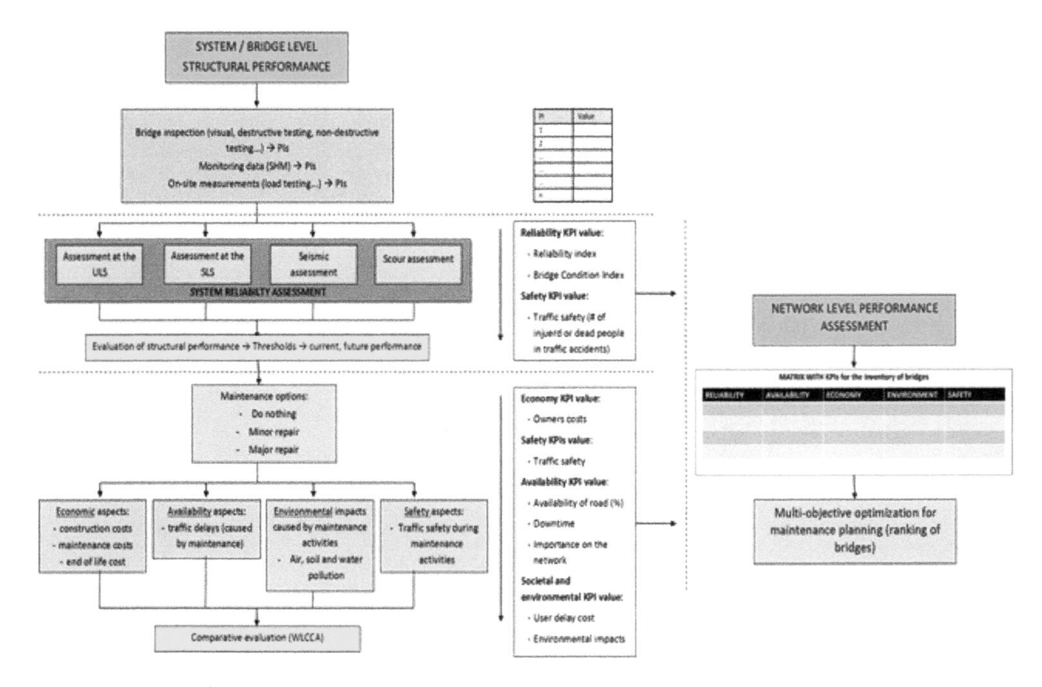

Figure 6. Overview of the process for performance goals assessment and transition from system to network level.

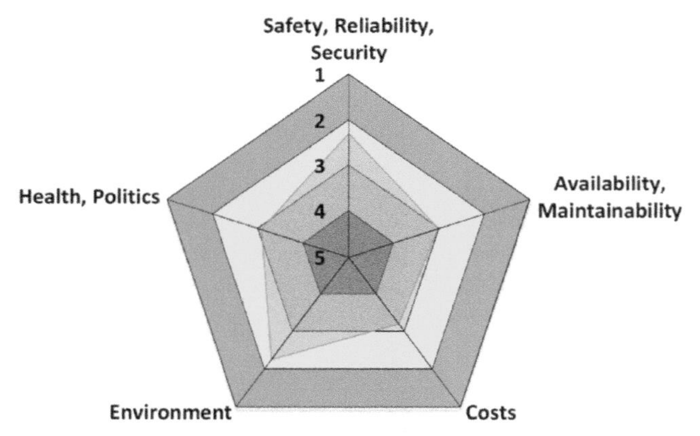

Figure 7. Spider diagram for multi-criteria decision making.

goals assessment is presented, also showing how the transition from the system (bridge) to the network level is achieved.

In order to get the optimum quality control plan, Multi-criteria decision-making (MCDM) provides a systematic approach to combine these inputs with benefit/ cost information and decision-maker or stakeholder views to rank the alternatives. MCDM is used to identify and quantify decision-maker and stakeholder considerations about various (mostly) non-monetary factors in order to compare alternative courses of action. Hierarchy structure for linking multi-objective bridge performance goals, covering most of the previously mentioned aspects with performance indicators is required. Possible result of multi-criteria assessment of different alternatives for bridge maintenance, represented in a spider diagram as in Figure 7 for a defined point in time, can be finally used for a decision making about the optimal maintenance or design solution.

Alternatively, the multiple performance criteria can be combined into a so-called utility function, in which all the criteria are brought into a single scale. In order to transform the various out into a single (mostly monetary) scale it is necessary to establish weight factor for the individual types of criteria. Some of the weight factors are available in some countries (for example weight factor for traffic delays, noise, injuries etc.). Depending on the selection of criteria, some weight factor may still need to be developed. In WG2 of TU1406 COST Action, a web-based tool was developed to apply the multi-objective optimization. The developed tool has implemented one of the methods of MCDA, namely Multi-Attribute Utility Theory (MAUT) by using the R Utility package (Reichert et al. 2013). The tool as well as further information on the treatment of performance goals within the Action can be found in the Report of WG2 at www.tu1406.eu.

4 FRAMEWORK OF QUALITY CONTROL PLAN

The quality specification or QC framework aims to provide a methodology with detailed step-by-step explanations for the establishment of QC plans for different bridge types. According to the identified PGs, these plans relate user/society goals, such as:

- Reliability: including the probability of structural failure (structural safety) or operational failure (serviceability)
- Availability: the proportion of time a system is in a functioning condition. In our case is the additional travel time due to imposed traffic regime on the bridge
- Safety (not structural safety): minimize or eliminate people harm during the service life
- Economy: minimize life-cycle cost
- Sustainability: environmental friendliness

In Figure 8 is shown the structure that supports the QC plan for highway bridges. Based on the static or dynamic nature of the goals, quality control plans are also divided in 2 groups:

- Static (snap shot) control: to inspect and investigate the bridge and determine whether reliability (structural safety and serviceability) and safety are met. This is fundamentally the basis for the decision making on actions
- Dynamic control: based on the static control and including the plan and actions (maintenance scenarios) to execute in order to ensure the long term fulfilment of safety and serviceability

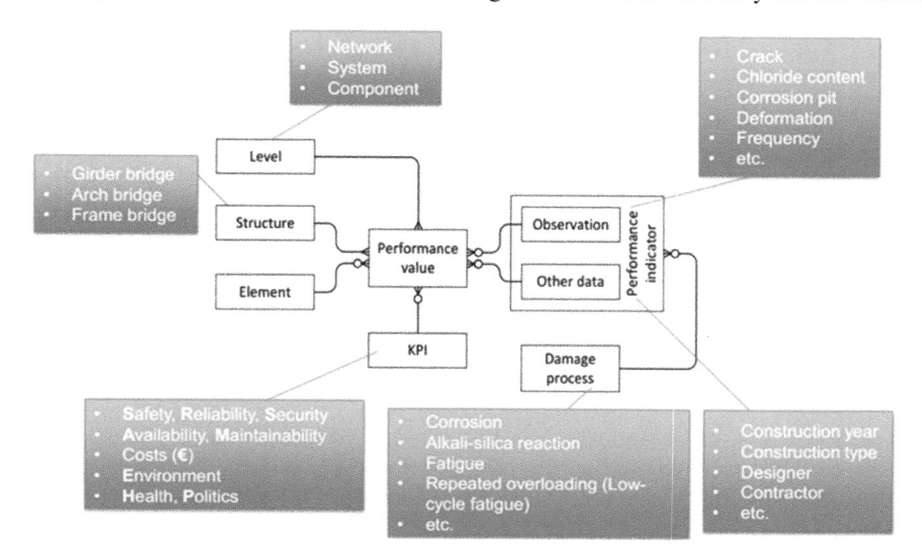

Figure 8. Quality control plan framework.

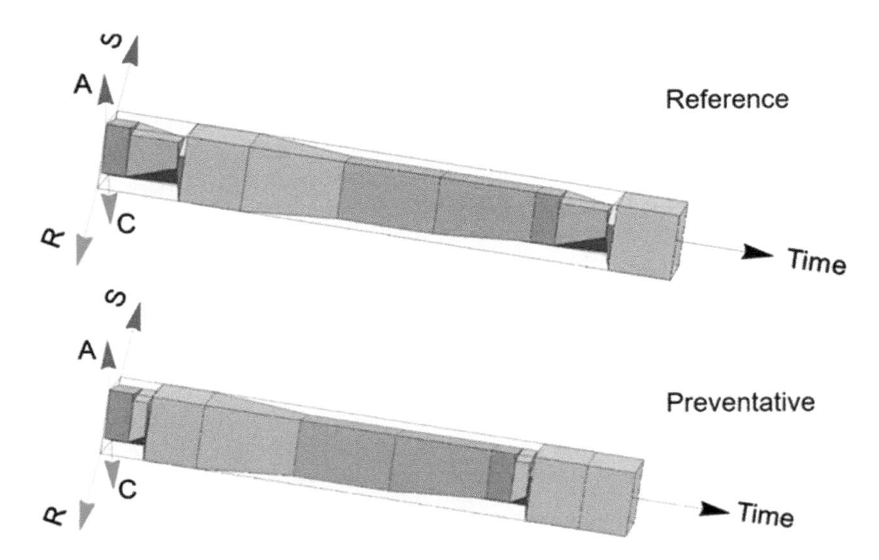

Figure 9. Spider diagram along time for two maintenance scenarios (reference: do nothing and preventative maintenance.

goals. The goals to achieve are related to availability, economy and sustainability as they include the feasible maintenance scenarios that define costs and availability over a certain time frame by using reliability and safety forecasts.

Because the reliability goal should be checked against feasible failures, the practical application of the method has divided the global group of bridges into the following bridge types: frame, arch and continuous beam. For each of these bridge types the most vulnerable zones can be identified and the corresponding PIs related to reliability observed and/or quantified. This division also helps in the process of selecting the best maintenance policy. The goals of availability, economics and sustainability are governed by maintenance scenarios. In fact, the snapshot assessment of availability and costs are of none or little interest. Therefore, the feasible maintenance scenarios (do nothing, preventive and corrective) are defined and the corresponding KPI's evaluated along time according to each maintenance scenario. In this way, the spider diagram of Figure 7 is developed along time as presented in Figure 9. Regarding the availability indicator, each maintenance intervention requires certain traffic regime, which may include closure for certain type of vehicles or lane closure or narrower lanes. The normal traffic regime can be assigned with the maximum performance value. The other traffic regimes can be ranked by the additional travel time they cause for the road users. This additional travel time can be also monetized according to the guidelines given in the WG2 report. The selected intervention scenario is obtained through MAUT based on the results presented in the diagrams as in Figure 9. Further information on the QC framework can be found in the WG3 report at www.tu1406.eu.

5 CASE STUDIES

The case studies developed are grouped according to the bridge types defined in the quality control plan: frame, arch and continuous beam. The proposed framework is thought for standard bridges, most present in the European highways, and not for landmark bridges. As an example, in the frame bridge of Figure 10, the vulnerable zones to failure in bending (yellow) and shear (read) are defined (Figure 11) and the corresponding performance indicators related to reliability observed and/or quantified. For instance, regarding a frame type bridge, the regions close to mid-span and

Figure 10. Example: Reinforced concrete frame bridge.

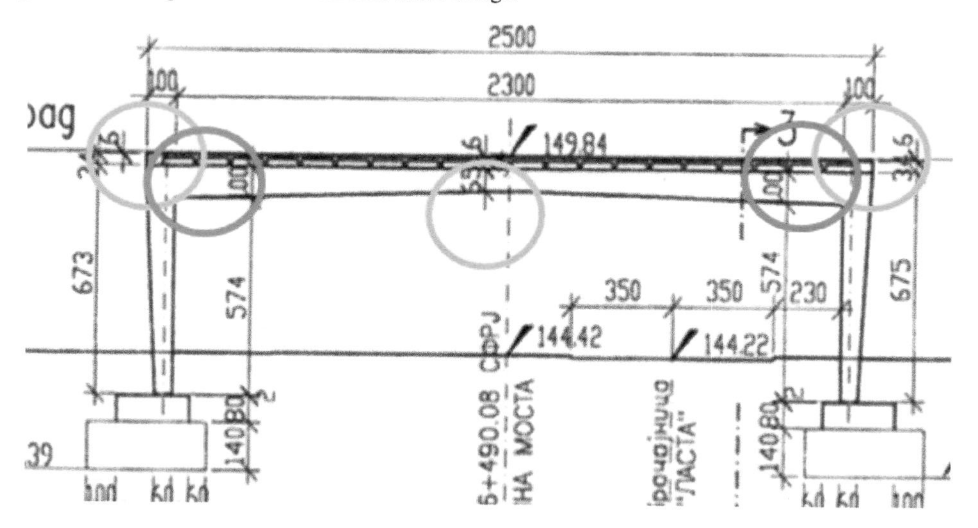

Figure 11. Vulnerable zones in bending and shear.

supports will be those selected for a reliability analysis. This division also helps in the process of selecting the best maintenance policy. For the frame bridge shown in Figure 10, the corresponding flowchart as proposed in the framework (Figure 8) and assigned values of reliability and safety based on the visual inspection is presented in Figure 12. The condition of the bridge observed in the visual inspection is shown in Figure 13. It becomes evident in Figure 13a the lack of safety due to condition of the bridge edge. Also the state of corrosion in some parts of the bridge (see Figure 13b) is used to fix the reliability level. The maximum value for both KPIs is 5. In this case, the values obtained are 3 for reliability and 2 for safety. In the report of WG4, available at www.tu1406.eu a complete description of 17 case studies, representative of several bridge types and countries, are presented as well as a Guideline with a very detailed flowchart and step-by-step process to apply

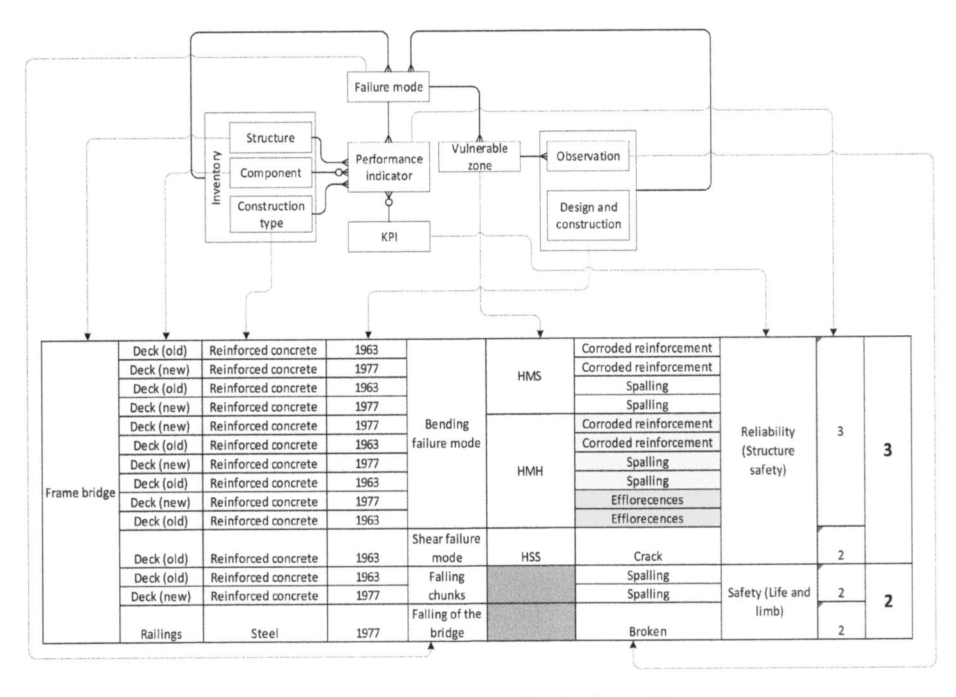

Figure 12. Flowchart and results corresponding to the bridge in Figure 10.

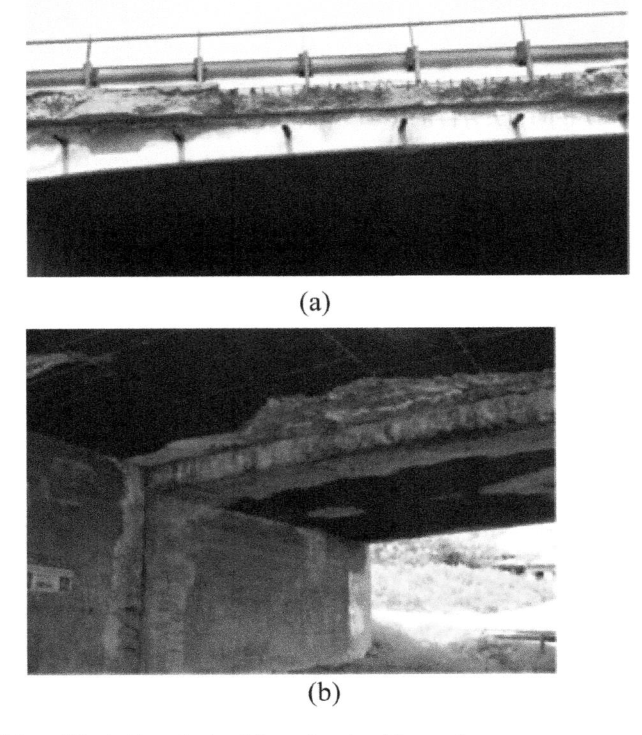

(a)

(b)

Figure 13. Condition of the bridge obtained from the visual inspection.

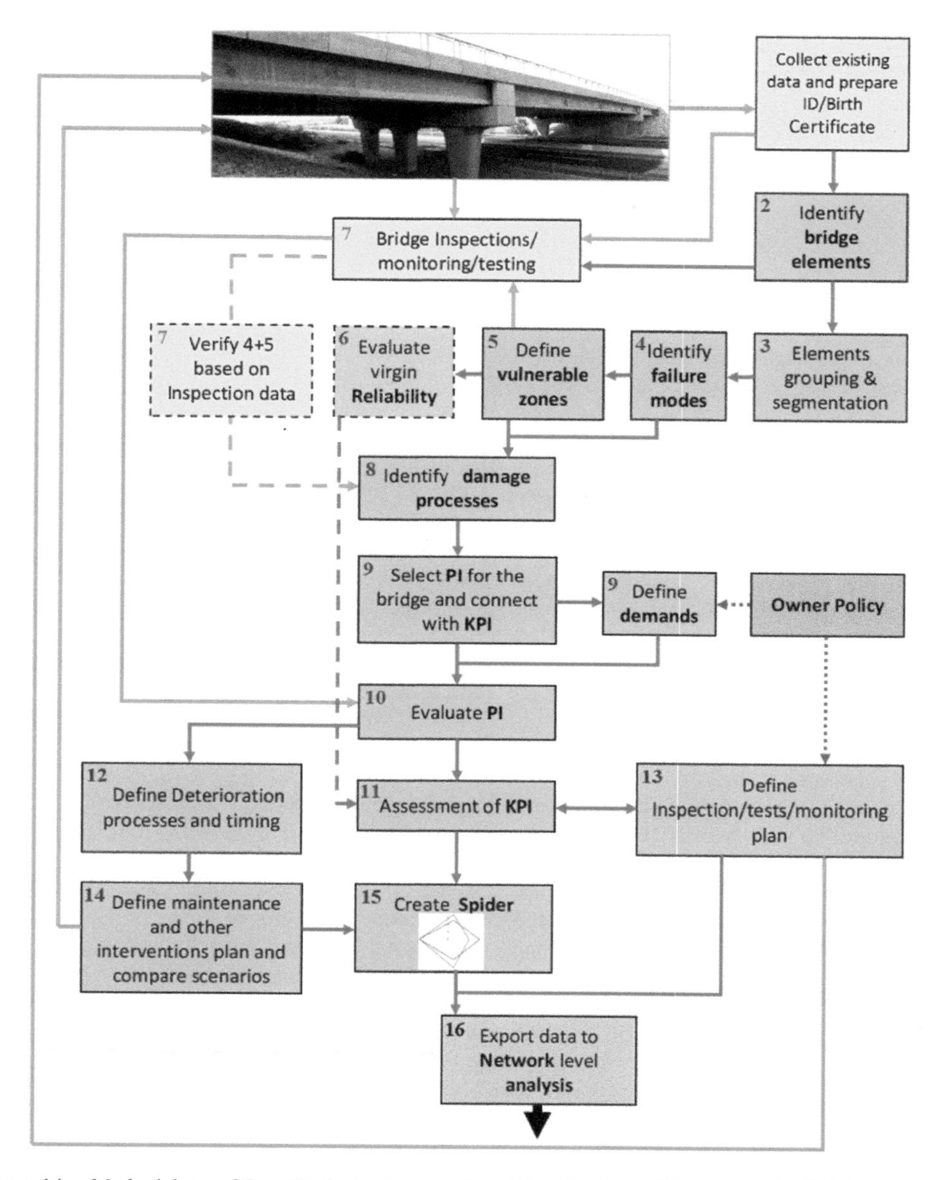

Figure 14. Methodology of Case Study Implementation of Quality Control Process in COST TU1406.

the quality control framework to a particular bridge (Figure 14). This facilitates the task for the bridge engineers and owners to apply the novel approach.

6 CONCLUSIONS

During the implementation of asset management strategies, maintenance actions are required in order to keep assets at a desired performance level. In case of roadway bridges, specific performance indicators are established for their components. These indicators can be qualitative or quantitative and can be obtained during principal inspections through a visual examination, non-destructive tests or a temporary or permanent monitoring system. Obtained indicators are compared with performance goals, in order to evaluate if the quality control plan is accomplished. It is verified that there is a large disparity in Europe regarding the way these indicators are quantified and how such goals are specified. Therefore, a discussion at a European networking level through Action COST TU1406, seeking to achieve a standardized approach in this subject, has brought significant benefits. The standardized approach unifies several formats of maintenance management in different networks and countries but allows them to be implemented in the format that they are already operational. Therefore, this Action achieved to bring together, for the first time, both research and practicing community in order to accelerate the establishment of a European guideline in this subject.

After an exhaustive analysis of the data obtained in the survey carried out among European countries, it was concluded that different countries have different definitions of PIs and how they are obtained. In some cases, simple observations are adopted as performance indicators. The methodology of the analysis was based on a deep analysis of the existing bridge management policies and available documents for inspection and evaluation existing in European countries and the main PIs used, with the objective to define a common group of quality specifications and control plans that can be assumed by all these countries. Due to the existence of different interpretations, an additional clustering and homogenization process was required. From this procedure it was possible to verify that all countries have a PI, named condition index, condition rating or deterioration index, mainly obtained through visual inspections. Although in some cases this is the only existing PI used, there are countries, like Denmark or The Netherlands, in which operators and bridge owners are currently using other relevant indicators. In Denmark, concepts like remaining service life, robustness, safety index, reliability and vulnerability are addressed. On the other hand, in The Netherlands, the performance is evaluated by the RAMSSHEEP (reliability, availability, maintainability, safety, security, health, environment, economics and politics) approach, where risk is used to obtain social, environmental, economic and political indicators.

TU1406 has a high societal relevance and brought together a collaborative network of several stakeholders, namely, partners from research and practicing community, aiming to joint efforts to build consensus on this subject. Multidisciplinary and complementary expertise covering a wide range of topics form visual inspection, on-site testing, numerical modelling, asset management and sustainability are considered. The collaborative dialogue developed in the process amongst researchers, engineers and owners and supported through networking, capacity building and training activities in COST TU1406 thus forms an invaluable reference point in the evolution of bridge management in EU.

To establish the suitability of the final reports and proposed methodologies along TU1406 for industrial application, an Industry Advisory Board (IAB) was formed by representatives of bridge owners and consultants. The industry Advisory Board has been asked by the COST TU1406 core group / chair to review the outcome and deliverables from the cost actions and comment based on the applicability in practice for the industry. The IAB uses web-meetings and subsequent individual contributions to arrive at comments and suggested improvements of the reports of relevant WGs. This is especially true where the exploitation of the outcome is relevant for several groups of stakeholders. The IAB has been able to create an impact in various reports, especially commenting on the need for harmonization of terms and definitions, data screening and access, consolidation of results, feasibility of scope, industry verification, standardization requirements and possibilities, scalability and ease of implementation in an industrial format.

Based on the results and recommendations of the reports from WG1 to WG4 (freely available at www.tu1406.eu) and the comments provided by the IAB, normative bodies and stakeholders, COST Action TU1406 has stated the following guidelines in the adoption of a common QC framework for highway bridges in the European countries:

1. Guideline on the definition of performance indicators for QC and collection of a European data base of observations and performance indicators

2. Guideline for the adoption of a common QC framework based on the adoption of 4 KPI (Key Performance Indicators: Reliability, Safety, Availability and Cost) and the 3-D spider tool for the practical implementation in a specific bridge in order to obtain the optimum maintenance scenario.
3. Guideline for practical implementation of the QC plans and definition of the 3-D spider tool for different bridge types.

In the guidelines, a multi-objective approach is recommended to address diverse PGs of a stock of bridges. Five performance aspects (Key Performance Indicators (KPI)) are selected in this regard: 1. Reliability; 2. Availability; 3. Safety; 4. Cost; 5. Environment.

A multi-criteria decision-making (MCDM) approach can systematically combine the inputs with cost-benefit models to rank available decision options about the bridges at component, system or network levels. A web-based Multi-Attribute Utility Theory (MAUT) and the corresponding software tool developed within the Action is recommended to be useful in this regard (https://maut.shinyapps.io/application_of_maut/).

For a simpler approach, the guidelines also recommend the use of a Spider Diagram for QC, by quantifying the overall performance related to the area in the diagram enclosed by KPI values, for a single or a collection of bridges. The KPI is qualitatively quantified between the ordinal scale of 1–5 (1 being the best and 5 being the worst). The most relevant KPIs should be selected for use in a particular situation. When assessed over time, the Spider Diagram forms a tube. The various KPIs can be expressed in their native units and then normalized to obtain their integer values.

Within the QC framework, the KPIs are evaluated for different maintenance scenarios (based on inspection/investigation or prediction), looking for the most feasible one. KPIs of Availability, Economy and Environment can be only reasonably applied as a function of time. Damage processes, defined as independent of combined actions having a detrimental effect on a bridge can be crucial for performance prediction, preventative maintenance and eventual rehabilitation. Information on damage can be obtained from inspection and testing. Impact of natural hazards on bridges is yet to be included in BMS but should be considered to understand consequences.

The QC framework has a a) static and b) dynamic stage. The steps for a static (snapshot) quality control comprise: 1. Preparatory work (inventory, conceptual weakness of design, material weakness, traffic load changes, identification of vulnerable zones, estimating à priori reliability; 2. Inspection on site (damage detection, material property measurement, sample collection); 3. Laboratory tests; 4. Assessment of the Reliability KPI (resistance reduction estimates, reliability estimates); 5. Assessment of the Safety KPI . The steps for a dynamic quality control comprise:

1. Assessment of remaining service life (damage speed and forecast, time dependent safety and reliability); 2. Maintenance scenario (reference scenario - end of service life, preventative scenario, long term cost, availability and reliability/safety estimates for scenarios); 3. Decision making (multi-objective/attribute optimization, monetize non-monetary KPIs, find optimal scenario).

The following challenges exist for the work carried out:

1. Training of inspectors are variable from country to country
2. For preliminary or approximate estimates of reliability, the experience and engineering judgement of the consultant is relied upon and this can have human variability in them.
3. For damage processes, the assumptions made about the type and rates of changes of damage need to be better calibrated and quantified by inspections and destructive/non-destructive testing.
4. Definitions around terminology, performance indicators and goals are still not entirely homogenized.
5. Understanding of uncertainty and reliability is further required in industrial scenarios.
6. There needs to be more demonstrations and direct benefits for the owners to implement the more holistic approach considered in this Action.

Finally, the following summary and related ongoing work can be considered:

1. The Action was extremely successful in a) understanding b) documenting and c) assessing the approaches taken for Road Bridge infrastructure in EU and around the world (e.g. USA, India, Australia, Russia etc.) and collating such information.
2. Experiences and limitations from bridge owners and managers from various countries were documented
3. The definitions and deep understanding of key governing terms were better clarified, key performance indicators and their measurements were better established and expanded through the action
4. For Operational Database, more work is still necessary to identify Key Performance Indicators (KPIs) for achieving Performance Goals for optimal Quality Control Plan and to allocate them with appropriate weights related to their respective levels of importance. The following steps are recommended to select the most important Performance Indicators:
   1. Define crucial Performance Goals
   2. Categorize Performance indicators in relation to Performance Goals
   3. Consider the following qualities for selecting a PI: a) measurability, b) quantifiability, c) availability of target value, d) validity for ranking purposes and e) applicability in making economic decisions
5. This consorted effort led to an overall approach developed in the Action for assessing road bridges, for which we now have a comprehensive set of case studies using several countries around EU as a demonstration and covering a wide range of road bridges to create a technical evidence base
6. At the end of the Action we are able to recommend, guidelines for best practice in maintenance and management of Road Bridges. The guidelines will harmonize (note: not homogenize) in EU the principles, approach and methodology based on which their maintenance and management will be carried out while retaining the disparate implementation and levels of data that is currently present in different countries in the presence of resource and funding constraints.
7. As a follow-up implementation, we have created EUROSTRUCT (www.eurostruct.org) as a first platform for the advancement of such actions through industrial leadership and academic collaborations to provide the bridge managers and owners a platform to discuss, decide and develop further activities in future for safer road bridges.
8. In order to diffuse the work done by the Action into normative documents we have been liaising closely with relevant bodies (ISO, CEN) to identify and follow-up the scope of contributing to their activities from our results and findings.

ACKNOWLEDGMENTS

This article is based upon the work from COST Action TU1406: Quality Specifications for Roadway Bridges. Standardization at a European Level, supported by COST (European Cooperation in Science and Technology).

REFERENCES

Casas, J.R. 2016a. European Standardization of Quality Specifications for Roadway Bridges: an Overview. Proceedings of IABMAS 2016. Foz do Iguaçu (Brasil): Taylor and Francis.

Casas, J.R. 2016b. Quality control plans and performance indicators for highway bridges across Europe. Proceedings of the 5th International Symposium on Lifecycle Civil Engineering. Delft (The Netherlands): Taylor and Francis

Matos, J.C., Casas, J.R., & Fernandes, S. 2016. COST Action TU1406 Quality Specifications for Roadway Bridges (BridgeSpec). Proceedings of IABMAS-2016, Foz do Iguaçu (Brasil): Taylor and Francis.

Reichert, P., Schuwirth, N., & Langhans, S. 2013. Constructing, Evaluating and Visualizing Value and Utility Functions for Decision Support. Environmental Modelling and Soft-ware,46:283–91.

Chapter 25

Multi-hazard financial risk assessment of a bridge-roadway-levee system

A. Nikellis, K. Sett, T. Wu & A.S. Whittaker
Department of Civil, Structural and Environmental Engineering, University at Buffalo, The State University of New York, Buffalo, NY, USA

ABSTRACT: This study deals with multi-hazard risk assessment of a spatially distributed civil infrastructure system with interdependent constituents. A regional system – consisting of six bridges, two roadway stretches and a levee – is selected and analyzed. During the analysis of this dynamic and relatively complex system, the following hazards are considered: (i) flood hazard due to extreme rainstorms which can cause internal erosion of the levee and (ii) seismic hazard and its triggering effects which can cause structural damage to the bridges and levee, and liquefaction-induced damage to the roadway stretches. The risk assessment of the system is conducted through a probabilistic event-based analysis. The results of this analysis are presented in terms of risk metrics commonly used in the field of financial engineering for portfolio optimization. An optimum risk mitigation strategy is also discussed.

1 INTRODUCTION

Risk assessment and performance evaluation of individual structures subjected to seismic hazard were pioneered by the Pacific Earthquake Engineering Research (PEER) Center (Cornell and Krawinkler 2000). A few recent studies (Bocchini and Frangopol 2012; Guidotti et al. 2016; Alipour and Shafei 2016; Gardoni et al. 2016; Sharma et al. 2017) have also explored the performance of civil infrastructure systems subjected to multiple hazards. These studies have mainly discussed the quantification of resiliency through the resilience metric defined by Bruneau et al (2003) or indirectly through other metrics (e.g., financial losses). A few recent review articles presented progresses and challenges regarding the risk assessment of civil infrastructure systems under multi-hazard conditions. An extensive literature review of community resilience studies was recently conducted by Koliou et al. (2018), concluding that more research is needed for the development of policies on risk-informed decision-making and optimization, prioritization of efficient retrofit solutions. Bruneau et al. (2017) qualitatively argued that retrofit strategies for the constituents of a system should be evaluated through a system-wide, multi-hazard analysis. Otherwise, these strategies might not be optimum for the maximization of the resilience of the system.

This paper quantitatively evaluates retrofit strategies for an interdependent civil infrastructure system subjected to multiple hazards, in terms of risk metrics commonly used in the field of financial engineering for portfolio optimization. A hypothetical, interdependent civil infrastructure system – consisting of bridges, roadway stretches and a levee - is subjected to seismic and high-water (storm surge) hazards. For this system different retrofit strategies are evaluated in terms of risk mitigation and financial loss reduction. To this end, risk metrics, including the annual frequency of exceedance of losses, average annual losses, value at risk, conditional value at risk, and worst case losses are utilized. Direct financial losses related to the potential damage of the constituents of the system, as well as indirect economic losses due to traffic disruption to the system are considered in this study. Cost-benefit analyses, utilizing the risk metrics, for the retrofit strategies of the bridges and the levee are also conducted and an optimum risk mitigation strategy is presented. The importance of conducting risk assessment at the system level and in a multi-hazard context, is underlined. Otherwise, the risk and the potential financial losses can be significantly underestimated. Different conclusions, regarding the cost-benefit evaluation of the optimum retrofit strategy, are drawn for different risk metrics.

2 THE TESTBED SYSTEM

The testbed is a hypothetical system, assumed to be part of the Bayshore Freeway, near the San Jose International Airport in California. It is shown schematically in Figure 1. The system consists of 6 bridges, 2 freeway stretches and part of a levee along the Guadalupe river. The bridges are located at 3 sites: (37.382 N, 121.964 W), (37.377 105 N, 121.942 W), and (37.375 N, 121.933 W). At each site there are 2 bridges (one bridge each for the southbound and northbound freeways). Between the bridges there are freeway stretches with four lanes and a shoulder at each directional carriageway. The length of the levee considered in this study is 0.65 km. The structural characteristics of the constituents of the system are assumed. Two-span reinforced concrete box-girder bridges supported on single circular piers and diaphragm abutments are selected. The details of the bridges selected in this study are presented in Table 1. The levee cross section and structural properties are assumed to be the same as those analyzed by Zimmaro et al. (2018).

Steel jacketing of the pier of the bridges and improvement of the foundation of the levee through jet grouting are the selected retrofit strategies for the constituents of this system.

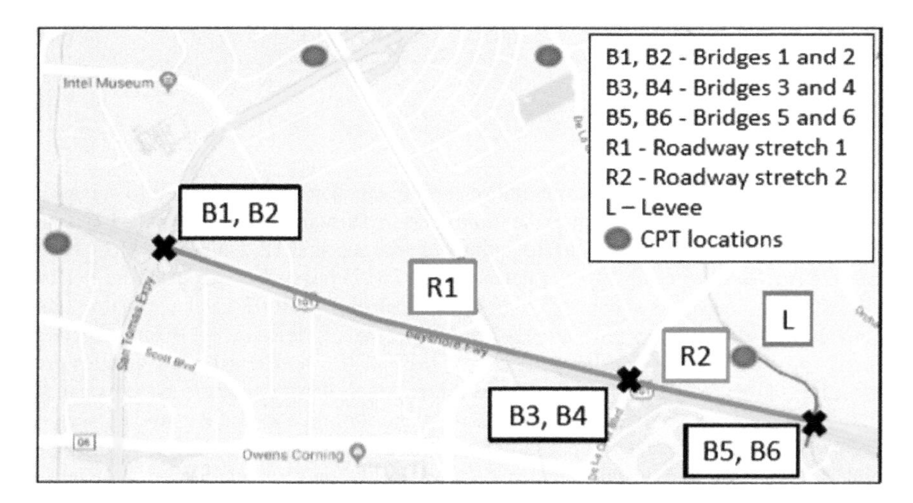

Figure 1. The testbed system.

Table 1. Details of the bridges of the system.

	Bridges 1 and 2	Bridges 3 and 4	Bridges 4 and 5
Span length (m)	25	30.48	36.58
Pier height (m)	5.49	7.62	10.36
Pier diameter (m)	1.53	1.83	2.74
Pier foundation	25-pile group	25-pile group	25-pile group
Deck width (m)	10	10.67	12.19
Box-girder height (m)	1.38	1.68	2.01
Abutment backwall height (m)	3.35	2.97	4.5
Number of piles per abutment	7	8	9

3 LOSS ANALYSIS

The risk assessment of the system and the risk-informed evaluation of its retrofit strategies are conducted through an event-based approach and consider both structural and downtime losses. This approach relies on the performance-based earthquake engineering framework (Moehle & Deierlein 2004), extended here in a multi-hazard context. The hazards considered during the analysis of this system are: (i) high-water (storm surge) hazard and (ii) seismic hazard and its triggering effects. The high water hazard can cause internal erosion of the levee. The seismic hazard can cause structural damages to the bridges and levee, and liquefaction-induced damages to the roadway stretches. Analysis of historical gage height (water elevation) measurements available for the river and probabilistic seismic hazard analysis for the entire system are performed for the quantification of the high-water and seismic hazards, respectively. For the vulnerability analysis of the bridges under seismic excitation, the technique of incremented dynamic analysis (Vamvatsikos and Cornell 2002) is employed. Typical fragility curves for both an original and retrofitted bridge are presented in Figure 2. The vulnerability of the roadway stretches against liquefaction is evaluated following HAZUS (National Institute of Building Sciences 1999). For the estimation of the probability of liquefaction site-specific cone penetration test (CPT) data are analyzed according to Boulanger & Idriss (2016). The vulnerability of the levee system against both hazards is quantified with fragility curves provided by Zimmaro et al. (2018).

Exceedance probability (EP) curves are utilized in the field of earthquake engineering to express the annualized losses of the system. This curve describes the probability that a level of loss will be exceeded annually (Miller & Baker 2015). The EP curves for the selected system with and without the mitigation of the retrofit strategies and for both the seismic and the flood hazards, are presented in Figure 3. The bridges produce lower losses than the roadway stretches for higher annual probabilities of exceedance, whereas the levee system produces higher losses for lower probabilities of exceedance, which are related to extreme, rare seismic events. These observations underline the importance of analyzing all three interdependent constituents of the system.

Furthermore, through a comparison of the EP curves of the original system while considering the seismic and the flood hazard independently, it is observed that even though the flood hazard is less than the seismic hazard, it is not negligible. This observation corroborates the importance of conducting risk assessment at the system level and in a multi-hazard framework.

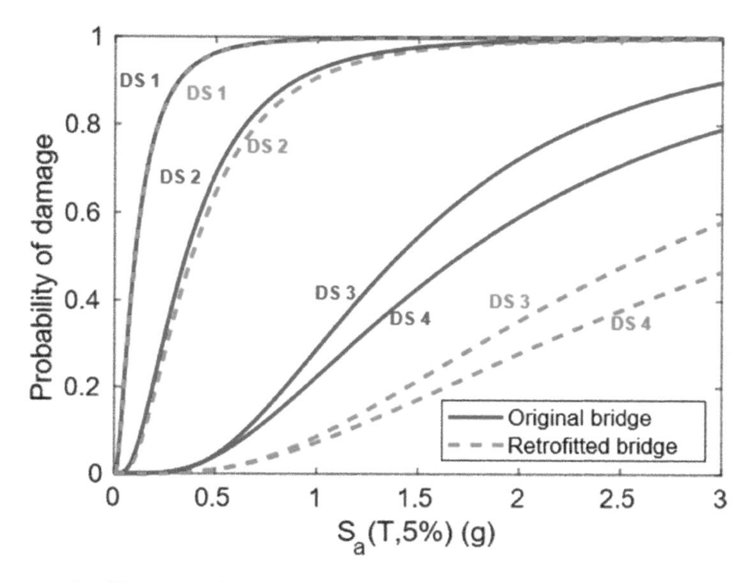

Figure 2. Damage fragility curves for Bridge 3.

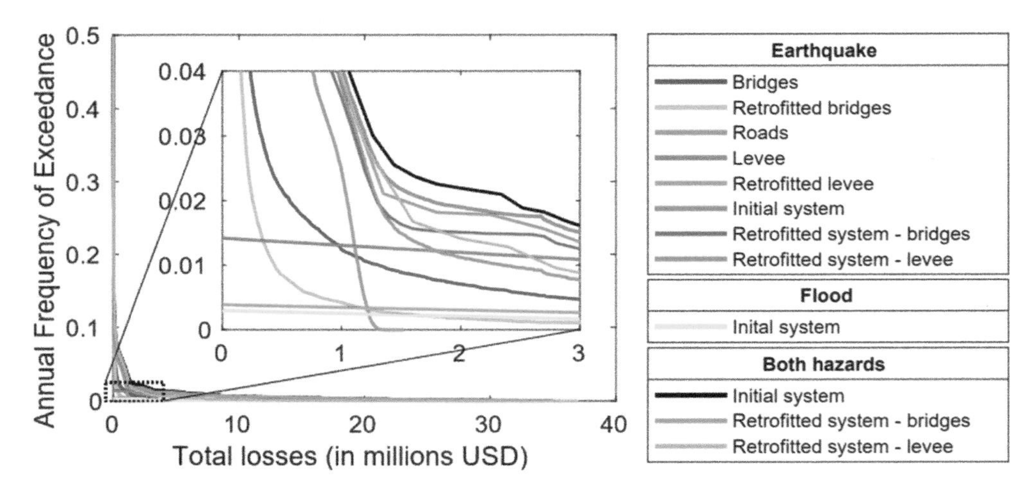

Figure 3. Exceedance probability curves for original and retrofitted systems along with their constituents.

3.1 *Risk metrics*

The risk related to extreme events is quantitatively assessed with different risk metrics. The selected metrics for this study are the Value at Risk (Var), the Conditional Value at Risk (CVaR), the Annual Average Loss (AAL) and the Worst Case Loss (WCL). These metrics are extensively used in the insurance and reinsurance sectors for assessing catastrophe risk and in financial engineering for conducting portfolio optimization.

3.2 *Average annual losses*

The Average Annual Loss (AAL) represents the expected loss per year and is equal to the area under the EP curve. In the insurance sector, the pricing of insurance premiums is based on the AAL.

As shown in Figure 4(a), at the constituent level, the retrofit of the bridges leads to a reduction of the total AAL equal to 57%, whereas the retrofit of the levee reduces the total AAL due to seismic events by 82%. At the system level, the total AAL for the seismic, the flood and both hazards are $0.26 million, $0.015 million and $0.27 million, respectively. Considering only the seismic or the flood hazard would lead to a misrepresented, lower total AAL of the system by 3.7% and 94%, respectively.

Furthermore, while considering only the seismic hazard, the total AAL reduces by 9.0% and 46.6% when the bridges and the levee are retrofitted. When both hazards are considered, the retrofit of the bridges reduces the total AAL of the system by 8.5%, whereas the retrofit of the levee reduces the total AAL of the system by 44.1%. These observations underline the importance of conducting risk assessment at the system level and under multi-hazard conditions.

Based on the total AAL of the selected system, retrofitting the levee for the seismic hazard results to the highest reduction of the total AAL both at the constituent and the system level while considering only the seismic or both hazards. Thus, such a retrofit strategy could be of the benefit of the owner of the system in order to negotiate better insurance premiums.

3.3 *Value at risk*

Value at Risk (VaR) is a risk metric which indicates the minimum loss that it will be reached or exceeded annually with a given probability. The selected annual probability of exceedance for the calculation of VaR is equal to 0.4% corresponding to events with a return period equal to 250 years. The VaRs of the system and its constituents, for both hazards are presented in Figure 4(b).

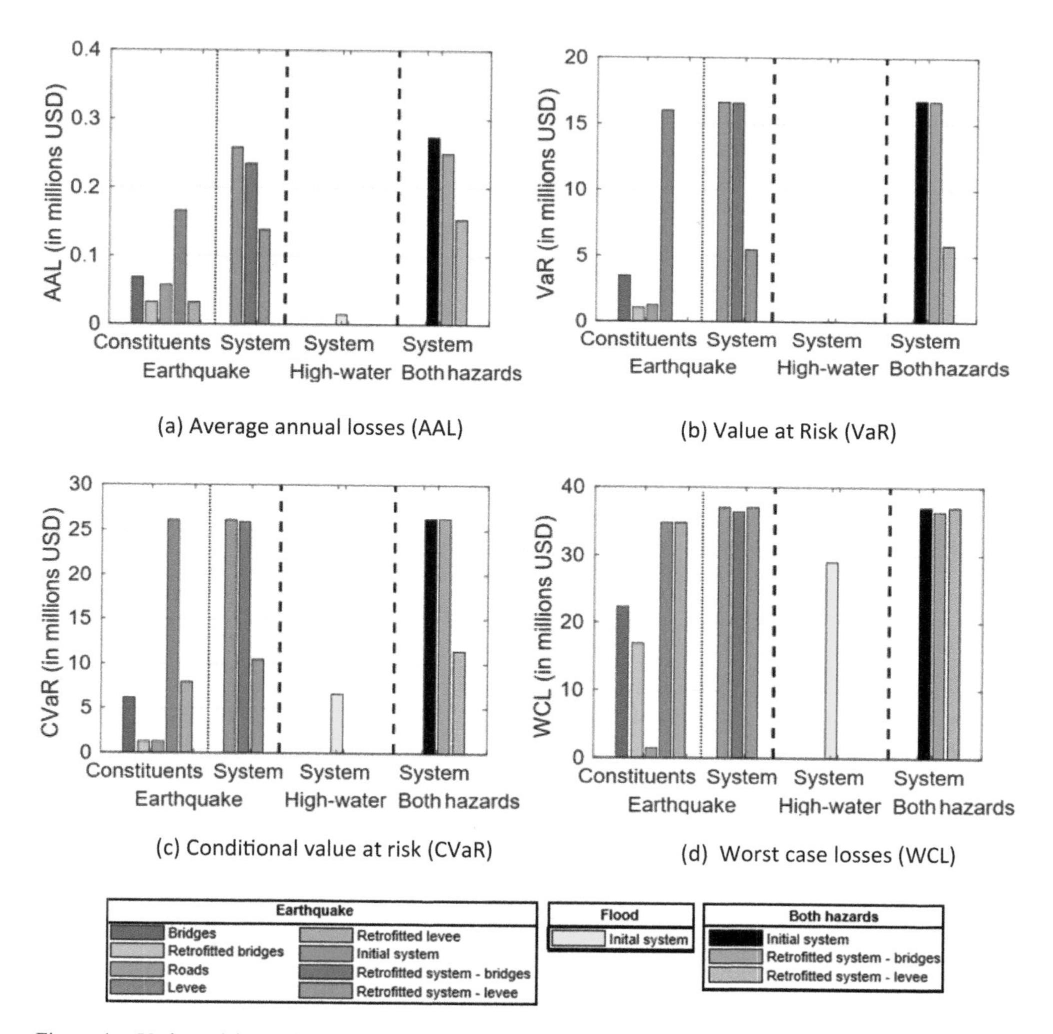

Figure 4. Various risk metrics for the original and retrofitted systems along with their constituents.

Insurance companies utilize the VaR in order to decide whether to hedge their risk through reinsurance and financial instruments such as insurance linked securities (e.g. catastrophe bonds). Furthermore, the VaR could be utilized by the policy maker or the owner of the system, to quantify the exposure of the system to a certain amount of risk related to a certain value of financial losses.

Regarding the seismic hazard, the total VaR of the levee contributes the most to the total VaR of the system, followed by the total VaR of the bridges and the roadway stretches. The total VaR of the bridges and of the roadway stretches are 22% and 7.5% of that of the levee, respectively. The total VaR of the original system, considering only the seismic or the flood hazard, is 0.6% and 100% less than that considering both hazards.

At the system level, considering only the seismic hazard, the retrofit of the levee system reduces the total VaR of the original system by 67%, whereas the retrofit of the bridges by 0.4%. If the same comparison is made, considering both hazards, then the retrofit of the levee system reduces the total VaR of the original system by 65% and the retrofit of the bridges by 0.5%. These observations confirm that the levee system is a better retrofit strategy for mitigating risk at events with a return period equal to 250 years.

3.4 *Conditional value at risk*

Even though the VaR is a quite simple risk metric to calculate, once the EP curve is constructed, it only provides information regarding a single loss related to an annual probability of exceedance. Further information can be extracted from the EP curve and better conclusions can be made for the exposure of the system to catastrophe and financial risk. The Conditional Value at Risk (CVaR) is a measure which quantifies the average value of losses which exceed a specific VAR. This is a better risk metric than the VaR, in making conclusions regarding the tail risk of the EP curve as it considers all losses at the tail of the EP curve under the condition that they exceed a certain VaR value.

Figure 4(c) shows the CVaR values of the system conditioned at the selected VaR value for this study (related to an annual probability of exceedance of 0.4%). For the original system the total CVaR related to flood hazard is 25% of that related to the seismic hazard. This observation confirms that these two hazards compete each other in terms of potential financial losses, which are related to the tail of the risk curve and corroborates the importance of conducting risk assessment under multi-hazard conditions.

Furthermore, the difference of the total CVaR between the original and retrofitted bridges at the system level is negligible, whereas that is not the case for the total CVaR related to the retrofit of the levee. Thus, retrofitting the levee is a better risk management strategy for mitigating risk related to extreme events with low probability of occurrence, but high financial consequences.

3.5 *Worst case losses*

This risk measure is defined in this study as the maximum losses that the system or its constituents could experience. These losses are related to the most extreme case hazard scenarios. It is related to the value of losses at the point where the EP has an annual probability of exceedance equal to zero. For the owner of the system or a policy maker the Worst Case Loss (WCL) could be of great interest since such losses are related with the absolute failure of some constituents or the system. The WCL of the constituents of the system are presented in Figure 4(d).

Regarding the roadway stretches of the system the total WCL is 95% lower than the total WCL of the bridges and the levee. This large difference between the total WCL of the roadway stretches and the other constituents of the system is attributed to the probability of liquefaction of the soil at the selected location. After the retrofit of the bridges of the system the seismic related total WCL is reduced by 24%. On the contrary, the retrofit of the levee does not affect the seismic related total WCL. This means that there are potential extreme seismic events that can cause the total collapse of the levee even after its retrofit is implemented. The retrofit of the bridges reduces the total WCL of the system, due to both hazards, by 3% whereas the retrofit of the levee does not affect it. This observation leads to the conclusion that retrofitting the bridges is a better option for mitigating risk associated with extreme case seismic events with very large return periods.

4 COST-BENEFIT ANALYSIS

The risk measures presented in the previous sections characterize the EP curve. Based on these metrics conclusions are made regarding the assessment of the catastrophe risk of the system and its constituents. Reducing the risk of a system has social impact and could be of great interest for both the owner and the policy maker. At the same time, for an owner it is important to identify whether reducing the catastrophe risk of the system, by retrofitting certain constituents, would lead to an economically beneficial investment strategy. To this end, a cost-benefit analysis for each retrofit strategy is conducted while utilizing each risk metric. The capital cost of the retrofit strategies is compared with the losses related to each risk metric prior and after the mitigation of the retrofit strategy. Based on the results of these cost-benefit analyses conclusions regarding an optimum retrofit strategy are made.

Table 2. Capital costs for retrofitting the constituents of the system

	Capital cost ($)
Total Bridges	2,171,432
Levee	7,932,600

A retrofit strategy is economically viable if it produces a reduction in the losses of the system higher than the capital cost spent for the implementation of the strategy. The retrofit strategy for the bridges (steel jacketing of the pier) is related to the cost of the structural steel used for it. The cost related to the steel jacketing of the pier of the bridge is estimated to be $39.7/kg ($18/lb) based on the bidding prices provided by Caltrans (2007). The retrofit strategy of the levee is related to the improvement of its foundation through jet grouting. The jet grout price is estimated to be $130.8 per cubic meter of levee foundation (Department of Water Resources 2012). Based on this price the cost for retrofitting the levee is estimated to be equal to $12,204 per meter. The capital costs of the retrofit strategies are presented in the Table 2.

4.1 *Value at risk*

If losses associated with a specific annual probability of exceedance are of primary interest, then the VaR could be selected as an appropriate risk metric for conducting a cost-benefit analysis for the retrofit strategies of the system. Thus, if the purpose of retrofitting constituents of the system is the reduction of risk and potential financial losses with an annual probability of exceedance equal to 0.4% then the following observations are made by comparing the VaR, shown in Figure 4(b), and the capital cost of the retrofit strategies, presented in the Table 2.

While considering the seismic induced total VaR of the levee before and after the implementation of the retrofit strategy, it is evident that by investing $7.93 million for retrofitting the levee, the losses whose annual probability to be exceeded is 0.4%, are reduced by $16.05 million. In addition, losses equal to $16.05 million for the retrofitted levee system have an annual probability of exceedance equal to 0.06%. Thus, there is a reduction of 85% at the annual probability of seismic-induced losses of the levee to exceed $16.05 million.

The reduction of the total VaR of the system under multi-hazard conditions and while retrofitting the levee is equal to $10.95 million. Thus, investing $7.93 million for this retrofit strategy would lead to an economically beneficial strategy for mitigating financial losses related to events with a return period equal to 250 years.

The reduction of the total VaR of the bridges at the constituent level is $0.29 million higher than the capital cost of the retrofit strategy whereas at the system level, both under single and multi-hazard conditions, it is negligible.

4.2 *Conditional value at risk*

If the purpose of retrofitting systems of the system is the reduction of potential average financial losses beyond a 0.4% VaR, then a comparison between the CVaR shown in Figure 4(c), and the capital cost of the retrofit strategies, presented in the Table 2 makes sense.

At the system level, under seismic events, the total CVaR related to the retrofit of the levee system is reduced by $15.56 million and outweighs the capital cost of this retrofit strategy. If both hazards are included in the cost-benefit analysis then the difference of the total CVAR between the retrofitted and original system is $14.88 million and is 100% larger than the capital cost for the retrofit of the levee.

The reduction of the total CVaR, due to retrofitting the bridges, is negligible at the system level, both under seismic and multi-hazard conditions, even though at the constituent level while considering the seismic hazard only, the total CVAR of the original bridges is reduced by 80%.

4.3 *Worst case losses*

If retrofitting the constituents of the system is conducted in order to minimize the potential worst-case losses of the system and its constituents, then the cost-benefit analysis should be based on the comparison of the total WCL shown in Figure 4(d), and the capital cost of the retrofit strategies, presented in the Table 2.

By investing in the retrofit of the levee the total WCL is not affected at the constituent and system level, both under seismic-only and multi-hazard conditions.

At the constituent level, the total WCL of the bridges is reduced by $5.4 million, which is approximately 2.5 times higher than the capital cost of the retrofit of the bridges. This reduction is 7.9 times larger than the reduction of the total WCL, associated with the retrofit of the bridges, at the system level under multi-hazard conditions. This observation underlines that by neglecting the flood hazard and while considering only one constituent of the system the error in the analysis would be equal to 790%.

5 CONCLUSIONS

This study explores the use of risk metrics, broadly used in the field of financial engineering, for risk assessment of a spatially distributed civil infrastructure system subjected to multiple hazards.

The results of this study are presented in terms of direct economic losses due to structural damage of the constituents of the system and indirect economic losses due to traffic disruption of the transportation network. The results presented in this study underline the importance of conducting risk assessment and cost-benefit analysis at the system level and under multi-hazard conditions. Otherwise, the error in the risk and loss estimation can be very high.

Furthermore, it is also shown that the retrofit strategies for a system should always be evaluated using a multi-hazard framework while considering all the interdependent constituents. Otherwise, a risk-informed decision-making process could result in erroneous conclusions.

Finally, the optimum decision depends upon the person who is making it. An insurance company or a policy maker could potentially perceive risk from different perspectives. Thus, different risk metrics should be utilized for better risk communication, in an attempt to provide more resilient civil infrastructure systems to society.

ACKNOWLEDGEMENTS

This work is partly supported by USDOT through Region 2 University Transportation Center and by the Institute of Bridge Engineering at the University at Buffalo.

REFERENCES

Alipour, A. & Shafei, B. 2016. "Seismic resilience of transportation networks with deteriorating components." *Journal of Structural Engineering*, 142(8), C4015015.

Boulanger, R. W. & Idriss, I. M. 2016. CPT-based liquefaction triggering procedure. *Journal of Geotechnical and Geoenvironmental Engineering* 142(2), 04015065.

Bocchini, P. and Frangopol, D. M. 2012. "Optimal resilience- and cost-based postdisaster intervention prioritization for bridges along a highway segment." *Journal of Bridge Engineering*, 17(1), 117–129.

Bruneau, M., Barbato, M., Padgett, J. E., Zaghi, A. E., Mitrani-Reiser, J., and Li, Y. 2017. "State of the art of multihazard design." *Journal of Structural Engineering*, 143(10), 03117002.

Caltrans 2007. Contract cost data. *Department of Transportation*. Sacramento, CA. http:// 655 sv08data.dot.ca.gov/contractcost/, accessed 2019-03-01.

Cornell, C. A. and Krawinkler, H. 2000. "Progress and challenges in seismic performance assessment." *PEER Center News*, Pacific Earthquake Engineering Research Center, University of California, Berkeley, CA. http://peer.berkeley.edu/news/2000spring/index.html.

Department of Water Resources 2012. Central valley flood management planning program. *Attachment* textit674 8J: Cost estimates, The Natural Resources Agency, CA.

Gardoni, P., Guevara-Lopez, F., and Contento, A. 2016. "The life profitability method (LPM): A financial approach to engineering decisions." *Structural Safety*, 63, 11–20.

Guidotti, R., Chmielewski, H., Unnikrishnan, V., Gardoni, P., McAllister, T., and van de Lindt, J. 2016. "Modeling the resilience of critical infrastructure: the role of network dependencies." *Sustainable and Resilient Infrastructure*, 1(3-4), 153–168.

Koliou, M., van de Lindt, J., Mcallister, T., Ellingwood, B., Dillard, M., and Cutler, H. 2018. "State of the research in community resilience: progress and challenges." *Sustainable and Resilient Infrastruc ture*, 1–21, DOI: 10.1080/23789689.2017.1418547.

Miller, M. & Baker, J. 2015. Ground-motion intensity and damage map selection for probabilistic infrastructure network risk assessment using optimization. *Earthquake Engineering & Structural Dynamics* 44(7): 1139–1156.

Moehle, J. & Deierlein, G. 2004. A framework methodology for performance-based earthquake engineering. *Proceedings of the 13th World Conference on Earthquake Engineering (15WCEE)*. Vancouver, 715 B.C., Canada (August 1-6). Paper No. 679.

National Institute of Building Sciences 1999. Multi-hazard loss estimation methodology. *HAZUS-MH 717 2.1, technical manual*. Technical report, National Institute of Building Sciences. Washington D.C.

Sharma, N., Tabandeh, A., and Gardoni, P. (2017). "Resilience analysis: a mathematical formulation to model resilience of engineering systems." *Sustainable and Resilient Infrastructure*, 1–19.

Vamvatsikos, D. & Cornell, C. A. 2002. Incremental dynamic analysis. *Earthquake Engineering and Structural Dynamics* 31(3): 491–514.

Zimmaro, P., Stewart, J., J. Brandenberg, S., Youp Kwak, D., & Jongejan, R. 2018. Multi-hazard system reliability of flood control levees. *Soil Dynamics and Earthquake Engineering*, in print, DOI: https://doi.org/10.1016/j.soildyn.2018.04.043.

Chapter 26

Benefits of using isolation bearing and seismic analysis on Bridge (MA-14) with Class F Soil

V.M. Liang, H.K. Lee & B.P. McFadden
Greenman-Pedersen, Inc., Lebanon, New Jersey, USA

I.S. Oweis
Oweis Engineering Inc., Cedar Knolls, New Jersey, USA

ABSTRACT: Monmouth County of New Jersey retained Greenman-Pedersen, Inc. to perform preliminary and final design of a new four-span continuous (90′-100′-80′-80′) steel multi-girder horizontally curved bridge that would replace a nearly 100-year old existing thru-girder bridge over Matawan Creek. Organic clay soils throughout the project site resulted in a Site Class F soil classification and the need for a site-specific ground motion response spectrum analysis. Due to the highly compressible soil of about 50 to 70 feet thick in some areas, deep foundations were required and extended into the stiff clay and dense sand layers below, ending nearly 120 feet below mudline. To minimize the size of the deep foundations, seismic isolation bearings were used to reduce the seismic demand on substructures. Due to the unique nature of the isolated system involving tall piers, a 3D time history seismic analysis was performed in addition to response spectrum analysis. The focus of this paper is to discuss how isolators can reduce the foundation size of a bridge situated in a weak soil site. It also describes the subsurface conditions, the development of a site-specific response spectrum for a 1000-year return period earthquake, isolation bearing preliminary design, the benefits of using isolation bearings and seismic analysis.

1 PROJECT OVERVIEW

Monmouth County retained Greenman-Pedersen, Inc. (GPI) to perform preliminary and final design of a new bridge to replace an existing four span thru-girder bridge, built in 1915, over Matawan Creek in the Borough of Keyport and Township of Aberdeen, New Jersey. GPI retained Oweis Engineering Inc. (OEI) as a subconsultant to perform the geotechnical analysis and foundation design of the bridge. The existing structure was in poor condition due to advanced deterioration of the superstructure, which resulted in a reduced live load carrying capacity and the need for replacement. The proposed replacement bridge, shown in Figure 1, is a horizontally curved four-span continuous steel multi-girder structure which is approximately 47 feet wide and 350 feet long. The bridge was constructed along the existing alignment to avoid impacts to the adjacent marinas

Figure 1. Photo of the four-span continuous steel bridge MA-14 over Matawan Creek.

located on each corner of the bridge. The bridge was widened to provide standard lane widths, shoulders, and sidewalks.

The vertical profile was raised six feet to improve navigational clearance. Retaining walls were constructed along the approaches to limit property impacts and were supported by a column supported embankment system. Drilled shafts extending into pier caps were utilized to minimize potential conflicts with the existing bridge substructure, to minimize environmental impacts, and to avoid large costly cofferdams.

2 SURFACE CONDITIONS AND SEISMIC RESPONSE SPECTRUM

The soil profile throughout the site consisted of layers of fill and alluvial sand, underlain by as much as 75 feet of very soft to soft organic clay, followed by stiff to very stiff clay and dense sand. Rock was not encountered through the depth of exploration of 130 feet. Geologic literature established rock at anywhere between elevations −550 to elevation −830.

Shear wave velocity data was collected and used. For the top 130 feet of soil, the shear wave velocity was approximated based on the site-specific SPT N values and the undrained shear strength of the clay. For lower layers of soil, published values were used. To account for variability, seven wave velocity profiles were used as shown in Figure 2; this analytical profile used for the shallow soil profile turned out to produce a more conservative seismic response.

The soil conditions in the project site are classified as Site Class F per AASHTO BDS (2014). Per Section 38 of NJDOT (2016) and 3.10.2 of AASHTO BDS (2014), a site-specific procedure is required due to the site classification. To assess the ground modification for site response, a rock motion that matches the AASHTO Response Spectrum for Rock Class B, which is a reference site category for the USGS and NEHPRP MCE ground shaking map, needs to be generated. The seed motion used to conduct the spectrum matching was sourced from events described in Table 1. Event 1 is a synthetic record from NYCDOT Bridges with a 1500-year event record, which was scaled down to a 1000-year event. Events 2 and 3 are recorded motions of events with magnitudes generally believed to be appropriate for New York City and its surroundings. The selection of these three events aimed to simulate the site seismic setting; the magnitudes ranged from 6 to 6.4, and the distance to the source ranged from 16 km to 24 km to simulate a local event. The postulated rock motion was specified at rock outcrop and modified by the site soils as it propagates through the soil profile (AASHTO class F soil profile).

The soil profile (deep and shallow) was subjected to rock motion derived from the three events in Table 1 by spectra matching and specified at outcrop using Pro-Shake software. Pro-Shake is an equivalent linear model. The program's default relationships were used for the modulus degradation and damping increase changes with increasing shear strain.

With the three generated rock motions, the deep and shallow soil profiles were shaken by the three events. Although the basement rock was deep based on the literature, the depth to rock also varied to as shallow as −120 feet. The shallow soil profile (basement rock at −530 feet) turned out to be more critical than the deep profile. The computed spectra are depicted in Figure 3 together with the recommended spectrum of this bridge.

3 HOW ISOLATION BEARINGS CAN REDUCE PIER SIZE AND COST

3.1 *Pier type selection*

During preliminary design, three pier type alternatives were considered as noted below.

Alternative 1: Pile foundation with both vertical and battered piles supporting conventional pier columns and cap is illustrated in Figure 4. Due to proximity to the existing substructure, battered piles may conflict with existing substructure elements. In addition, constructing a footing in the

EL. (ft)	Layer No.	Soil Type	Unit Weigth (pcf)	SPT (N)	Su (tsf)	Shear Wave Velocity (ft/s)[4]	Shear Wave Velocity (ft/s)[5]	Shear Wave Velocity Used (ft/s) Profile S1*	Profile S2*	Profile S3*	Profile S4*	Profile S5*	Profile S6*	Profile S7*
+4 / -3	1	Fill	105	3~36				345	260	430	345	345	430	260
-3 / -7	2	Alluvial Sand	105	2~7				300	240	360	300	300	360	240
-7 / -55	3	Orangic Clay	94	–	0.06 ~0.37			270	165	375	270	270	375	165
-55 / -80	4	Stream Terraced Sand	110	4~34			778, 632,1179, 771, 448	560	390	725	560	560	725	390
-80 / -100	5	Varred Clay	121	28~36				655	490	820	655	655	820	490
-100 / -120	6	Dense Sand (Navesink)	125	25~185		928 ~1279	723, 2087, 674,1172	885	650	1120	885	885	1120	650
-120 / -200	7	Clay (Merchantville_ Woodbury)	125					900	900	900	675	1125	1125	675
-200 / -300	8	Sand (Magothy)	125					1370	1370	1370	1028	1713	1713	1028
-300 / -460	9	Clay	125				1220	1400	1400	1400	1050	1750	1750	1050
-460 / -550	10	Sand	125					1700	1700	1700	1275	2125	2125	1275
-550	11	Basement Rock	135					3500	3500	3500	3500	3500	3500	3500

*Profile S1: Medium- Medium; S2: Low-Medium; S3: High-Medium; S4: Medium-Low; S5: Medium-High; S6: High-High; S7: Low-Low

Figure 2. Shear wave velocity profiles.

Table 1. Event seed motion used for spectrum matching.

Event No.	Event Used as Seed Motion	Station	Magnitude	Distance to Source-km
1	NYCDOT (Scaled to 1000-year event)	Artificial	6–6.4 (assumed based on actual records used to generate it)	16–24 (assumed based on actual records used to generate it)
2	1984 Morgan Hill	57007 Corralitos	6.2	22.7
3	1986 N. Palm Springs	5069 Fun Valley	6	15.8

waterway requires cofferdams, which increase construction duration, cost and disturbance. The presence of battered piles will further increase the cofferdam size and footprint of disturbance.

Alternative 2: Vertical and battered piles extending up to a pier cap that directly supports the superstructure without temporary cofferdams is a more cost-effective alternative as illustrated in Figure 5. Preliminary analysis determined that the unbraced length of the piles due to the poor soil was too long for this to be a feasible option. Furthermore, proposed battered piles may interfere with piles from the existing bridge.

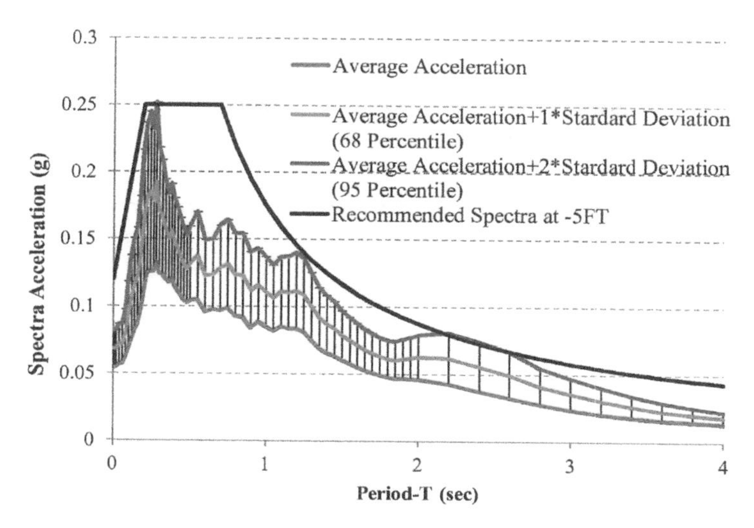

Figure 3. Recommended Response Spectrum for Bridge MA-14.

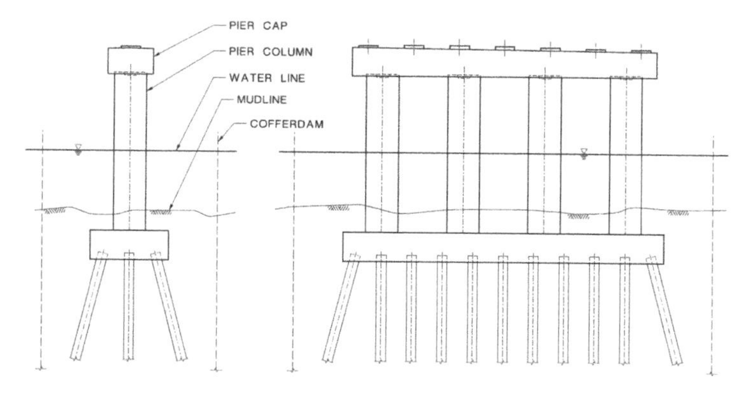

Figure 4. Section and elevation view of a hypothetical bridge pier in water, supported on a pile foundation with large cofferdams due to battered piles.

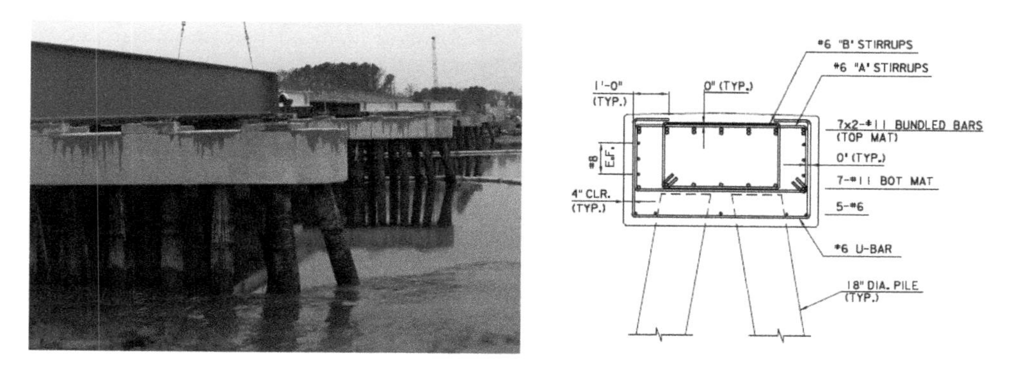

Figure 5. Photo of a bridge pier with battered piles extending up to a concrete pier cap of a previous GPI project (Left); Cross-section view of pier cap (Right).

Figure 6. Photo of Bridge MA-14 pier during construction. Note drilled shafts extend directly to pier cap.

Alternative 3: Drilled shaft extending up to a pier cap that directly supports the superstructure without cofferdams is another cost-effective alternative. Since drilled shafts cannot be battered, lateral resistance comes from the size of the shaft. To reduce shaft diameter, isolation bearings are proposed to reduce seismic load. During the selection process, both isolated and non-isolated conditions were analyzed.

- 3A: Isolated condition using seismic isolation bearings at all supports.
- 3B: Non-isolated condition using conventional HLMR bearings with fixed bearings assigned at Piers 1 and 2, and expansion bearings at the remainder of the supports.

Analysis results discussed in Section 3.2 demonstrate that seismic demand per 3A is about half of 3B and hence is the preferred alternative. Figure 6 below shows this (3A) alternative, a pier with four 5′ diameter drilled shafts having a 3/4″ steel casing.

3.2 *Isolated versus non-isolated comparison*

Preliminary seismic response spectrum analyses were performed to compute the seismic demand on the bridge, particularly the drilled shafts, for both isolated (Alternative 3A) and non-isolated (Alternative 3B) conditions, with both alternatives using the same shaft section and point of fixity for analysis. For the drilled shafts, resultant of bending in both transverse and longitudinal directions at point of fixity increased from 2700 k-ft (isolated) to 4500 k-ft (non-isolated), which is a 70% increase; the deflection at the top of shaft also increased 70%. Since the point of fixity computed based on the isolated model is still located in the weak soil zone, increasing demand will lower the point of fixity and increase the cantilever arm. In addition, as the demand increases, more concrete inside the shaft will crack and reduce the shaft stiffness. As the shaft deflects more, p-delta effect will further increase the bending and deflection. After considering these effects, it is estimated that the (non-isolated) seismic displacement will be about three times that of the isolated condition at fixed Piers 1 and 2.

FHWA (2010) recommends the maximum deflection of a drilled shaft to be 10% of its diameter for geotechnical design against lateral loads. It is estimated that a 6 ft (minimum) diameter drilled shaft, which is about one foot larger in diameter than the isolated condition, is required to satisfy this recommendation. This one-foot increase in diameter (44% increase in area and 107% increase in moment of inertia) is required to increase the shaft bending stiffness to reduce deflection as well as to increase the allowable deflection limit.

Further stiffness increase can be achieved by using thicker steel casing. While there is no need to use a larger shaft at the expansion supports, two different sets of equipment will be mobilized to the site to install different sizes of shaft. Having two different diameter shafts in the same bridge may not be cost-effective nor visually pleasing. To maintain proper spacing between larger shafts, the overall pier size at fixed piers will become much larger, increasing the area of disturbance and negatively affecting the visual appearance of the bridge. Based on bid prices of the isolated bridge,

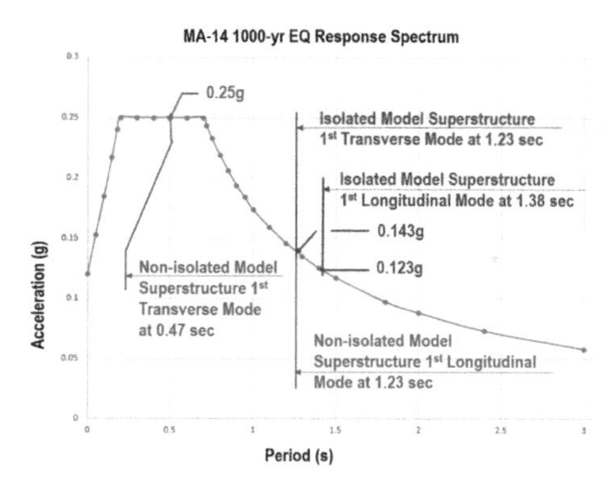

Figure 7. MA-14 Response Spectrum indicating first longitudinal and first transverse modes of the isolated and non-isolated conditions.

it is estimated that the non-isolated alternative may cost another \$1.5 to \$2 million due to the size increase of these 140 feet long drilled shafts, which is about 20% to 30% increase in drilled shaft related items cost; whereas the estimated additional cost of the 35 isolators is about \$70,000 as compared to conventional HLMR bearings, reflecting a 25% to 40% cost increase.

3.3 *Load reduction on piers through use of isolation bearings*

For a 4-span continuous steel curved girder bridge with five supports, conventional design illustrated under alternative 3B is to make the two middle supports fixed and allow the remainder of the bridge to expand/contract under thermal load. In this case, longitudinal lateral loads will only go to the fixed supports leading to larger foundations, whereas transverse lateral loads can be distributed to all supports. Using isolation bearings at all supports allows lateral loads to be distributed to all supports, instead of only fixed supports, resulting in a more economical pier design. Additionally, isolators will displace together with the superstructure relative to the substructure during a seismic event. This displacement increases the predominant period of superstructure vibration to lower acceleration due to its inertia as illustrated in Figure 7, as well as induces additional structural damping which was conservatively ignored for the analysis of this bridge. However, isolators cannot be made too flexible since they need to remain stable under service loads.

Figure 7 above shows the spectrum of MA-14, first mode periods and corresponding accelerations under both longitudinal and transverse directions for both isolated and non-isolated conditions. In the longitudinal direction, while the period of the first isolated mode is only slightly longer than the non-isolated, seismic load is distributed to all supports under isolated condition as compared to concentrated to the two fixed piers under non-isolated; in the transverse direction, the period of the first isolated mode is much longer than the non-isolated which corresponding acceleration is about twice of the isolated one. Please note that the periods of non-isolated model indicated in Figure 7 are computed using the same shaft section and point of fixity as the isolated model.

3.4 *Three common types of isolation bearings*

There are three common types of isolators, employing different mechanisms to make their isolators flexible under seismic load while remaining stable under service loads. They are:

- Lead-rubber isolator as shown in Figure 8 (Left).
- Eradiquake isolator as shown in Figure 8 (Right).
- Friction pendulum isolator as shown in Figure 9, which was ultimately used for MA-14.

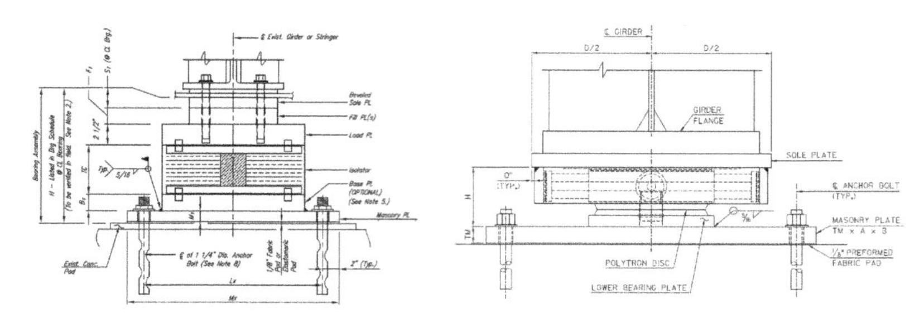

Figure 8. Section view of a lead-rubber isolator (Left); Section view of an Eradiquake isolator (Right).

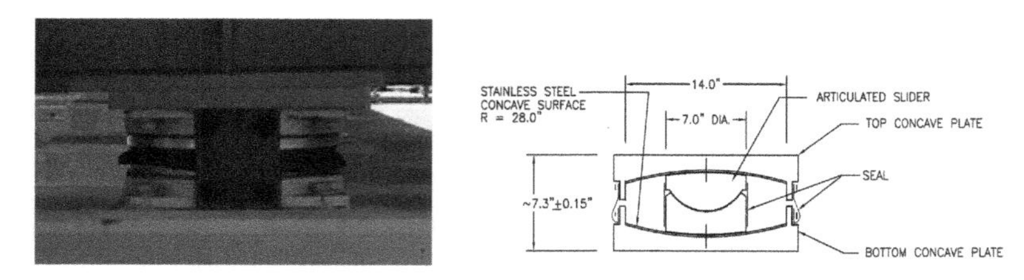

Figure 9. Photo of the friction pendulum isolator used on MA-14 (Left); Section of the friction pendulum isolator from MA-14 (Right).

For friction type isolators (friction pendulum and Eradiquake), the material friction coefficient under fast motion (seismic load) is higher than slow to moderate motion (thermal, wind and braking). For lead-rubber isolators, the responses under different speeds are similar. This characteristic makes isolators stiffer under seismic load and less stiff under non-seismic loads. While we want a flexible isolator under seismic load and a stable isolator under non-seismic loads, isolators can only be efficiently used when anticipated seismic load is much larger than non-seismic loads.

3.5 *Preliminary isolation bearing design*

All three common types of isolators are proprietary bearings which will be designed by the manufacturer to be selected by the Contractor. The designer only needs to determine the isolation bearing design parameters based on the results of seismic analysis, in accordance with AASHTO GSSID (2014). Table 2 shows the MA-14 isolation bearing design parameters for two types of isolators for seismic load (high velocity motion), non-thermal service loads such as wind load and braking forces (moderate velocity motion) and thermal load (low velocity motion). The third type did not work well for this project without modifying the isolator, which made it less cost competitive.

Isolation bearing design parameters include characteristic strength (Q_d), post-elastic stiffness (k_d), maximum bearing displacement (d_{max}), effective stiffness at d_{max} (K_{eff}), energy dissipated per cycle at d_{max} (EDC) and damping ratio. Since isolator behavior under seismic load is different than under wind and thermal loads, different sets of parameters are required. The designer needs to coordinate with at least one bearing manufacturer or ideally three to make sure the manufacturer can fabricate isolators meeting these requirements and determine the preliminary size of the isolators so that bridge seat can accommodate the bearings.

As shown in Table 2, bearing design parameters for the piers are different from the abutments. As the end supports are located further away from the center of the bridge, isolators at the ends need to be more flexible to accommodate higher thermal movement than those near the center of bridge subject to less thermal movement. Even though isolators at abutments are more flexible, they still can distribute a portion of the seismic load to the abutment lessening the demand at the center

Table 2. Bearing characteristic table for Bridge MA-14.

TYPE OF ISOLATORS	LOCATIONS	BEARING CHACTERISTICS – SEISMIC													
		LONGITUDINAL							TRANSVERSE						
		Q_d Range (Kip)	Weighted Avg. Q_d (Kip)	K_d (k/in)	d_{max} (in)	k_{eff} (k/in)	EDC (K-in)	DAMPING RATIO, ξ	Q_d Range (Kip)	Weighted Avg. Q_d (Kip)	K_d (k/in)	d_{max} (in)	k_{eff} (k/in)	EDC (K-in)	DAMPING RATIO, ξ
Type E	W & E Abuts.	1.0 to 1.0	1.0	0.0	1.7	0.6	6.7	0.27	1.0 to 1.0	1.0	3.0	1.3	3.8	5.3	0.31
Type E	Piers 1 to 3	9.2 to 13.8	11.5	3.5	1.0	15.2	45.1		9.2 to 13.8	11.5	3.5	1.1	14.3	48.8	
Type F	W & E Abuts.	2.9 to 4.7	3.5	1.0	1.5	3.3	21.1	0.27	2.9 to 4.7	3.5	1.0	1.4	3.6	19.0	0.36
Type F	Piers 1 to 3	6.8 to 12.9	9.4	2.5	1.2	10.2	45.9		6.8 to 12.9	9.4	2.5	1.2	10.7	43.2	

TYPE OF ISOLATORS	LOCATIONS	– SERVICE (NON-THERMAL)							– SERVICE (THERMAL)			
		AVG. Q_d (Kip)	LONGITUDINAL			TRANSVERSE			AVG. Q_d (Kip)	K_d (k/in)	LONGITUDINAL	
			K_d (k/in)	d_{max} (in)	SERVICE FORCE (k)	K_d (k/in)	d_{max} (in)	SERVICE FORCE (k)			d_{max} (in)	THERMAL FORCE (k)
Type E	W & E Abuts.	2.4	0.0	0.0	2.0	6.0	0.2	4.5	2.4	0.0	1.37	2.4
Type E	Piers 1 to 3	7.6	7.0	0.0	3.0	7.0	0.0	5.5	7.6	3.5	0.74	10.2
Type F	W & E Abuts.	3.5	1.0	0.0	2.0	1.0	0.5	4.5	2.3	1.0	1.37	3.7
Type F	Piers 1 to 3	9.4	2.5	0.0	3.0	2.5	0.0	5.5	6.3	2.5	0.74	8.2

piers. Structural analysis shall be performed for dead, live, seismic, wind, braking and thermal loads under various AASHTO Limit States to optimize the pier design and ensure bridge stability.

3.6 *Drilled shaft pier design based on isolated condition*

For this bridge situated in a "Soil Class F" site having about 50 to 70 ft deep of weak soils, the design of the drilled shaft and isolation bearing is an iterative process. When lateral load is small, it demands less geotechnical lateral resistance from the weak soil; shaft point of fixity may still be located within the weak soil layer. When lateral load increases, it demands more resistance and lowers the point of fixity. This increases shaft cantilever length leading to higher bending at the base of the shaft and deflection at the shaft top. It may also require larger shaft diameter. This process requires multiple rounds of modification to the 3D bridge model and seismic analysis, as well as rounds of drilled shaft design and geotechnical analysis. The depth to fixity as defined here refers to a length of shaft fixed at some equivalent length below the top of shaft without any soil restraint above tip, and top rotations and deflection are identical to the real pile with soil restraint. The average of four lengths are calculated considering deflections due to applied shear and moment at top of the shaft, and top rotations due to applied shear and applied moment. The four quantities and equivalent length are explained in Table 3.

Soil exhibits non-linear behavior, and the response in terms of deflection or rotation is not proportional to shear and moment. For example, the moment and shear increase by 50%, and the rotations and deflection increase by more than 50% because the stiffness of the soil degrades with increasing moments. This non-linearity produces increased fixed length as readily apparent from the relationships in Table 3.

As a starting point, lateral load is typically assumed based on a certain percentage of dead load to determine the shaft diameter and point of fixity. Based on the initial information, a 3D bridge structural model is developed, and seismic analysis is performed. Based on the initial results from the model, the next round of geotechnical analysis is performed to refine the shaft diameter and point of fixity. It is essential to consult with at least one isolation bearing manufacturer to come up with preliminary bearing stiffness based on the preliminary seismic force and displacement at the bearing. Continual revisions to the 3D bridge model, with the updated bearing stiffness, shaft diameter and point of fixity, as well as drilled shaft design and geotechnical analysis, are required until the design converges. At the end of this process for bridge MA-14, the optimal pier design was to use four 60-inch diameter drilled shafts with a 3/4 inch thick steel casing at each pier, Figure 10.

Table 3. Definition of equivalent depth of fixity. *(Source: Caltrans, 1990.)*

Equivalent Length L_{δ_F} based on deflection on top of shaft due to shear	$L_{\delta_F} = (\frac{3(EI)\delta_F}{F})^{\frac{1}{3}}$
Equivalent Length based on deflection on top of shaft due to Moment L_{δ_M}	$L_{\delta_M} = (\frac{3(EI)\delta_M}{M})^{\frac{1}{2}}$
Equivalent Length L_{θ_F} based on rotation of shaft on top due to shear	$L_{\delta_F} = (\frac{2(EI)\theta_F}{F})^{\frac{1}{2}}$
Equivalent LengthL_{θ_M} based on rotation on top of shaft due to moment	$L_{\delta_F} = (\frac{(EI)\theta_M}{M})$
Notation	EI: Flexural rigidity, F = Shear, M = Moment, δ_F = Deflection due to shear, δ_M = Deflection due to moment, θ_F = Rotation due to shear, θ_M = Rotations due to shear
Depth to Fixity	$\frac{1}{4}(L_{\delta_F} + L_{\delta_M} + L_{\theta_F} + L_{\theta_M})$

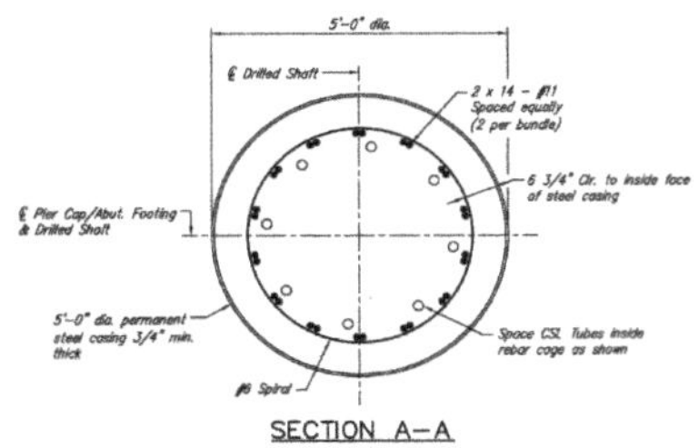

Figure 10. Photo of four 60-inch diameter steel casings during construction of Bridge MA-14 (Left); Reinforcement section of the MA-14 drilled shaft (Right).

4 SEISMIC ANALYSIS

Section 7 of AASHTO BDS (2014) mentions four different types of analysis procedures for isolated bridges and their applicability. Out of these four, Multimodal Spectral Method (Response Spectrum Analysis – RSA) and Time History Method (THA) are considered applicable for the project. Section 7 of AASHTO BDS (2014) further states that "a nonlinear time-history analysis shall be required for structures with effective periods greater than 3 seconds"; "for isolation systems where the equivalent viscous damping ratio, ξ, exceeds 30 percent, either a nonlinear time-history analysis shall be performed utilizing the hysteresis curves of the system or the damping coefficient, BL, shall be taken as 1.7"; "this spectrum may be scaled by the damping coefficient (BL), as defined in Article 7.1, to include the effective damping of the isolation system for the isolated modes." Based on the response spectrum, spectral acceleration at 3 seconds is only 0.063 g, and it is most likely that the effective period of the structure will be less than 3 seconds. Therefore, the RSA method is adequate for isolation bearing design, and time history analysis is not necessary. In addition, AASHTO GSSID (2014) allows single or multi-modal analysis for all bridges in Zone 2, for all bridges in Zone 3 other than irregular critical bridges, and for all bridges in Zones 3 and 4 other than "critical bridges". Even if MA-14 is a critical bridge, AASHTO GSSID (2014) allows modal analysis because the bridge is in Zone 2 ($0.15 < SD1 < 0.3$ per Table 3.10.6.1). Consequently, a time history analysis of the bridge, although it may be required for other reasons, is not required based on 4.7.4.3.1-1 of AASHTO GSSID (2014).

However, the depth/length of drilled shaft from the tip of shaft to the pier cap is expected to be more than 140 feet long, with a point of fixity 40 to 50 ft below pier cap. Having flexible bearings (isolators) on top of flexible substructure units, the dynamic structural behavior may become complex, due to 1) substructure displacement not being in sync with bearing displacement; 2) difference in damping and response due to isolators under different magnitude of vertical loads over different substructure; and 3) damping stemming from isolators being different from that of the flexible substructure, making RSA less reliable. Therefore, both RSA and THA were performed with RSA being used as preliminary analysis to design the substructure and isolator and THA as a final analysis to verify and calibrate the results from RSA. RSA was conducted in accordance with AASHTO GSSID (2014).

According to AASHTO BDS (2014), THA must be performed with at least three appropriate sets of acceleration time histories for the bridge site. Each set shall be comprised of three orthogonal components selected. In most cases, however, vertical time history is not necessary as it has little overall impact, which requires very careful and accurate modeling and adds complexity in combining the results. The omission of vertical ground motion in seismic analysis of a bridge is permitted by several studies (FHWA, 2006; AASHTO GSSID, 2014). In particular, FHWA (2006) states "the impact of vertical ground motion may be ignored if the bridge is greater than 30 miles from an active fault". AASHTO SBD (2011) C4.2.2 does not state such requirements for structures outside of 6 miles of an active fault. As the structure is located under Seismic Zone 2 and does not fall under any of above categories, vertical motion was not considered. Nevertheless, axial loads on the isolator generated by the lateral motion were considered in the analysis.

Therefore, only two horizontal orthogonal directions were considered. For a minimum of three distinct time histories, E1, E2, and E3, vector excitation for three runs R1, R2, and R3 can be combined as follows:

$$R1 = 0.919\ E1\ L + 0.919\ E2\ T$$

$$R2 = 0.919\ E2\ L + 0.919\ E3\ T$$

$$R3 = 0.919\ E3\ L + 0.919\ E1\ T$$

where L and T stands for longitudinal and transverse directions, respectively, in relation to the structure's alignment. A factor of 0.919 is used to account for directionality and correlation to response spectrum analysis.

It was determined that use of three distinct time histories were necessary to avoid "in-phase" movement that can potentially overestimate structure response. Three sets of distinct time histories compatible with the design spectrum were generated from three distinct seed movements from one synthetic source retrieved from the NYC Bridges database and two actual records from the Caltech database. The maximum displacement and responses of the isolation system were calculated from the vectoral sum of the orthogonal displacements at each time step.

CSiBridge, a computer structural analysis program used for the project, offers Fast Non-linear Analysis (FNA). FNA is a modal analysis method for dynamic evaluation that is computationally efficient and well-suited for time-history analysis; however, to correctly utilize FNA, the structural model under analysis must have primarily linear-elastic behavior and have a limited number of predefined nonlinear members in forms of link elements; the efficiency of FNA formulation is largely due to the separation of the nonlinear-object force vector from the elastic stiffness matrix and the damped equations of motion, according to CSI (2010). From a preliminary RSA, it was determined that about 95% of the drilled shaft concrete core cross section remain uncracked and assumed elastic during the design 1000-yr earthquake; isolation bearings are the only type of inelastic elements in the model, making FNA suitable for the structure. Use of FNA was able to reduce running time of each THA analysis by about 90% compared to a more traditional direct-integration method.

In general, results from THA were consistent with those of RSA. THA yielded lower bearing displacement and higher seismic demand on substructure than RSA, while yielding similar demands on bearings and superstructure as RSA. One reason for THA yielding higher demand on the substructure (particularly the abutment) than RSA is due to the significantly different damping effect from isolators and substructure. THA can account for different damping effects on different elements in the model, whereas RSA analysis is based on one single damping factor. The RSA analysis damping factor is approximated through averaging the different damping factors coming from different isolators and estimating the overall superstructure and substructure damping based on displacement and force. In addition, from THA, it was evident that superstructure and substructure move in a synchronized manner.

5 BENEFITS OF USING ISOLATION BEARING

Benefits of using isolation bearings for this multi-span curved bridges include:

- Lower seismic demand and reduced substructure size and cost.
- Lower substructure seismic demand implies smaller damage and lower repair cost after earthquake.
- Lower thermal load and stress as well as reduced size and numbers of cracks on substructure to extend service life and reduce future maintenance cost.
- No jammed bearings for curved girders due to the use of flexible bearings.

6 CONCLUSIONS

Through MA-14 it was demonstrated that a bridge located in a relatively minor seismic zone can also benefit from the use of seismic isolation bearings when subsurface conditions are poor. The thick layer of weak soils below the mudline required the use of drilled shafts having an average length of 140 feet with limited lateral resistance from soil. With the use of isolation bearings, which are incrementally more expensive than conventional bearings, shaft deflection was reduced by more than 50%, leading to a much smaller shaft cross-sectional area and reduction in construction cost.

ACKNOWLEDGEMENTS

The authors would like to thank the Monmouth County Division of Engineering for giving us the opportunity to work on the design of this unique structure and for allowing us to develop and present this paper. We would also like to thank Paul Bradford of PB Engineering for his guidance in time-history analysis; Greg Johnson, Judy Bowen, Peter Mahally, Lenny Lembersky and Jason Bell of GPI for their help during the design and construction phases of the project; and Ashley Kocsis of GPI for her help reviewing this paper.

REFERENCES

AASHTO BDS 2014. *AASHTO LRFD Bridge Design Specifications*, 7th Edition, American Association of State and Highway Transportation Officials (AASHTO), Washington, D.C.

AASHTO GSSID 2014. *AASHTO Guide Specifications for Seismic Isolation Design*, 4th Edition, American Association of State and Highway Transportation Officials (AASHTO), Washington, D.C.

AASHTO SBD 2011. *AASHTO Guide Specifications for LRFD Seismic Bridge Design*, 2nd Edition, American Association of State and Highway Transportation Officials (AASHTO), Washington, D.C.

Caltrans 1990. *Bridge Design Aids,* Caltrans, Sacramento, CA

CSI 2010. *CSI Analysis Reference Manual,* Computer and Structures, Inc, Berkeley, CA.

FHWA-HRT-06-032 2006. *Seismic Retrofitting Manual for Highway Structures: Part 1 – Bridges,* U.S. Department of Transportation Federal Highway Administration (FHWA), McLean, VA.

FHWA-NHI-10-016 2010 (NHI Course No. 132014). *Drilled Shafts: Construction Procedures and LRFD Design Methods,* U.S. Department of Transportation Federal Highway Administration (FHWA), Washington, D.C.

NJDOT 2016. *NJDOT Design Manual for Bridges & Structures,* 6th Edition, New Jersey Department of Transportation, Trenton, NJ.

Author index